LONDON MATHEMATICAL SOCIETY STUDENT TEXTS

Singular Points of Plane Curves

C. T. C. Wall
University of Liverpool

PUBLISHED BY THE PRESS SYNDICATE OF THE UNIVERSITY OF CAMBRIDGE
The Pitt Building, Trumpington Street, Cambridge, United Kingdom

CAMBRIDGE UNIVERSITY PRESS
The Edinburgh Building, Cambridge CB2 2RU, UK
40 West 20th Street, New York, NY 10011–4211, USA
477 Williamstown Road, Port Melbourne, VIC 3207, Australia
Ruiz de Alarcón 13, 28014 Madrid, Spain
Dock House, The Waterfront, Cape Town 8001, South Africa

http://www.cambridge.org

First published 2004

Printed in the United Kingdom at the University Press, Cambridge

Typeface Computer Modern 10/13pt *System* LaTeX 2ε [AUTHOR]

A catalogue record for this book is available from the British Library

Library of Congress Cataloguing in Publication data

Wall, C. T. C. (Charles Terence Clegg)
Singular points of plane curves/C. T. C Wall.
p. cm. – (London Mathematical Society student texts; 63)
Includes bibliographical references and index.
ISBN 0 521 83904 1 (hardback) – ISBN 0 521 54774 1 (paperback)
1. Singularities (Mathematics) – Congresses. 2. Curves, Plane – Congress.
I. Title II. Series.

QA614.58.W35 2004
512′.22 – dc22 2004052121

ISBN 0 521 83904 1 hardback

ISBN 0 521 54774 1 paperback

Contents

Preface

The study of singular points of algebraic curves in the complex plane is a meeting point for many different areas of mathematics. The beginnings of the study go back to Newton. During the nineteenth and early twentieth century algebraic geometers working on plane curves developed methods which allowed them to deal with singular curves: see e.g. [168], [84], and several articles in volume III of the mathematical encyclopaedia published 1906–1914. A notable achievement was the resolution of singularities of such curves. In the late 1920s results in the then new area of topology were applied to the knots and links in the 3-sphere obtained by looking at the neighbourhood of such a singularity. There was a resurgence of interest about 1970 due to the interaction with newly developing ideas from singularity theory in higher dimensions, most importantly, the fibration theorem which Milnor had just discovered, in the context of functions of several complex variables. There has been continuous development since then, a particular point of interest being the application of Thurston's (circa 1980) decomposition theorems for 3-manifolds and for homeomorphisms of 2-manifolds.

The interaction between ideas from these different sources makes the study of curve singularities particularly fruitful and exciting. Equisingularity is an equivalence relation which admits characterisations from numerous differing points of view. The development of the ideas leading up to this is the leitmotif of the first half of this book. I thus emphasise the equivalence of different approaches, and feel that many results gain in clarity from appearing in an integrated account.

This book is based on an M.Sc. course given a number of times at the University of Liverpool. On the first such occasion (Autumn 1975) the course was given jointly by myself and two colleagues: Hugh Morton and Peter Newstead. It is a pleasure to acknowledge the insights derived from our collaboration on that occasion, and conversations subsequently.

The chapter on preliminaries is included to define a starting point: the topics mentioned here are not covered in detail. The core of the book is contained in Chapters 2–5, which lead to the equivalence of a range of conditions defining equisingularity. Here the level of exposition has been kept to that of the M.Sc. course (though I have added the decomposition theorem for general polar curves). We begin in Chapter 2 with a proof that a curve given by an equation may be represented by a particular type of parametrisation. Some foundational material on complex analysis is included to enable questions of convergence to be dealt with. It is then easy to proceed in Chapter 3 to a proof of resolution, and this in turn leads on naturally to a discussion of the invariants and configurations arising in the resolution process. The key concept in Chapter 4 is that of order of contact: this is developed to a flexible tool for giving the relation with intersection numbers and for answering questions that arise in the case of a curve with several branches. In Chapter 5 we begin the discussion of topology with a detailed geometrical picture of the knot, and proceed to calculate its Alexander polynomial, which suffices for the application to equisingularity. This fits well for students attending a parallel course on knot theory.

The later chapters are written at a more sophisticated level, and include introductions to a number of topics of recent research. The next two chapters deal with topics only briefly mentioned in the M.Sc. course. Chapter 6 contains proofs of Milnor's fibration theorems, first remarks about the Milnor fibre, and several calculations of Milnor numbers, emphasising the use of the Euler characteristic. Then we treat curves in the complex projective plane, with proofs of the general form of Plücker's theorems, Viro's proof of Klein's equation (using Euler characteristics of constructible functions), and an analysis of singularities of dual curves; and conclude with a survey of known results about curves whose singularities are maximal in some sense.

The next three chapters lead up to the calculation of the monodromy of the Milnor fibration. Chapter 8 introduces calculations and notation for later results in the form of several numerical invariants and their representation using exceptional cycles on a resolution tree. We include an introduction to the topological zeta function. Chapter 9 analyses the Thurston decomposition of the Milnor fibre and the JSJ decomposition of the link complement (following Le–Michel–Weber and Eisenbud–Neumann). Students attending a suitable parallel course, for example following [36], will find this an interesting application. A plumbing description of the neighbourhood of the resolution graph is given in sufficient detail for application to the study of the monodromy: it follows that the monodromy can be chosen to have no fixed points, and we obtain a direct relation of the Eggers tree and the resolution tree. We discuss how to calculate the E–N invariants, and give several necessary and sufficient conditions for the monodromy to have (pointwise) finite order. Chapter 10 opens with the definition of the Seifert matrix, and its interrelation with monodromy and intersection form, and proceeds to a detailed calculation of the monodromy map in homology, using the Thurston decomposition to define the weight filtration. It is shown how to classify Seifert forms over a field, and some of the invariants required for classification over $\mathbb{Q}$ are calculated.

In Chapter 11 we discuss ideals in relation to resolution, and a representation by exceptional cycles. There is a relation between ideals and clusters of infinitely near points, which takes the form of a Galois connection, in which a cluster is closed if and only if it satisfies the proximity inequalities and an ideal is closed if and only if it is valuatively closed. Enriques' unloading process is seen to be intimately related to the Zariski decomposition of cycles. There is a neat formula for the codimension of a closed ideal. The equivalence of valuative and integral closure is proved as an application of Noether's $Af + Bg$ theorem. We conclude with brief treatments of determinacy and of differential forms.

The later chapters can also be viewed as forming two parallel but interdependent developments; the geometry of the link complement and the Milnor fibration being studied in Chapters 6, 9 and 10; and the algebraic and combinatorial information being developed in Chapters 4, 8 and 11.

Each chapter is concluded by sections on 'Notes' and 'Exercises'. The notes include historical remarks, references – which we do not in general include in the main text – comments on related material (for example, characteristic p and the real as opposed to the complex case), and some references for further developments. The exercises include routine exercises on applying the results in the text to specific examples, and problems related to an alternative approach to a topic treated in the text.

1

Preliminaries

1.1 What is a plane curve?

There are many ways to give a precise interpretation of the informal notion of a plane curve as something drawn on a sheet of paper. Here we will explain just what will be meant in this book, and indicate why this is a fruitful choice.

We begin with thinking of the plane as consisting of points, which may be described by two coordinates (x, y). Then one basic idea is that a curve is a set of points whose coordinates satisfy some equation $f(x, y) = 0$. What kinds of function f are to be allowed? The simplest kind, where f is just a polynomial function – expressible, say, as $f(x, y) = \sum_{0 \leq i,j; i+j \leq d} a_{i,j} x^i y^j$ for some numerical coefficients $a_{i,j}$ – already leads to an extensive theory of curves, mostly developed in the nineteenth century. We take this as our starting point.

The next thing to decide is what sort of numbers are to be allowed for the coefficients $a_{i,j}$ and indeed for the values of x and y. For the kind of curve one may draw and picture most easily, it is natural to choose to allow arbitrary real numbers: these permit the kind of continuity one expects from a sketch of a curve. However, it turns out that a much richer structure is obtained if we allow complex coefficients, and although the geometry involved when there are two complex variables is harder to picture, there have been major advances in this type of geometry in recent years. One of our main concerns will be understanding the interrelation between these geometrical aspects of our curves with invariants defined from a more algebraic viewpoint. We will occasionally restrict to real coefficients; results over other fields will be mentioned sometimes in the notes at the end of chapters.

The topic of this book is not so much the study of entire curves, though we will obtain a number of important results valid for the curve as a whole, as a very detailed study of what may happen near a singular point of a curve. Thus we will mainly study curves just in a neighbourhood of the origin O (with coordinates $(0,0)$) in the complex plane $\mathbb{C}^2$. For this local study it is not important that the equation $f(x,y)=\sum_{0\leq i,j} a_{i,j}x^iy^j$ defining the curve should be given by a finite sum. So we need to consider more carefully just what kind of expression f is to be: let us give some precise definitions.

A *polynomial* in one variable t is a sum of finitely many terms a_nt^n, where the *coefficients* a_n are complex numbers and the *exponents* n are non-negative integers. The *degree* of the polynomial is the largest number n such that a_n is non-zero. Polynomials may be added and multiplied in the usual way, and form a ring, denoted $\mathbb{C}[t]$. An immediate and important consequence of our choice of complex numbers as coefficients is that polynomials can be factorised: if $f=\sum_0^n a_rx^r$ has degree n, then the *roots* α_i of $f(x)=0$ are such that $f(x)\equiv a_n\prod_1^n(x-\alpha_i)$.

An expression $\sum_0^\infty a_rt^r$, where infinitely many non-zero terms are allowed, is called a *formal power series.* The usual rules still allow us to add and multiply such expressions, yielding a much larger ring, denoted $\mathbb{C}[[t]]$. The *order* of the formal power series is the smallest number n such that a_n is non-zero.

It is easy to write down a formal power series such that if any non-zero complex number is substituted for t, the resulting series of complex numbers fails to converge. A simple example is $\sum_0^\infty r!t^r$. We recall from complex variable theory that if $\sum_0^\infty a_rt^r$ converges at $t=v$, then the terms $|a_n|R^n$ (where $R=|v|$) are bounded (indeed, they tend to 0), and that if conversely this condition holds then the series converges for all values of t such that $|t|<R$. If this condition holds for some value of $R>0$, we call the series a *convergent power series.*

A function of one (or several) complex variable(s) defined on some region $U\subset\mathbb{C}$ (or $U\subset\mathbb{C}^n$) and which possesses a derivative on U is said to be *holomorphic* on U. Standard complex variable theory tells us that any function of t which is holomorphic on some neighbourhood of O can be expanded as a convergent power series (some books use the word *analytic* to describe functions defined by convergent power series, so the result can be stated as 'analytic = holomorphic'). Given two holomorphic functions, each defined on some neighbourhood of O, we can add and multiply them (on a smaller neighbourhood of O); it follows that the convergent power series form a subring of $\mathbb{C}[[t]]$. It is denoted by $\mathbb{C}\{t\}$.

Correspondingly for functions of two variables we have the polynomials, which are sums of finitely many terms $\sum a_{i,j}x^iy^j$ (with i and j non-negative integers and the coefficients $a_{i,j}$ complex numbers), and form a ring denoted $\mathbb{C}[x, y]$. We have the formal power series, which are expressions $\sum_{i=0}^{\infty} \sum_{j=0}^{\infty} a_{i,j} x^i y^j$ and form a ring, denoted $\mathbb{C}[[x, y]]$. The *degree* of a polynomial is the largest number n such that $a_{i,j}$ is non-zero for some i, j with $i + j = n$. The *order* of a power series is the smallest number n such that $a_{i,j}$ is non-zero for some i, j with $i + j = n$.

A power series in two variables is said to be *convergent* if there exist positive real numbers R, S such that the numbers $|a_{m,n}|R^m S^n$ are bounded. In this case, the series is convergent on the region $|x| < R$, $|y| < S$, and summing it defines a holomorphic function on this region. Conversely, a function which is holomorphic on a neighbourhood of O is said to be holomorphic at O and can be expanded as a convergent power series. Such series form a ring, denoted $\mathbb{C}\{x, y\}$.

Each convergent power series converges on some neighbourhood of $t = 0$. Formal power series do not in general converge at any $t \neq 0$: a compensating advantage is that they can be constructed term-by-term. Thus a sequence $\{f_k(t)\}$ of polynomials (or power series) such that, for each n, the coefficient of t^n in $f_k(t)$ is the same for all large enough values of k, defines a formal power series $f_\infty(t)$ having these coefficients. We say that f_k converges to f_∞ in the $\mathfrak{m}_t$*-adic sense.*

As a simple example, observe that a series $1 - a(t)$ with constant term 1 has an inverse, since the series $\sum_0^\infty a(t)^r$ converges in the $\mathfrak{m}_t$-adic sense, and its sum is the desired inverse. Thus any series with non-zero constant term also has an inverse in the ring. This result is also true in the ring $\mathbb{C}\{x, y\}$ for the simpler reason that if the function $f(x, y)$ is differentiable in a neighbourhood of O and $f \neq 0$ at – and hence in some neighbourhood of – O, then the usual rule allows us to differentiate also $\frac{1}{f(x,y)}$ in such a neighbourhood.

Thus in the 1-variable cases $\mathbb{C}[[x]]$ and $\mathbb{C}\{x\}$, any element of order m is equal to x^m multiplied by a power series with non-zero constant term, which has an inverse in the ring. This gives a complete description of factorisation in these rings.

Another way to approach the idea of plane curves is by parametrisations. The most familiar example is that of a graph, where y is expressed as a function of x: in general we have a parameter t, with each of x and y expressed in terms of t – say $x = \phi(t)$, $y = \psi(t)$. As above, the functions ϕ and ψ may be taken as polynomials or power series (preferably convergent).

The starting point for the analysis of singular points is the solution of a holomorphic equation $f(x, y) = 0$ to express y as a function of x. This will be discussed in the next chapter. It follows from the implicit function theorem 1.4.2 that if $\partial f/\partial y$ is non-zero at O, we can solve for

y as a holomorphic function of x near O. In this case, the curve Γ given by the equation $f(x,y)=0$ is said to be *non-singular* or (more briefly) *smooth* at the origin; its tangent there is given by $x\frac{\partial f}{\partial x}(O)+y\frac{\partial f}{\partial y}(O)=0$. The example

$$f(x,y)\equiv x+y-xy; \quad y=-\sum_{1}^{\infty} x^r$$

shows that even if f is a polynomial we will need power series, not just polynomials, to express y as a function of x. The example

$$f(x,y)\equiv x^2-y^3; \quad y=x^{\frac{2}{3}}$$

shows that in general fractional powers of x will be involved.

This last example may be represented by a parametrisation $(x,y)=(t^3,t^2)$. We also have the parametrisation $(x,y)=(u^6,u^4)$, obtained by substituting $t=u^2$. Clearly the latter is less satisfactory: each point of the curve is represented by two values of u, differing in sign. We will say that a parametrisation $(x,y)=(\phi(t),\psi(t))$ is *good* (an alternative term is 'primitive') if a general point of the curve corresponds to just one value of the parameter, i.e. the map $t\to(\phi(t),\psi(t))$ is injective on some region $|t|<\epsilon$.

Thus an equation $f(x,y)=0$ with $f\in\mathbb{C}[x,y]$, or a parametrisation with $\phi,\psi\in\mathbb{C}[t]$, defines a curve Γ as a subset of the plane $\mathbb{C}^2$. If we merely have $f\in\mathbb{C}\{x,y\}$ or, respectively, $\phi,\psi\in\mathbb{C}\{t\}$, then there is a neighbourhood U of O in $\mathbb{C}^2$ on which the series converges, so the equation $f(x,y)=0$ defines a curve in U. We will see in Chapter 2 that we obtain the same class of holomorphic curves whether we use equations or parametrisations.

It is convenient to introduce some terminology for this situation. Two functions $f_i:U_i\to\mathbb{C}$ $(i=1,2)$, defined on neighbourhoods U_1,U_2 of O in $\mathbb{C}^2$ are said to define the same *germ* at O if they coincide on some neighbourhood $U\subset U_1\cap U_2$ of O. In the case of holomorphic functions, this is the case if and only if the power series expansions of f_1 and f_2 coincide. Correspondingly, subsets $X_i\subset U_i$ of neighbourhoods U_1,U_2 of O define the same *germ* at O if for some neighbourhood $U\subset U_1\cap U_2$ of O we have $X_1\cap U=X_2\cap U$. In practice, rather than use the word 'germ', we will speak of curves defined in some neighbourhood of a point, usually O, and always be prepared to pass to smaller neighbourhoods.

The above discussion is concentrated on what happens in a small neighbourhood of the origin in the plane $\mathbb{C}^2$. Sometimes we wish to think of curves in the large, and then it is more convenient to work in the projective plane (we may refer to $\mathbb{C}^2$, when we wish to emphasise the distinction, as the *affine plane*).

In general we define n-dimensional *projective space* $P^n(\mathbb{C})$ to be the set of lines through the origin in $\mathbb{C}^{n+1}$. We may take coordinates $\mathbf{x} = (x_0, \ldots, x_n)$ in $\mathbb{C}^{n+1}$; then any point other than the origin (so with coordinate vector $\mathbf{x} \neq \mathbf{0}$ in $\mathbb{C}^{n+1}$) is joined to the origin by a unique line, and if $\mathbf{x}$ and $\mathbf{y}$ lie on the same such line, then $\mathbf{y} = \lambda\mathbf{x}$ for some $\lambda \neq 0$. Thus the point is determined by the *ratios* of the coordinates: $(x_0 : x_1 : \ldots : x_n)$. Note also that we may (if it is convenient to do so) change coordinates by any linear transformation of $\mathbb{C}^{n+1}$ onto itself.

If $f(x_0, \ldots, x_n)$ is some function, then when we replace $\mathbf{x}$ by $\mathbf{y} = \lambda\mathbf{x}$ we will get a different function, so the condition $f(x_0, \ldots, x_n) = 0$ is not in general well defined on $P^n(\mathbb{C})$. It is so is f is *homogeneous*, that is, if $f(\lambda x_0, \ldots, \lambda x_n) \equiv \lambda^k f(x_0, \ldots, x_n)$ for some value of k, the degree of f. In general, when working in projective space, only homogeneous polynomial functions f are considered.

Ratios are not always convenient to work with. Observe that the subsets U_r of $P^n(\mathbb{C})$ given (for $0 \leq r \leq n$) by $x_r \neq 0$ are well defined. Since we had $\mathbf{x} \neq \mathbf{0}$ above, any point in $P^n(\mathbb{C})$ lies in at least one of these subsets. In the subset U_r we may omit x_r and take the ratios $z_s = x_s/x_r \quad (s \neq r)$ as coordinates in the usual sense, or equivalently, fix $x_r = 1$. Thus each U_r is isomorphic to the affine space $\mathbb{C}^n$. Observe that if f is any polynomial function on U_r, and d is the highest degree of any term in f, then we may define a homogeneous polynomial function on $P^n(\mathbb{C})$ by

$$F(x_0, \ldots, x_n) = x_r^d f(z_0, \ldots, z_{r-1}, \uparrow^r, z_{r+1}, \ldots, z_n).$$

A projective algebraic variety is a subset of $P^n(\mathbb{C})$ defined by some homogeneous polynomial equations. We will be particularly interested in subsets of $P^2(\mathbb{C})$ defined by a single such equation: *projective plane curves.* The solutions of $f = 0$ coincide with the solutions of $f^2 = 0$. It will usually be convenient to insist that we consider only equations with no squared factor (these are called *reduced*). If $f = 0$ is a reduced equation, homogeneous of degree d, for a curve Γ, then Γ is said to be of degree d, and we write $d = \deg \Gamma$.

A useful background reference for the first three sections (and some later ones) of this book is the student text [99] by Kirwan, which assumes less background than we do.

1.2 Intersection numbers

We will have frequent occasion to consider intersections of curves, and to count intersection numbers. If the curves Γ_1 and Γ_2 are both smooth at the origin and have distinct tangents there, we will say that they have intersection number 1 at O. In general, we seek to deform one or both the curves a small amount so that at each point of intersection of the resulting curves, both are smooth and they have distinct tangents. Thus for example, for the curves given by $y = 0$ and $y = x^2$, we deform the latter to $y = x^2 - t^2$, giving two intersection points $(t, 0)$ and $(-t, 0)$, so the intersection number is 2.

More generally, consider the curves $y = 0$ and $y = f(x)$ in $\mathbb{C}^2$. If f is a polynomial, we may factorise it as $A \prod_1^n (x - t_i)$, where the roots t_i need not all be distinct. Since a small deformation will make them so, if just r of the t_i take a given value T, the intersection number at $(T, 0)$ is equal to r. If f is not a polynomial, but can be expressed as a power series $f(x) = \sum_0^\infty a_r x^r$ of order m, then we may write $f(x) = x^m g(x)$ with $g(0) \neq 0$, and then the intersection number at O is equal to m. Sometimes we will also refer to the order m as the *multiplicity* of 0 as a root of f.

To count intersections of $y = 0$ with a general curve Γ given by an equation $g(x, y) = 0$, we substitute $y = 0$ in the equation to obtain $g(x, 0)$ and proceed as above with $g(x, 0)$ in place of $f(x)$. Unless g has a repeated factor, Γ will intersect a general line $y = \epsilon$ in distinct points, which (for small ϵ) provide a deformation of the intersection $\Gamma \cap \{y = 0\}$ as before.

Suppose we have a curve Γ_1 given by an equation $g(x, y) = 0$ and a curve Γ_2 given by a good parametrisation $(x, y) = (\phi(t), \psi(t))$ such that $\phi(0) = \psi(0) = 0$. Then the intersection number of Γ_1 and Γ_2 at O is equal to the order of $g(\phi(t), \psi(t))$. For a first perturbation of g will ensure that g does not vanish at singular points of Γ_2, and for intersections at non-singular points we can take local coordinates in which Γ_2 is given by $y = 0$ with parameter $t = x$, and then argue as above. In future we will often find it convenient to write the parametrisation as a single map $\gamma : \mathbb{C} \to \mathbb{C}^2$, with $\gamma(0) = O$.

We will denote the intersection number of Γ_1 and Γ_2 at a point P by $(\Gamma_1.\Gamma_2)_P$ or, if P is understood, simply by $\Gamma_1.\Gamma_2$. The above unsymmetrical rule for calculating intersection numbers is very convenient, but does not cover all needs. For a general discussion of intersection numbers, including precise definitions, the reader may refer to [188], [99] or, for a more advanced treatment, [74]. We now list some key properties.

Lemma 1.2.1

(i) *Let C_1, C_2 be germs of holomorphic curves at a point $P \in \mathbb{C}^2$, with P an isolated point of $C_1 \cap C_2$, then there is a well defined intersection number $C_1.C_2$, which is a positive integer.*

(ii) *If C_1 has equation $g_1(x, y) = 0$ and C_2 has a good parametrisation $\gamma : \mathbb{C} \to \mathbb{C}^2$, with $\gamma(0) = O$, then $C_1.C_2$ is equal to the order of $g(\gamma(t))$.*

(iii) *If g_1 factorises as $g_3 g_4$ so that $C_1 = C_3 \cup C_4$, then $C_1.C_2 = C_3.C_2 + C_4.C_2$.*

(iv) *If C_i has equation $g_i(x, y) = 0$ for $i = 1, 2$ and $P = O$, then $C_1.C_2$ is equal to the dimension of the quotient ring $\mathbb{C}\{x, y\}/\langle g_1, g_2\rangle$.*

(v) *Intersection numbers are symmetric: $C_1.C_2 = C_2.C_1$.*

(vi) *Suppose given a holomorphic map $F : \mathbb{C}^2 \to \mathbb{C}^2$, defined near O, with $F(O) = O$, a good parametrisation $\gamma_2 : \mathbb{C} \to \mathbb{C}^2$ with $\gamma(0) = O$ of a curve C_2 such that $F \circ \gamma_2$ is a good parametrisation of $F(C_2)$, and an equation g_1 of a curve C_1, so that we can define a curve $F^{-1}C_1$ by the equation $g_1 \circ F$. Then $C_1.FC_2 = F^{-1}C_1.C_2$.*

Proof We omit (or defer) the proof that (ii) and (iv) give the same result, which is independent of all choices. Then (i) follows from either version of the definition, (iii) follows from (ii) and (v) from (iv).

(vi) is an immediate consequence of (ii), since each of the intersection numbers is equal to the order of the composite function $g_1 \circ F \circ \gamma_2$. □

Consider a curve Γ of degree d in the projective plane $P^2(\mathbb{C})$. A general line in the plane intersects Γ in d points, and (unless Γ contains the whole line as a subset) any line has intersections with Γ whose intersection numbers add up to d. For we may choose coordinates $(x : y : z)$ so that the line is given by $y = 0$ and Γ does not pass through $(1, 0, 0)$. Then if $f(x, y, z) = 0$ is the equation of Γ, the intersections are given by the vanishing of $f(x, 0, 1)$, which is a polynomial of degree d, and the sum of the multiplicities of the roots of such a polynomial is equal to d.

More generally, suppose Γ_1, Γ_2 are two projective plane curves, of respective degrees d_1, d_2, and such that their intersection does not contain a curve. Then the sum of the intersection multiplicities at all points of $\Gamma_1 \cap \Gamma_2$ is equal to $d_1 d_2$. This is known as Bézout's theorem; a proof may be found in [99]; see also the following section.

1.3 Resultants and discriminants

Suppose $f(x,y), g(x,y)$ are homogeneous polynomials of respective degrees m,n. Then

$$x^{n-s-1}y^s f(x,y)\ (0 \le s < n), \quad x^{m-r-1}y^r g(x,y)\ (0 \le r < m)$$

are $m+n$ homogeneous polynomials of degree $m+n-1$, so their coefficients form a square matrix. Its determinant is denoted $R(f,g)$ and called the *resultant* of f and g. If f and g have a common factor (e.g. $(x-\lambda y)$), then this divides all the above polynomials, so these are linearly dependent, the matrix is singular, and $R(f,g) = 0$. In fact we have the following well-known results: see e.g. [18]1.4.4; see also [182] Sections 26–28 for this section.

Lemma 1.3.1 *For polynomials* $f(x,y) = c\prod_1^m(x - a_i y)$, $g(x,y) = c'\prod_1^n(x - b_j y)$ *in factorised form, with* $c, c' \neq 0$, *we have:*

(i) $R(f,g) = c^n c'^m \prod_{i,j}(a_i - b_j)$;
(ii) $R(g,f) = (-1)^{mn} R(f,g)$;
(iii) $R(f,gh) = R(f,g)R(f,h)$;
(iv) $R(f,g) = c^n \prod_1^m g(a_i, 1)$;
(v) *if* $\deg\phi = \deg f - \deg g$, $R(f + g\phi, g) = R(f,g)$;
(vi) $R(f,g) = 0$ *if and only if* f *and* g *have a common factor.*

We have used the fact that we are working over the field $\mathbb{C}$ to factorise our polynomials. But the definition and the formulae (ii), (iii), (v) are valid if the coefficients are taken in any commutative ring. The conclusion (vi) also holds, provided that we work over a unique factorisation domain. General results about unique factorisation in commutative rings may be found in elementary algebra texts, e.g. in [182] Section 19.

For a single homogeneous polynomial $f(x,y)$, we can form the resultant of $\partial f/\partial x$ and $\partial f/\partial y$: the result $D(f) = R(\partial f/\partial x, \partial f/\partial y)$ is called the *discriminant* of f. This also has important properties.

Lemma 1.3.2 *If* $f(x,y) = c\prod_1^m(x - a_i y)$, *with* $c \neq 0$, *we have*

(i) $D(f) = m^{m-2}c^{2m-2}\prod_{i\neq j}(a_i - a_j)$, *and hence*
(ii) $D(f) = 0$ *if and only if* f *has a repeated factor.*

Indeed since $mf = x\partial f/\partial x + y\partial f/\partial y$, we have

$$R(mf, \partial f/\partial x) = R(y\partial f/\partial y, \partial f/\partial x) = (-1)^{m-1}R(y, \partial f/\partial x)D(f),$$

and as $R(y, \partial f/\partial x) = mc$, this reduces to $(-1)^{m-1}mcD(f)$.

By substituting $y = 1$ we can regard these as results about polynomials in a single variable. Although this version is more familiar, there is a certain ambiguity since, for example, the function $f = bx^2 + cx + d$ may be regarded as a quadratic, or as a special case of a cubic $f = ax^3+bx^2+cx+d$ where the coefficient a happens to be 0. This ambiguity disappears if we insist, as we often will, that the coefficient of the highest power of x is 1. Such polynomials are called *monic*. Thus for monic polynomials, $D(f) = (-1)^{m-1}m^{m-2}R(f, df/dx)$.

We will sometimes be interested in the situation where the coefficients of f and g depend on a further parameter. Consider for example two homogeneous polynomials $f(x, y, z)$, $g(x, y, z)$ of respective degrees p and q. Substituting yt for y and zt for z makes them homogeneous in the two variables x and t; forming the resultant as above gives a homogeneous polynomial $P(y, z)$ of degree pq in y and z. The roots of $P(y, 1) = 0$ are those values y_0 of y for which the polynomials $f(x, y_0, 1)$, $g(x, y_0, 1)$ have a common root, and thus correspond to the intersections of the curves $f(x, y, z) = 0 = g(x, y, z)$. One proof of Bézout's theorem consists in counting these intersections carefully to see that indeed the multiplicities correspond to those of the roots of P.

1.4 Manifolds and the Implicit Function Theorem

One of the main objectives of the book is to explore the topology of plane curve singularities, so it will be necessary from Chapter 5 on to assume that the reader knows the rudiments of topology. There are numerous textbooks on this subject: for example the beginner's text [9] and the rather detailed exposition [169]. Although we need only elementary algebraic topology, the concept of manifold is important to us, and we now recall some basic facts.

A space X is an n-dimensional manifold if every point of X has a neighbourhood U_α such that there is a homeomorphism (a 'chart') $\phi_\alpha : U_\alpha \to V_\alpha$ where V_α is an open set in $\mathbb{R}^n$. We have coordinate transformations defined on the overlaps: if all of these are differentiable, more precisely, C^∞, then X is a differentiable manifold. If Y_1 and Y_2 are subsets of X which are differentiable manifolds, and the identity map of $Y_1 \cap Y_2$ is smooth in terms of charts of Y_1 on one side and of Y_2 on the other, then the collection of all the charts gives X the structure of differentiable manifold. This construction is known as glueing.

A differentiable n-manifold X has a well defined tangent space $T_P(X)$ at each point P, which is an n-dimensional vector space; there are several equivalent definitions. The notion of differentiable, or smooth, map is defined using charts and requiring differentiability in the local coordinate systems. A smooth map $f : X \to Y$ induces a linear map $T_P f : T_P(X) \to T_{f(P)}(Y)$ of corresponding tangent spaces. In local coordinates, if $f : \mathbb{R}^m \to \mathbb{R}^n$ is defined and differentiable at O, its partial derivatives $\partial f_i / \partial x_j$ form an $m \times n$ matrix, the *Jacobian matrix* of f, which we denote Df_O, which is the matrix of the map $T_O f$.

A bijection $f : X \to Y$ such that both f and f^{-1} are smooth maps is called a diffeomorphism of X to Y. Diffeomorphism is the basic equivalence relation between smooth manifolds.

A smooth map $f : X \to Y$ is called a smooth embedding if it is injective and, for each $P \in X$, the induced map $T_P(X) \to T_{f(P)}(Y)$ is also injective (strictly speaking, if X is non-compact one adds a further requirement to ensure that at each point of $f(X)$ there is a chart of Y in which X corresponds to a linear subspace). A smooth embedding $S^1 \to S^3$ is called a (smooth) knot. A link is a finite collection of knots with disjoint images, so can be taken as a smooth embedding $A \times S^1 \to S^3$ with A a finite set.

A manifold with boundary is defined in the same way as a manifold except that charts may map to open subsets of $\mathbb{R}^n_+ := \{(x_1, \ldots, x_n) \in \mathbb{R}^n \mid x_n \geq 0\}$. The boundary is the part corresponding in these charts to the subset where $x_n = 0$: it is an $(n-1)$ dimensional manifold. All the above extend naturally to this case.

The key to discussing changes of coordinates is the Inverse Function Theorem, which is proved in many textbooks, e.g. [52].

Theorem 1.4.1 *Let U be a neighbourhood of $O \in \mathbb{R}^n$; let $f : U \to \mathbb{R}^n$ be differentiable, and suppose $T_O f$ an isomorphism. Then there is a neighbourhood $U_1 \subset U$ of O such that $f|_{U_1}$ is a bijection of U_1 with a neighbourhood V_1 of $f(O)$ and its inverse $f^{-1} : V_1 \to U_1$ is again differentiable.*

An important application is the Implicit Function Theorem, which gives a first general result for proceeding from a subset of Euclidean space defined by equations to one given by a parametrisation.

Theorem 1.4.2 *Let $f_j(x_1, \dots, x_m, y_1, \dots, y_n)$ $(1 \leq j \leq n)$ be smooth functions of the variables indicated, such that at the origin $O = (0, \dots, 0)$ we have each $f_j(O) = 0$ and the matrix $J_y f := \left(\frac{\partial f_j}{\partial y_k}(O)\right)$ is non-singular. Then, in some neighbourhood of O, there are unique functions $h_j(x_1, \dots, x_m)$ with $h_j(0, \dots, 0) = 0$ such that $f_j = 0$ for each j if and only if $y_j = h_j(x)$ for each j.*

In particular, if $f(x, y)$ is such that $f(0,0) = 0$ and $\frac{\partial f}{\partial y}(0,0) \neq 0$, there is a unique function $h(x)$ with $h(0) = 0$ such that, in some neighbourhood of $(0,0)$, $f(x, y) = 0$ if and only if $y = h(x)$.

Proof Define $F : \mathbb{R}^{m+n} \to \mathbb{R}^{m+n}$ (in some neighbourhood of O) to have coordinates $x_1, \dots, x_m, f_1, \dots, f_n$. The Jacobian matrix of F can be expressed in block form $\begin{pmatrix} I & J_x(f) \\ 0 & J_y(f) \end{pmatrix}$, so is non-singular if and only if $J_y(f)$ is which, by hypothesis, holds at O.

By the Inverse Function Theorem, F has an inverse function G in a (smaller) neighbourhood. Since F preserves the coordinates x_i, we can write

$$G(x_1, \dots, x_m, z_1, \dots, z_n) = (x_1, \dots, x_m, g_1(x, z), \dots, g_n(x, z)).$$

The functions $h_j(x_1, \dots, x_m) = g_j(x_1, \dots, x_m, 0, \dots, 0)$ have the desired property. □

There is also a complex analytic version of the notion of manifold. We can identify $\mathbb{C}^n$ with $\mathbb{R}^{2n}$: if $(z_1, \dots, z_n)$ are complex coordinates in the former, write $z_r = x_r + iy_r$ and take the x_r and y_r as real coordinates. If U is an open subset of $\mathbb{C}^m$, and $f : U \to \mathbb{C}^n$ a differentiable map, f is called holomorphic if, at each $P \in U$, the linear map $T_P f$ is a linear map of the $\mathbb{C}$-vector spaces $\mathbb{C}^m \to \mathbb{C}^n$. The notion of complex n-manifold is now defined as above, with charts mapping to open sets in $\mathbb{C}^n$ and coordinate transformations required to be holomorphic. We have already given the example of projective space $P^n(\mathbb{C})$ with charts U_r $(0 \leq r \leq n)$.

The Inverse Function Theorem in the complex case is formally the same as the real version, of which it is an immediate consequence, since if $T_P f$ is an isomorphism for some P, we know that f has a local inverse f^{-1}, and this inverse is holomorphic, since each $T_{f(P)} f^{-1}$ is the inverse map to $T_P f$ and hence is complex linear. As in the real case, the Implicit Function Theorem follows at once.

The Inverse Function Theorem also gives us facility in changing local coordinates. Given holomorphic functions $f(x,y)$, $g(x,y)$ both defined near O, then provided the Jacobian determinant

$$\frac{\partial(f,g)}{\partial(x,y)} := \frac{\partial f}{\partial x}\frac{\partial g}{\partial y} - \frac{\partial f}{\partial y}\frac{\partial g}{\partial x}$$

does not vanish at O, the map $(x,y) \to (f(x,y), g(x,y))$ gives a bijection of a neighbourhood of O in $\mathbb{C}^2$ with another such neighbourhood, whose inverse is also holomorphic. We may thus also take (f,g) as holomorphic coordinates on a neighbourhood of O in $\mathbb{C}^2$.

Another way to construct functions which we will find useful is the following.

Lemma 1.4.3 *Suppose f a smooth function, defined on a neighbourhood of $O \in \mathbb{R}^n$, which vanishes on the linear subspace where $x_1 = \ldots = x_k = 0$. Then there are smooth functions $g_1, \ldots, g_k$, defined on a (perhaps smaller) neighbourhood of $O \in \mathbb{R}^n$, such that $f(x_1, \ldots, x_n) = \sum_1^k x_i g_i(x_1, \ldots, x_n)$.*

The same applies if we replace $\mathbb{R}^n$ by $\mathbb{C}^n$ and 'smooth' by 'holomorphic'.

Proof Consider the behaviour of f at a point running along the straight line segment from $(0, \ldots, 0, x_{k+1}, \ldots, x_n)$ to $(x_1, \ldots, x_n)$. Since f vanishes at the former point, we may write

$$\begin{aligned} f(x_1, \ldots, x_n) &= \int_0^1 \frac{d}{dt} f(tx_1, \ldots, tx_k, x_{k+1}, \ldots, x_n)dt \\ &= \int_0^1 \sum_1^k x_i \frac{\partial f}{\partial x_i}(tx_1, \ldots, tx_k, x_{k+1}, \ldots, x_n)dt. \end{aligned}$$

Thus we can take $g_i(x_1, \ldots, x_n) := \int_0^1 \frac{\partial f}{\partial x_i}(tx_1, \ldots, tx_k, x_{k+1}, \ldots, x_n)dt$. The argument works in both real and complex cases. □

1.5 Polar curves and inflections

If $f(x,y,x) = 0$ is a homogeneous equation of degree d, defining a curve C in the projective plane $P^2(\mathbb{C})$, its tangent at a (non-singular) point $P_0 \in C$ with coordinates (x_0, y_0, z_0) is given by the equation $\frac{\partial f}{\partial x}(P_0)x + \frac{\partial f}{\partial y}(P_0)y + \frac{\partial f}{\partial x}(P_0)z = 0$. For this is a linear equation; if we substitute $(x,y,z) = (x_0, y_0, z_0)$ it is satisfied since by Euler's theorem the left hand side gives $df(P_0)$, which vanishes by hypothesis; and it has the correct slope, as we can see by taking affine coordinates $z = 1$ and regarding x as a parameter, since then we have $\partial f/\partial x + (\partial f/\partial y)dy/dx = 0$.

Changing our viewpoint, we can think of the point (x, y, z) as fixed and the point P_0 as variable. The equation is homogeneous of degree $d-1$ in the coordinates of P_0, and the above discussion shows that it is satisfied at any point P_0 of C such that the tangent to C at P_0 passes through P. The locus of points P_0 satisfying this equation is called the *polar curve* of the point P with respect to the curve C. We often just say that the locus is a polar of C.

The above definition is global, but the concept is of use more generally. If C is a curve-germ at $O \in \mathbb{C}^2$ defined by an equation $f(x, y) = 0$, we define a polar curve of C to be any curve with equation $P\frac{\partial f}{\partial x} + Q\frac{\partial f}{\partial y} + Rf = 0$, where P and Q are functions at least one of which is non-vanishing at the point O.

This definition is invariant under multiplying the equation either of f or of the polar curve by a function non-vanishing at O. It is also invariant under holomorphic coordinate change, for if $(u(x, y), v(x, y))$ represents such a change, $P\frac{\partial f}{\partial x} + Q\frac{\partial f}{\partial y} + Rf = P_1\frac{\partial f}{\partial u} + Q_1\frac{\partial f}{\partial v} + Rf$, where $P_1 = P\frac{\partial u}{\partial x} + Q\frac{\partial u}{\partial y}$ and $Q_1 = P\frac{\partial v}{\partial x} + Q\frac{\partial v}{\partial y}$.

We may however normalise.

Lemma 1.5.1 *For any polar curve* $P\frac{\partial f}{\partial x} + Q\frac{\partial f}{\partial y} + Rf = 0$ *of* $f = 0$, *there exist holomorphic germs* g, u, v, *with* $u(O) \neq 0$ *and* $v(O) \neq 0$ *such that*

$$P\frac{\partial f}{\partial x} + Q\frac{\partial f}{\partial y} + Rf \equiv v\left\{\frac{\partial(uf)}{\partial x}\frac{\partial g}{\partial y} - \frac{\partial(uf)}{\partial y}\frac{\partial g}{\partial x}\right\}.$$

Proof Equating coefficients, we see that we need to solve

$$P = vu\frac{\partial g}{\partial y}, \quad Q = vu\frac{\partial g}{\partial x}, \quad R = \frac{\partial(u, g)}{\partial(x, y)}$$

for g, u and v. The key is the integration theorem for vector fields (see Theorem 5.1.1), applied to the vector field $\xi := Q\partial/\partial y - P\partial/\partial x$.

Since P and Q do not both vanish at O we may suppose that $Q(O) \neq 0$. We first integrate $\xi(g) = 0$, with the initial condition $g(x, 0) = x$ (note that this prescribes g along the x-axis, which is transverse to ξ). This gives a unique function-germ g, and $\partial g/\partial x$ takes the value 1 at O, so is non-zero in a neighbourhood.

Next we integrate $\xi(u) = -QR/(\partial g/\partial x)$, with the initial condition $u(x, 0) = 1$. This gives a function u with $u(O) = 1$ and $\frac{\partial(u,g)}{\partial(x,y)} = R$. Finally define $v := Q/(u\partial g/\partial x)$, and all the desired equations hold. □

Thus, writing $f' = uf$, we may take C to be given by $f' = 0$ and the polar curve by $0 = \frac{\partial(f',g)}{\partial(x,y)}$ for some function g whose first derivative does not vanish at O. We may then take new local coordinates $(x', y') = (g(x, y), y)$, and in these the polar becomes just $\partial f'/\partial y = 0$.

Each polar curve is associated to a direction at the point O. In the projective case, this is the line joining our point to $(x_0 : y_0 : z_0)$; in the general case it is $Q(O)x = P(O)y$; and for $\partial(f, g)/\partial(x, y) = 0$ we have the tangent at O to $g(x, y) = g(O)$. The polar is called a *transverse polar* if this line is transverse to C at O.

A point of inflection on a curve C is usually characterised by the condition $\partial^2 y/\partial x^2 = 0$. If C is given implicitly by $f(x, y) = 0$, a short calculation gives

$$\partial^2 y/\partial x^2 = (f_y)^{-3} \det \begin{pmatrix} f_{xx} & f_{xy} & f_x \\ f_{yx} & f_{yy} & f_y \\ f_x & f_y & 0 \end{pmatrix}.$$

For a curve C in $P^2(\mathbb{C})$ given by a homogeneous equation $f(x, y, z) = 0$ of degree d we define the *Hessian* $H(f)$ to be the determinant of the (Hessian) matrix $\begin{pmatrix} f_{xx} & f_{xy} & f_{xz} \\ f_{yx} & f_{yy} & f_{yz} \\ f_{zx} & f_{zy} & f_{zz} \end{pmatrix}$. If we multiply the final row of this matrix by z, and add x times the first row and y times the second, the row reduces (using Euler's formula for homogeneous functions) to $(d-1)(f_x, f_y, f_z)$. Operating in the same way with the columns now gives $z^2 H(f) = (d-1)^2 \det \begin{pmatrix} f_{xx} & f_{xy} & f_x \\ f_{yx} & f_{yy} & f_y \\ f_x & f_y & \frac{d}{d-1} f \end{pmatrix}$. At points where $z \neq 0$, we can take affine coordinates $z = 1$, and so see that the Hessian curve $H(f) = 0$ intersects C at exactly the points of inflection on C.

2

Puiseux' Theorem

The theorem of Puiseux states that a polynomial equation $f(x, y) = 0$ has a solution in which y is expressed as a power series in fractional powers of x. In this chapter we will give several versions of this theorem, of increasing sharpness. In the first section we present the classical algorithm for calculating the successive terms in the power series, and show that this does yield a solution. However, to obtain a convergent power series requires more work, and in the second section we give a different approach giving an introduction to the geometry of the situation and an existence proof for convergent power series solutions.

The next short section collects the results describing the relations between curves, their branches, tangents and multiplicities, which are basic for later chapters.

The fourth section establishes some basic properties of the rings of power series, in particular that they are unique factorisation domains, and deduces that the solutions obtained in the preceding sections must all be the same.

2.1 Solution in power series

We want to solve a polynomial equation $f(x, y) = 0$. There are several ways to find a solution for y in terms of x, but we begin with one which gives an effective method of calculation. For this, it will make no difference if we allow f to be a formal power series. The basis of the method of proof goes back to Newton [142].

Theorem 2.1.1 *Any equation $f(x, y) = 0$, where f is a polynomial with $f(O) = 0$ or more generally $f \in \mathbb{C}[[x, y]]$ with zero constant term, admits* *at least one solution in formal power series of the form*

$$x = t^n, \quad y = \sum_1^\infty a_r t^r, \quad (\text{some } n \in \mathbb{N}).$$

Proof The basic idea for constructing a power series to solve a problem is always to solve for one term at a time. Often – as here – the hardest part is knowing where to start. So we try writing

$$y = c_0 x^\alpha + \text{terms of higher order } (\alpha \in \mathbb{Q}).$$

Now substitute for y in $f(x, y)$. Each term $a_{r,s}x^r y^s$ in f contributes $a_{r,s}c_0^s x^{r+s\alpha}$, plus terms of higher order. When we add these, the terms of least order $r + s\alpha$ in x will have to cancel.

To see what this entails, we consider a (real) plane with coordinates (r, s) and mark those points (r, s) for which the coefficient $a_{r,s}$ is non-zero. The lines $r + s\alpha = C$ (where C is constant) are all of the same slope. Start with $C = 0$ and steadily increase C until the first time the corresponding line goes through one of the points we have marked. In order for the terms in x^C to cancel, there must be at least two of them. So this line must pass through two marked points. This leads us to the following definition.

The *Newton diagram* of f is the convex hull of the regions above and to the right of the marked points (r, s) (that is, points with $a_{r,s} \neq 0$). Its boundary is made up of straight line segments. The union of those segments which do not lie on the coordinate axes is the *Newton polygon* of f.

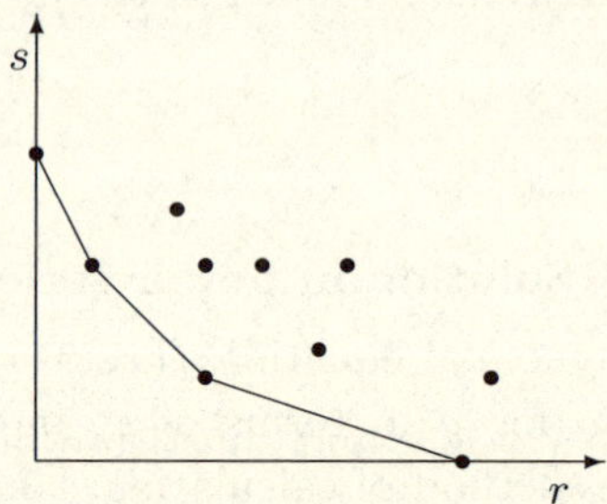

Fig. 2.1. A Newton polygon

It follows from the above discussion that we need to choose a line $r + s\alpha = C$ lying along an edge of this polygon. We can write the coordinates of the two points at the ends of this edge as (r_0, s_0) and (r_0+ka, s_0-kb) where a, b and k are positive and a and b are coprime, so that the points on the edge with integer coordinates are the (r_0+la, s_0-lb) with $0 \leq l \leq k$. When we make the above substitution in f, no powers of x lower than x^C appear, and the coefficient of x^C is

$$\sum_{l=0}^{k} a_{r_0+la,s_0-lb} c_0^{s_0-lb}.$$

To simplify the notation let us set

$$u_l = a_{r_0+la,s_0-lb} \quad \text{for } 0 \leq l \leq k,$$

and

$$\phi(T) = \sum_{l=0}^{k} u_l T^{k-l}.$$

Then the coefficient of x^C may be written as $c_0^{s_0-kb}\phi(c_0^b)$. We have chosen notation so that $u_0 \neq 0, u_k \neq 0$.

With these preparations made, we are ready to start the constructive part of the proof. First we check whether f is divisible by x: if so, then of course one solution is parametrised by $(x, y) = (0, t)$. In general, write f as a power of x multiplied by f', where f' is not divisible by x, and from now on deal only with f'. (It is necessary to discuss this point in the argument since it will appear at each step of the inductive procedure below.)

It follows that the Newton polygon of f intersects the s-axis; the point of intersection is determined by the order of $f(0, y)$, which we denote from now on by m_0.

We choose an edge of the Newton polygon, and a solution T_0 of the equation $\phi(T) = 0$, and set $c_0 = T_0^{1/b}$. Since ϕ has non-zero constant term, T_0 and c_0 cannot be zero. We take $y = c_0 x^{a/b}$ as a first approximation to a solution of $f(x, y) = 0$, and will show that there is a solution of the desired kind with this as first term.

Set $x = x_1^b$, $y = x_1^a(c_0 + y_1)$, and substitute in $f(x, y) = 0$. By the above choices,

$$f(x_1^b, x_1^a(c_0 + y_1))$$

is divisible by $x_1^{br_0+as_0}$: dividing by this gives $f_1(x_1, y_1)$, say. Now repeat the above procedure, but replacing f, x, y by f_1, x_1, y_1. We write m_1 for the order of $f_1(0, y_1)$. Note that $f_1(0, y_1)$ is obtained from the terms on the edge of the previous Newton polygon by dividing by $x_1^{br_0+as_0}$, and hence is given by

$$(c_0 + y_1)^{s_0 - kb}\phi((c_0 + y_1)^b).$$

Since ϕ has degree k, we deduce that $m_1 \leq kb \leq m_0$.

Continually repeating this procedure leads after r steps to an expression of the form

$$Y_r := x^{\frac{a_0}{b_0}}(c_0 + x^{\frac{a_1}{b_0 b_1}}(c_1 + x^{\frac{a_2}{b_0 b_1 b_2}}(c_2 + \ldots (c_r + x^{\frac{a_r}{b_0 b_1 \cdots b_r}})\ldots))), \quad (2.1)$$

which multiplies out to an expansion in increasing fractional powers of x. Observe that Y_r differs from Y_{r-1} only in coefficients of $x^{\frac{a_r}{b_0 b_1 \cdots b_r}}$ and higher terms. Also, if the lowest non-vanishing coefficient in $f(x, Y_{r-1})$ is the coefficient of $x^{M_{r-1}}$, then since the procedure amounts to killing this coefficient, we have $M_r > M_{r-1}$.

It remains to prove – and this is a key point – that the denominators $b_0 b_1 \ldots b_r$ are bounded, or equivalently, that $b_r = 1$ for all large enough r. To see that this concern is not without foundation, consider the expression $R(x) = \sum_{k=0}^{\infty} x^{2-2^{-k}}$: if our approximations behaved like this, it would not be clear how to deal with powers of x higher than the second. We will show

Lemma 2.1.2 *If $b_r > 1$, then $m_r > m_{r+1}$.*

Since m_r is a monotone decreasing sequence of positive integers, it must be constant from some point onwards, so the lemma will imply that $b_r = 1$ from some point onwards. Thus all the Y_r are power series in $x_d = x^{1/d}$ for some fixed d, and since they agree up to increasingly high powers of x_d, they converge in the $\mathfrak{m}$-adic sense to a power series Y_∞. The numbers dM_r are also integers, forming a strictly increasing sequence, and so tending to infinity. Thus $f(x, Y_\infty) = 0$, so we have an exact solution, completing the proof of the theorem. □

Proof [of the lemma] Since we repeated the same procedure at each step, it suffices to consider the first step. Thus suppose $m_1 = m_0$. Then the least power of y_1 in

$$(c_0 + y_1)^{m_0 - kb}\phi((c_0 + y_1)^b),$$

is that of degree m_0. This polynomial is thus a constant multiple of $y_1^{m_0}$. Now $c_0 \neq 0$, so we must have $m_0 = kb$, so in fact the Newton polygon has only one edge. Substitute $z = y_1 + c_0$. This produces $(z - c_0)^{bk}$ equal to a polynomial in z^b. Since the coefficient of z^1 in $(z-c_0)^{bk}$ is non-zero, it follows that $b=1$. □

The final argument fails if, instead of working over $\mathbb{C}$, we are working over a field of prime characteristic p, for then the crucial coefficient may well vanish. And the main result also fails. Indeed, consider the function $R(x)=\sum_{k=0}^{\infty} x^{2-2^{-k}}$ cited earlier as an example. Over a field of characteristic 2, we have $R(x)^2=\sum_{k=0}^{\infty} x^{4-2^{1-k}}$; substituting $k=j+1$ gives

$$R(x)^2 = x^2 + \sum_{j=0}^{\infty} x^{4-2^{-j}} = x^2 + x^2 R(x),$$

so that $y=R(x)$ gives a parametrisation of the curve $y^2+x^2y+x^2=0$.

We illustrate the effective nature of the proof with one simple example (further examples may be found in [23]pp 501–5).

Example 2.1.1 Consider the equation

$$y^4 - 2y^2x^3 - 4yx^5 + x^6 - x^7 = 0.$$

The Newton polygon consists of a single edge joining $(6,0)$ to $(0,4)$, and the terms on it factorise as $(y^2-x^3)^2$. We thus set $x=x_1^2, y=x_1^3(1+y_1)$. When we substitute, the resulting equation has a factor x_1^{12}. Removing this, the result simplifies to

$$y_1^4 + 4y_1^3 + 4y_1^2 - 4y_1x_1 - 4x_1 - x_1^2.$$

Again the Newton polygon has a single edge, and the terms on it are $4(y_1^2-x_1)$. We thus next substitute $x_1=x_2^2, y_1=x_2(1+y_2)$ in the equation, and again simplify. The result has a factor x_2^2, but also has a factor y_2. Thus one solution is given by $y_2=0$, so $y_1=x_2$ and $y=x_2^6+x_2^7$, with $x=x_2^4$.

2.2 Convergent power series

In this section, we will show that if the equation $f(x,y)=0$ in Theorem 2.1.1 is holomorphic (e.g. if it is polynomial), then the solutions obtained are all convergent power series. There are several possible routes to achieve this. First, it is possible to give a direct verification that the series converge. Second, we may begin with the observation that if the order m of $f(0,y)$ is equal to 1, then the solution obtained in Theorem 2.1.1 is holomorphic, in virtue of the Implicit Function

Theorem 1.4.2, which applies since the hypothesis $m = 1$ is equivalent to the non-vanishing of $\partial f/\partial y$ at O. Since the proof of Theorem 2.1.1 proceeds by reducing eventually to this case, we may use this and then seek to reverse the argument. See the Exercise 2.6.6 for hints on this proof. We now, however, give a direct proof of convergence in one simple case.

Lemma 2.2.1 *If $f \in \mathbb{C}\{x\}$ has order n, there exists $g \in \mathbb{C}\{x\}$ with $g^n = f$.*

Proof Let the lowest order term in f be ax^n. Write $f = ax^n(1+E)$, and choose $b \in \mathbb{C}$ with $b^n = a$. We expand $E = \sum_1^\infty E_r x^r$, and substitute in the binomial expansion

$$\begin{aligned}(1+E)^{n^{-1}} &= 1 + n^{-1}E + \frac{n^{-1}(n^{-1}-1)}{1.2}E^2 \\ &+ \frac{n^{-1}(n^{-1}-1)(n^{-1}-2)}{1.2.3}E^3 + \cdots .\end{aligned}$$

Since there are only finitely many terms contributing to the coefficient of any particular power x^r, we obtain a well defined power series $G(x)$.

To prove convergence, use the fact that since $E \in \mathbb{C}\{x\}$, there exists $R' > 0$ such that the terms $|E_r|R'^r$ are bounded. We can thus find a number $R > 0$ such that $\sum_1^\infty |E_r|R^r < 1$. Since the binomial series has radius of convergence equal to 1, it follows that for $|x| < R$ the series obtained by substitution converges absolutely, so all rearrangements are permissible. Hence the power series $G(x)$ converges for $|x| < R$, and we can set $g := bG$. □

For the general case, we present a more geometrical approach, which is independent of the preceding. It is not constructive, so complements rather than replacing the arguments of Section 2.1. We need to prepare for this with an important result which reduces many arguments about holomorphic functions to results about polynomials.

Geometrically, we are thinking of projecting the curve $f(x, y) = 0$ on the x-axis. Above the origin $x = 0$ we have the part $f(0, y) = 0$ of the curve. We recall that (unless $f(0, y) \equiv 0$) we may write $f(0, y) = y^m A(y)$ where $A(0) \neq 0$ so that $\frac{1}{A(y)}$ is also holomorphic. In this situation we say that $f(x, y)$ is *regular in y of order m.*

Theorem 2.2.2 [Weierstrass Preparation Theorem] *Let $G \in \mathbb{C}\{x, y\}$ be regular of order s in y. Then there exist $U \in \mathbb{C}\{x, y\}$ with non-zero* *constant term and* $A_r \in C\{x\}$ *for* $0 \le r < s$ *such that*

$$G(x,y) = U(x,y)\left\{y^s + \sum_{r=0}^{s-1} A_r(x)y^r\right\}.$$

The proof we give depends on first establishing a type of division algorithm. It is convenient, in fact, to divide by a universal polynomial $P_s(y,\lambda) = y^s + \sum_1^s \lambda_i y^{s-i}$, where the coefficients λ_i are further independent variables.

Theorem 2.2.3 [Division Theorem] *If* $F(x,y,\lambda)$ *is holomorphic at* $(0,0,\mathbf{0})$, *there are unique holomorphic functions* $Q(x,y,\lambda)$, $A_i(x,\lambda)$ *for* $1 \le i \le s$, *such that*

$$F(x,y,\lambda) = P_s(y,\lambda)Q(x,y,\lambda) + R(x,y,\lambda),$$

where $R(x,y,\lambda) = \sum_1^s A_i(x,\lambda)y^{s-i}$.

Proof A straightforward calculation yields the identity

$$P_s(z,\lambda) - P_s(y,\lambda) = (z-y)\left(\sum_1^s P_{i-1}(z,\lambda)y^{s-i}\right),$$

and hence

$$\frac{1}{z-y} = \frac{P_s(y,\lambda)}{(z-y)P_s(z,\lambda)} + \sum_1^s \frac{P_{i-1}(z,\lambda)}{P_s(z,\lambda)}y^{s-i}.$$

We are concerned only with small values of y. If λ is taken to be small, the roots of $P_s(*,\lambda) = 0$ are also small, so we may choose a loop round 0 containing them all in its interior. Multiply by $F(x,z,\lambda)$ and integrate round this loop. Since, by Cauchy's Residue Theorem,

$$F(x,y,\lambda) = \frac{1}{2\pi i}\oint \frac{F(x,z,\lambda)}{z-y}dz,$$

we obtain an equation of the desired form with

$$Q(x,y,\lambda) = \frac{1}{2\pi i}\oint \frac{F(x,z,\lambda)}{(z-y)P_s(z,\lambda)}dz,$$

$$A_i(x,\lambda) = \frac{1}{2\pi i}\oint \frac{P_{i-1}(z,\lambda)F(x,z,\lambda)}{P_s(z,\lambda)}dz.$$

Uniqueness is immediate, since P_s cannot divide a non-zero polynomial of degree less than s. □

Proof [of Weierstrass Preparation Theorem]

Since $G(x,y)$ is regular of order s, the coefficient of y^s is $c \neq 0$. We apply the Division Theorem to $F(x,y,\lambda) = G(x,y)$. Setting $x = 0$ then yields

$$G(0,y) = P_s(y,\lambda)Q(0,y,\lambda) + \sum_1^s A_i(0,\lambda)y^{s-i}.$$

First set $\lambda = \mathbf{0}$, giving $G(0,y) = y^s Q(0,0,\mathbf{0}) + \sum_1^s A_i(0,\mathbf{0})y^{s-i}$. It follows by comparing powers of y^{s-i} for $i \geq 0$ that $Q(0,0,\mathbf{0}) = c$ and $A_i(0,\mathbf{0}) = 0$. Next, differentiate both sides with respect to λ_j and then again set $\lambda = \mathbf{0}$. We obtain

$$0 = y^{s-j}Q(0,y,\mathbf{0}) + y^s \frac{\partial Q}{\partial \lambda_j}(0,y,\mathbf{0}) + \sum_1^s \frac{\partial A_i}{\partial \lambda_j}(0,\mathbf{0})y^{s-i}.$$

Comparing coefficients of powers of y leads here to

$$\partial A_i/\partial \lambda_j = 0 \text{ if } j < i \text{ and } = -c \text{ if } j = i.$$

Since the entries $\{\partial A_i/\partial \lambda_j \mid 1 \leq i,j \leq n\}$ define a non-singular matrix, we may apply the Inverse Function Theorem 1.4.2 in a neighbourhood of $(0,\mathbf{0})$ to obtain (uniquely determined) holomorphic functions $\lambda_j = H_j(x)$ such that $A_i(x,\mathbf{H}(x)) = 0$ for each i. Now substituting $H_j(x)$ for λ_j in the equation given by the Division Theorem gives

$$G(x,y) = Q(x,y,\mathbf{H}(x))P_s(y,\mathbf{H}(x)),$$

which is of the desired form since $Q(0,0,\mathbf{0}) = c \neq 0$, while $P_s(y,\mathbf{H}(x)) = y^s + \sum_1^s H_i(x)y^{s-i}$. □

We note for later reference that these two results do allow us to perform division in essentially the usual way.

Corollary 2.2.4 *Let $F, G \in \mathbb{C}\{x,y\}$ with G regular of order s in y. Then we can write*

$$F(x,y) = G(x,y)D(x,y) + \sum_1^s C_i(y)x^{s-i}.$$

Proof By Theorem 2.2.2 we may write

$$G(x,y) = U(x,y)\left\{y^s + \sum_{r=0}^{s-1} A_r(x)y^r\right\},$$

with $U(0,0) \neq 0$, and from Theorem 2.2.3 (with the roles of x and y interchanged) we obtain

$$F(x,y) = P_s(y,\lambda)Q(x,y,\lambda) + \sum_1^s B_i(x,\lambda)y^{s-i}.$$

Now substituting $\lambda_i = A_{s-r}(x)$ in $P_s(x,\lambda)$ produces $U(x,y)^{-1}G(x,y)$, so the result follows with $D(x,y) = U(x,y)^{-1}Q(x,y,A(x))$ and $C_i(y) = B_i(x,A(x))$. □

The Weierstrass Preparation Theorem is the key to proving properties of $\mathbb{C}\{x,y\}$ considered as an abstract ring. One says, for short, that an element of a (commutative) ring is a *unit* if it has a multiplicative inverse. Since the function $U(x,y)$ of Theorem 2.2.2 has non-zero constant term, it follows that U is a unit in $\mathbb{C}\{x,y\}$. We next establish unique factorisation.

Theorem 2.2.5 *The ring $\mathbb{C}\{x,y\}$ is a unique factorisation domain.*

Proof The two essential assertions are that any element can be expressed as a product of factors that are *irreducible*, i.e. cannot be non-trivially factorised (a factorisation with one factor a unit is trivial), and that if such an irreducible element divides a product, it must divide one (at least) of the factors: 'irreducibles are prime'.

The 1-variable case is trivial: for $\mathbb{C}\{x\}$, the only primes are the products of x by a unit. For the case of polynomials, unique factorisation follows inductively from Gauss' lemma, which states that if A is a unique factorisation domain, so is $A[t]$. Thus $\mathbb{C}\{x\}[y]$ is a unique factorisation domain.

Now for any $f \in \mathbb{C}\{x,y\}$, suppose its image $\overline{f}$ in $\mathbb{C}\{y\}$ of order s, and write $f = Ug$ as in Theorem 2.2.2. Then g belongs to the unique factorisation domain $\mathbb{C}\{x\}[y]$, so can be expressed as a product of irreducible factors g_i. It follows (adjusting by constant factors if necessary) that $\overline{g}_i = y^{s_i}$ with $\sum s_i = s$.

Each g_i is irreducible in $\mathbb{C}\{x,y\}$. For if, say, $g_i = hk$ then applying the preparation theorem to h and to k shows that we can write $h = U_h h'$, $k = U_k k'$ with U_h, U_k units and h', k' monic polynomials in $\mathbb{C}\{x\}[y]$. Now $g_i = (U_h U_k)h'k'$, and by the uniqueness clause of the theorem, it follows that $g_i = h'k'$. But unless h' or k' is a constant, this contradicts the fact that g_i is irreducible in $\mathbb{C}\{x\}[y]$.

Finally we show that any g which is irreducible in $\mathbb{C}\{x, y\}$ is prime. Suppose g divides $h_1 h_2$. As above we may take g as an element of $\mathbb{C}\{x\}[y]$ of degree s with $\overline{g} = y^s$. By Corollary 2.2.4, we can write

$$h_1 = gQ_1 + R_1, \quad h_2 = gQ_2 + R_2,$$

with R_1, R_2 polynomials of degree $< s$. Now since g divides $h_1 h_2$, it divides $R_1 R_2$, which belongs to $\mathbb{C}\{x\}[y]$. Since this is a unique factorisation ring, and g is irreducible in it, g is also prime; since it divides the product, it divides one of the factors. But if g divides R_i, it also divides h_i. □

We begin our second approach to Puiseux' Theorem by supposing $f(x, y)$ regular in y of order m and applying the Preparation Theorem 2.2.2 to express $f(x, y)$ as $U(x, y)A(x, y)$ as above. Since $U(0, 0) \neq 0$, there exists a neighbourhood of O on which U does not vanish. We can thus find a number $\epsilon > 0$ such that, whenever $|x| < \epsilon$, $|y| < \epsilon$, all the above series are convergent, the above formula is true when evaluated at (x, y), and $U(x, y) \neq 0$. Thus the zero locus of $f(x, y)$ in this neighbourhood is the same as that of $A(x, y)$.

The function $A(x, y)$ is a polynomial in y, so has a discriminant $D(A) \in \mathbb{C}\{x\}$. If $D(A)$ is identically zero, A has a repeated factor: this follows from Lemma 1.3.2 since $\mathbb{C}\{x, y\}$ is a unique factorisation domain. Apply the euclidean algorithm to find the highest common factor h of A and $\partial A/\partial y$. Then A/h has the same set of zeros and no repeated factor. Thus we may replace A by A/h and for the new A, $D(A)$ is not identically zero. Now choosing ϵ smaller (if necessary), we may suppose that $D(A) \neq 0$ for $0 < |x| < 2\epsilon$.

Thus for each value of x in the region $0 < |x| < 2\epsilon$ we have a polynomial equation $A(x, y) = 0$ for y with m distinct roots $y_1(x), \ldots,$ $y_m(x)$, say. At any one of the points $(x, y_r(x))$ we have $A(x, y) = 0$ and $D(A)(x) \neq 0$, and hence $\partial A/\partial y \neq 0$. We may thus apply the Implicit Function Theorem to infer that in a neighbourhood of such a point there exists a holomorphic function $y = y_i(x)$ such that $y = y_i(x)$ if and only if $A(x, y) = 0$.

Each function y_i is well defined on a small neighbourhood of the originally chosen point x, and so if we move x along a path in the region $0 < |x| < 2\epsilon$ we can continue to make sense of $y_i(x)$ by insisting that it varies continuously. If the path returns to its starting point, we must certainly end up with a value of y such that $A(x, y) = 0$, but this need not be the same value as y_i. But since it does give a solution to the equation, it must be y_j for some j with $1 \leq j \leq m$.

Suppose we start at the point $x = \epsilon$, and define functions y_i on a neighbourhood of this point. It follows from the above discussion that we may regard all these as well defined functions of x in the region obtained from $0 < |x| < 2\epsilon$ by cutting along the negative real axis. However, if we proceed round a path encircling the origin – say $x = \epsilon e^{i\theta}$ for $0 \leq \theta \leq 2\pi$ – then the y_i may return in a different order. So they are affected by some permutation σ, say: the solution $y = y_i(x)$ becomes $y = y_{\sigma(i)}(x)$.

The permutation σ of the finite set $\{1, \ldots, m\}$ can be decomposed as a product of disjoint cycles. Equivalently, by renumbering the objects permuted, we arrange that, say, $\sigma(1) = 2, \sigma(2) = 3, \ldots, \sigma(m_1 - 1) = m_1, \sigma(m_1) = 1$, a *cycle* of length m_1, and similarly for the remaining objects permuted, with cycles of lengths $m_1, m_2, \ldots$ adding up to m. In particular, for a path encircling the origin m_1 times, the function $y_1(x)$ returns to its original value. This is the same behaviour as for x^{1/m_1}. More precisely, define an m_1^{th} root z of x by starting at $x = \epsilon$ with z being real, and proceeding from there: we find that y_1 is a well-defined function of z in $0 < |z| < (2\epsilon)^{1/m_1}$.

By Laurent's Theorem, we can expand $y_1(x)$ as a Laurent series in z converging in this region. The coefficients of the Laurent expansion $y_1 = \sum_{r=-\infty}^{\infty} a_{1,r} z^r$ have integral representations

$$a_{1,r} = \frac{1}{2\pi i} \oint \frac{y_1(z)\, dz}{z^{r+1}},$$

and taking this over a circle of radius ϵ gives

$$|a_{1,r}| \leq \frac{1}{2\pi} . 2\pi\epsilon . \frac{\sup |y_1|}{\epsilon^{r+1}} = \epsilon^{-r} \sup |y_1|.$$

If $r \leq 0$, this tends to 0 as the radius of the circle does, since $y_1(z)$ tends to 0 with z. Thus $a_{1,r} = 0$ for $r < 0$ so the Laurent series is a power series, and we have a convergent power series expansion for y_1 in positive powers of $z = x^{1/m_1}$.

It follows from the way we have obtained this expansion that the corresponding expansion for $y_{\sigma^j(1)}$ is obtained from it by substituting $ze^{2\pi ij/m_1}$ for z.

We summarise these conclusions in the following.

Theorem 2.2.6

(i) *Any equation* $f(x, y) = 0$ *where* $f \in \mathbb{C}\{x, y\}$ *with* $f(O) = 0$, $f(0, y) \not\equiv 0$ *admits at least one solution of the form* $y = g(x^{1/m_1})$ *with* $g \in \mathbb{C}\{z\}$.

(ii) *If* f *is regular of order* m *in* y, *and we write* $f = UF$ *with* U *a unit and* F *a monic polynomial of degree* m *in* y, *there are* m

such solutions $g_j(x^{1/m_j})$, all distinct unless the discriminant of F vanishes identically, and

$$F(y) \equiv \prod_{j=1}^{m}(y - g_j(x^{1/m_j})).$$

We have seen that the m solutions of $f(x, y) = 0$ fall into groups corresponding to the orbits of σ: each such group admits a common parametrisation $(x, y) = (t^{m_j}, \sum_{r=1}^{\infty} a_{j,r}t^r)$. Such a group of solutions is called a *branch* of the curve $f(x, y) = 0$. We see from the parametrisation that geometrically this should be regarded as a single solution of the equation. It is simpler to describe the structure of a single branch than of a general curve, and for many topics considered below we will consider the case of a single branch first.

The Weierstrass preparation theorem allows us to give a detailed treatment of the simplest types of singularity.

Theorem 2.2.7 *Let C be a curve with equation $f(x, y) = 0$ with f of order 2. Then there exist coordinates in $\mathbb{C}^2$ in which C is given by $y = 0$ or $y^2 + x^k = 0$ for some $k \geq 2$.*

Proof Making a linear change of coordinates if necessary, we may suppose f regular in y of order 2. By Theorem 2.2.2, we can express f in the form $U(x, y)(y^2 + a(x)y + b(x))$ with $U(0, 0) \neq 0$, so C is given by $y^2 + a(x)y + b(x) = 0$. Making the coordinate change $y' = y + \frac{1}{2}a(x)$, we reduce to $y'^2 + b'(x) = 0$. Now either $b' = 0$ and C is given by $y = 0$, or b' has order k for some $k \geq 2$, so by Lemma 2.2.1 we have $b' = x'^k$ for some x' which is a convergent power series in x of order 1. It thus suffices to take coordinates (x', y'). □

For reasons which will appear later, a singularity which has equation $y^2 + x^{k+1} = 0$ in some local coordinates will be said to be of *type* A_k.

Although the concept of branch is the important one geometrically, there are occasions when it is convenient to consider the equation as having m distinct solutions. There are several ways to give a precise sense to this. Following the above line of thought, we may fix a value of x, say, real and positive (and small) and consider the m corresponding solutions for y. These can be expressed as series $\sum_{r=1}^{\infty} a_{j,r}x^{r/m_j}$ in fractional powers of x: such series are called *Puiseux series*. The fractional powers of x may be interpreted as those taking positive real values when

x is real and positive: thus the series can be regarded as defined in a sector $|\arg x| < \alpha$, $|x| < \delta$. We will refer to the distinct solutions obtained in this way as *pro-branches*. We will give a more careful discussion in Section 4.1.

2.3 Curves, branches, multiplicities and tangents

We begin by summarising informally the results we wish to establish in this section.

An equation $f(x, y) = 0$ determines a curve C; conversely each curve has an essentially unique equation.

A curve C has a unique decomposition as a finite union of branches.

A parametrisation $(x, y) = (\phi(t), \psi(t))$ determines a unique branch; conversely each branch has an essentially unique good parametrisation.

A curve has a well defined multiplicity, which is the sum of the multiplicities of its branches.

A branch has a unique tangent direction at O.

We proceed to details. By Theorem 2.2.5, the ring $\mathbb{C}\{x, y\}$ of holomorphic functions (each defined in a neighbourhood of O) is a unique factorisation domain. Its invertible elements, or units, are the series $U(x, y)$ with non-zero constant term, so that U does not vanish in a neighbourhood of O.

We now formalise our notion of curve. We say that a holomorphic function $f \in \mathbb{C}\{x, y\}$, with $f(0, 0) = 0$, defines a curve C. Thus on some neighbourhood of O, C consists of the points (x, y) with $f(x, y) = 0$. We shall regard $f = 0$ and $g = 0$ as the same curve only if $g = Uf$ for some unit U.

We can factorise f essentially uniquely as a finite product $f = \prod g_j^{a_j}$, where the factors g_j are distinct (so we cannot write $g_k = Ug_j$ for any unit U if $k \neq j$) and do not admit non-trivial factorisations, and the a_j are positive integers. Then the curves B_j defined by $g_j = 0$ are the branches of C. We may write $C = \sum a_j B_j$. The curve C is said to be *reduced* if each $a_j = 1$. We will usually suppose all curves occurring are reduced.

The curves $f = 0$ and $f^2 = 0$ have the same points. However, this is the only ambiguity of this type: if g_1 and g_2 are irreducible elements of $\mathbb{C}\{x, y\}$ which are not unit multiples of each other, then in some neighbourhood of O, O is the only common point of the corresponding branches B_1 and B_2. This justifies the above definition of curve as an equivalence class of defining equations. It is true since, as we are about to show, we can parametrise each branch by Puiseux series; the Puiseux series are different, and if the difference has order κ then it behaves for x small like x^κ, and so vanishes only at $x = 0$.

Consider a branch B with equation $g(x, y) = 0$. According to Theorem 2.2.6, unless the branch coincides with $x = 0$, the equation admits at least one solution of the form $y = \psi(x^{1/m})$ with $\psi \in \mathbb{C}\{z\}$. We can write this as a parametrisation $x = t^m$, $y = \psi(t)$. Provided m is the highest common factor of the denominators of the exponents occurring in the series $\psi(x^{1/m})$, this is a good parametrisation, since if t_1, t_2 give the same point, then $t_1^m = t_2^m$, so $t_2 = e^{2\pi ir/m}t_1$ for some $r \in \mathbb{Z}$, so $\psi(t_1) = \psi(e^{2\pi ir/m}t_1)$. It follows from our choice of m that if r is not divisible by m, the power series $\psi(t)$ and $\psi(e^{2\pi ir/m}t)$ do not have identical coefficients. Hence their difference is non-zero in some neighbourhood $0 < |t| < \epsilon$. Thus on a small enough neighbourhood, our parametrisation is injective, so it is good. We next show that all points of B in a small enough neighbourhood of O are given by the parametrisation.

Lemma 2.3.1 *Given a good parametrisation of a branch B, we can write down an irreducible equation such that a point (near enough to O) satisfies the equation if and only if it is given by the parametrisation.*

Proof Let the parametrisation be $y = \sum_1^\infty a_r t^r$, where $x = t^m$. We first collect all the terms in the expansion where r lies in a single congruence class modulo m, setting $r = mq + s$:

$$y = \sum_{s=0}^{m-1} t^s \left(\sum_{q=0}^{\infty} a_{mq+s} t^{mq} \right),$$

and thus define $\phi_s(x) = \sum_{q=0}^\infty a_{mq+s} x^q$. These are convergent power series: we leave the proof as Exercise 2.6.3. We may now regard the equations

$$t^a y = \sum_{s=0}^{m-a-1} t^{a+s}\phi_s(x) + \sum_{s=m-a}^{m-1} t^{a+s-m} x \phi_s(x),$$

for $0 \le a \le m-1$, as simultaneous linear equations for the unknowns t^a with coefficients in $\mathbb{C}\{x, y\}$. Since the values t^a provide non-zero solutions to these equations, the determinant $D(x, y)$ of the system vanishes, and we can take $D(x, y) = 0$ as an equation for the branch.

By construction, D is a monic polynomial of degree m in y. We have exhibited one solution and can infer the remaining $m - 1$: thus we can factorise

$$D(x,y) = \prod_{k=0}^{m-1} \left(y - \sum_{1}^{\infty} a_r e^{2\pi i k r/m} x^{r/m} \right).$$

It follows that (in a small enough neighbourhood of O) all solutions of $D(x,y) = 0$ are given by the vanishing of a factor, and hence are given by the parametrisation. It also follows that D is irreducible in $\mathbb{C}\{x,y\}$, for any element of this ring whose image in $\mathbb{C}\{x^{1/m}, y\}$ is divisible by one of the above factors must also be divisible by all the others. □

The same procedure works to obtain an equation from a parametrisation for formal power series.

We illustrate with Example 2.1.1. If we start with $x = t^4$, $y = t^6 + t^7$ then $\phi_0(x) = \phi_1(x) = 0$, $\phi_2(x) = \phi_3(x) = x$. The system of equations has the matrix

$$\begin{pmatrix} y & 0 & -x & -x \\ -x^2 & y & 0 & -x \\ -x^2 & -x^2 & y & 0 \\ 0 & -x^2 & -x^2 & y \end{pmatrix},$$

with determinant $y^4 - 2y^2x^3 - 4yx^5 + x^6 - x^7$.

Essential uniqueness of good parametrisations follows by an argument related to the proof of Theorem 2.2.6.

Lemma 2.3.2 *If t and u are the parameters for two good parametrisations of the same branch B, we can write $u = \omega(t)$ for some $\omega \in \mathbb{C}\{t\}$ such that $\omega(0) = 0$, $\omega'(0) \neq 0$.*

Proof Suppose B has multiplicity m, equation $f(x,y) = 0$ regular in y of order m, and good parametrisation $(x,y) = (\phi(t), \psi(t))$. Expand ϕ in powers of t, and let n be the least exponent occurring: $\phi(t) = at^n + \ldots$, with $a \neq 0$. Then we can write $\phi(t) = t'^n$ with $t' = t + \ldots$ a convergent power series.

We saw in the above proof that there is also a good parametrisation with parameter z such that $x = z^m$. Now since each of these defines a bijection between a neighbourhood of 0 in the t or z plane and a neighbourhood of O in B, to each non-zero value of x in such a neighbourhood correspond m distinct values of t. It follows that $m = n$ and $z^m = at'^m$, so $z = bt'$ for some m^{th} root b of a. This shows that the parameter t is equivalent to z; similarly, so is u. □

Similarly, given a (not necessarily good) parametrisation with parameter u, we can write $u = \omega(t)$ for some holomorphic ω: if this has order N, then $u = u'^N$ with u' a good parameter.

We next discuss multiplicities and tangents. If C is defined by the equation $f(x, y) = 0$ (with $f \in \mathbb{C}\{x, y\}$), the order of f is called the *multiplicity* of C at O. We denote it by $m_O(C)$, where the point O and the curve C may be omitted if clear from the context. Since the order of a product is the product of the orders, we have $m(C) = \sum a_j m(B_j)$. The multiplicity is an important invariant, and multiplicities will play a prominent role in this book.

Let $m = m_O(C)$; denote by f_m the sum of the terms of lowest degree m in the power series expansion of f. Since this is a homogeneous polynomial in x and y, we may factorise it into linear factors. The lines whose equations are these linear factors are called the *tangent lines* to C at O. There are at most m such lines.

If $y - ax$ is not a factor of f_m, substituting $y = ax$ in f gives a function of x of order m; if $y - ax$ is a factor we obtain a function of higher order. Hence the intersection number of C with a straight line L through O is equal to m if L is not a tangent, and greater than m if it is (the line $x = 0$ may be included by reversing the roles of x and y). This remark is important enough for us to record it.

Lemma 2.3.3 *A straight line L is tangent to B at O if and only if the intersection multiplicity $(L.B)_O > m(B)$.*

We defined 'f is regular in y of order M' if M is the order of $f(0, y)$, i.e. if M is the intersection number of C with the line $x = 0$. Thus provided $x = 0$ is not a tangent to C at O, f will be regular in y of order $m = m_O(C)$.

Given a branch B, choose coordinates with $y = 0$ not a tangent line. Then an equation for B will be regular in y of order m, and so B has a good parametrisation $x = t^m$, $y = \sum_{r=1}^{\infty} a_r t^r$. Substituting in the equation $y = ax$ of a line L gives a power series in t whose order is the intersection number $B.L$. This cannot be less than the multiplicity m of B, so a_r vanishes for $r < m$. We now see that there is a unique tangent $y = a_m x$. Moreover replacing t by $e^{2\pi i/m} t$ gives Puiseux series all with the same coefficient a_m of x. Multiplying these gives an equation for B whose terms of order m are just $(y - a_m x)^m$.

Differentiating gives $\frac{dx}{dt} = mt^{m-1}$, $\frac{dy}{dt} = ma_r t^{m-1} + \ldots$, so $\frac{dy}{dx} \to a_m$ as $t \to 0$. Thus the tangent to B at the point with parameter t converges to the line $y = a_m x$ as $t \to 0$. The straight line from O to this point also converges to the same limit. Thus the tangent defined algebraically above has the expected geometric properties.

The curve C is smooth at O if and only if $m_O(C) = 1$. It follows that C consists of a unique branch with multiplicity 1. If $x = 0$ is not tangent, we can take x itself as parameter on C. Since $y = \psi(x)$ for ψ holomorphic, we can take as coordinate system the pair (x, y') with $y' := y - \psi(x)$: in these coordinates, C becomes a straight line. We may sometimes say O is a *simple point* of C, or C is *non-singular* at O to mean the same thing.

2.4 Factorisation

In this section, which may be omitted at a first reading, we clarify various points raised by the above. We need to show that the procedures of the two preceding sections actually lead to the same Puiseux series: this will follow from a fuller discussion of factorisation in rings of formal power series, and for this we need a version of the preparation theorem valid for these.

We may regard Puiseux' Theorem as stating 'Every equation has a root': this statement for complex numbers can be stated in terms of the field of complex numbers being algebraically closed: there is a corresponding statement here, but we first need to construct the field.

There is also a useful criterion for factorisation in terms of the Newton polygon.

We have already seen that if R is a ring of formal power series or a ring of convergent power series, any element of R with non-zero constant term has an inverse. Associating to each element of R its constant term defines a ring homomorphism $\epsilon : R \to \mathbb{C}$, whose kernel $\mathfrak{m}$ is thus an ideal in R. Thus any element of R not lying in $\mathfrak{m}$ has an inverse. So any ideal in R not contained in $\mathfrak{m}$ contains an element with an inverse, and hence contains 1, so is the whole of R. So $\mathfrak{m}$ is the unique maximal (proper) ideal in R. In general a ring R is said to be a *local ring* if it has an ideal $\mathfrak{m}$ such that all elements of $R \setminus \mathfrak{m}$ have inverses. So our rings are local rings. The terminology refers to the fact that such rings were originally introduced for studying the behaviour of an algebraic variety near a particular point; and this is precisely what we are doing here.

We can also formalise the key property of formal power series, that they can be constructed term-by-term. A local ring R with maximal ideal $\mathfrak{m}$ is said to be *complete* if any sequence $a_i \in R$ such that for each i, $a_i - a_{i-1} \in \mathfrak{m}^i$ converges $\mathfrak{m}$-adically: there exists an element $a_\infty \in R$

with $a_\infty - a_i \in \mathfrak{m}^i$ for each i. Clearly this holds for formal power series in any number of variables. The preparation theorem for rings of formal power series can now be stated as:

Theorem 2.4.1 *Let R be a complete local ring with maximal ideal $\mathbf{m}$, so that $K = R/\mathbf{m}$ is a field. Let $f \in R[[y]]$ have image $\overline{f} \in K[[y]]$ of order s in y. Then f may be uniquely expressed as a product Ug, with U a unit and $g \in R[y]$ a monic polynomial of degree s.*

Proof Set $g_0 = y^s$; let U_0 be any lift to $R[[y]]$ of $\overline{f}/y^s \in K[[y]]$. Then we have $f - U_0 g_0 \in \mathbf{m}.R[[y]]$.

Suppose inductively we can find U_r with $U_r - U_{r-1} \in \mathbf{m}^r.R[[y]]$ and monic polynomials g_r of degree s with $g_r - g_{r-1} \in \mathbf{m}^r.R[y]$ such that $f - U_r g_r \in \mathbf{m}^{r+1}.R[[y]]$: then U_r converges in the $\mathbf{m}$-adic sense to a unit U and g_r converges to a monic polynomial g with $f = Ug$.

For the induction step, write

$$(U_0)^{-1}(f - U_r g_r) = \sum_{i=0}^{s-1} c_i y^i + y^s B(y),$$

with $c_i \in R$. Since the right hand side belongs to $\mathbf{m}^{r+1}.R[[y]]$, we have $c_i \in \mathbf{m}^{r+1}$, and set

$$U_{r+1} = U_r + U_0 B(y), \quad g_{r+1} = g_r + \sum_{i=0}^{s-1} c_i y^i.$$

Then since we have (writing $\sum_{i=0}^{s-1} c_i y^i = \Sigma$ for short)

$$(U_0)^{-1}(f - U_{r+1} g_{r+1}) = U_0^{-1}(U_r - U_0)\Sigma + B(y)(y^s - g_r) - B\Sigma,$$

with $(U_r - U_0) \in \mathbf{m}.R[[y]]$ and $(y^s - g_r) \in \mathbf{m}.R[[y]]$ (as follow by induction) the above conditions are easily checked. Also these choices are determined uniquely modulo $\mathbf{m}^{r+2}$. □

It follows, by the same proof as for Theorem 2.2.5, that $\mathbb{C}[[x, y]]$ is a unique factorisation domain.

Corollary 2.4.2 *For any $f \in \mathbb{C}[[x, y]]$, with $f(0, y)$ of order m, there exist N and a factorisation*

$$f(t^N, y) = U(t^N, y) \prod_{i=1}^{m} \left\{ y - \sum_{r=1}^{\infty} c_{r,i} t^r \right\}.$$

Proof By Theorem 2.1.1, the equation $f(x,y) = 0$ admits a solution as a Puiseux series: say $(x,y) = (t^n, \sum_1^\infty a_r t^r)$. If we substitute $x = x_1^r$ in f, we can than take out a factor $(y - \sum_1^\infty a_r x_1^r)$. The quotient has the order m in y decreased by 1. We may thus proceed by induction: when m such factors are taken out, the result has order 0 in y, so is a unit. □

This result corresponds to Theorem 2.2.6. Now observe that the power series in this result are uniquely determined, since unique factorisation holds, and $y - \alpha(x)$ can only divide $y - \beta(x)$ if $\alpha(x) \equiv \beta(x)$. Thus if in fact $f \in \mathbb{C}\{x,y\}$ the series obtained must be the same whichever method we have used to obtain them.

This resembles the usual factorisation of polynomials over $\mathbb{C}$, and indeed the result can be formulated in a similar manner. We first define a ring $\mathbb{P}[[x]]$ as the union of the rings $\mathbb{C}[[x^{1/n}]]$. (Strictly speaking we should introduce new variables x_n for $n \in \mathbb{N}$ subject to the relations $x_{mn}^m = x_n$ for $m, n \in \mathbb{N}$, and then interpret x_n as $x^{1/n}$.) Next, $\mathbb{C}[[x]]$ is made into a field $\mathbb{C}[[x]][x^{-1}]$ by adjoining x^{-1}. Similarly we make $\mathbb{P}[[x]]$ into a field $\mathbb{P}[[x]][x^{-1}]$ by adjoining x^{-1}. An element of this field is a series of the form $\sum_{r=\nu}^\infty a_r x^{r/n}$, where ν need not be positive.

Correspondingly we have the ring $\mathbb{P}\{x\}$, defined as the union of the $\mathbb{C}\{x^{1/n}\}$, for $n \in \mathbb{N}$, and the field $\mathbb{P}\{x\}[x^{-1}]$ obtained from it by adjoining x^{-1}.

Then Puiseux' Theorem can be stated as:

Theorem 2.4.3 *The fields $\mathbb{P}[[x]][x^{-1}]$, $\mathbb{P}\{x\}[x^{-1}]$ are algebraically closed.*

Proof The argument is the same for both cases: we present it for the formal power series. If y satisfies a polynomial equation $f(y) = 0$ over $\mathbb{P}[[x]][x^{-1}]$, we may suppose f monic. Replacing y by $z = x^k y$ for suitable k gives z satisfying an equation $g(z) = 0$ all of whose coefficients belong to $\mathbb{P}[[x]]$.

Choose a common denominator n for the exponents so that all the coefficients belong to $\mathbb{C}[[x_n]]$. We may now regard g itself as belonging to $\mathbb{C}[[x_n, z]]$, and apply the theorem to obtain a solution for z as a Puiseux series in x_n and hence in x, so as an element of $\mathbb{P}[[x]]$. Thus the equation for y has a solution in $\mathbb{P}[[x]][x^{-1}]$. □

Over an algebraically closed field, a polynomial equation of degree m always has m roots (allowing for multiplicities), which agrees with the conclusions above. The presence of multiple roots can, as usual, be detected by applying the euclidean algorithm to test for a common factor of f and $\partial f/\partial y$, or more simply by the identical vanishing of the discriminant.

Consideration of the Newton polygon of the equation f leads to a different type of result about factorisation.

Lemma 2.4.4 *Suppose the ith edge of the Newton polygon of f corresponds to an increment k_i in the first coordinate and a decrement of n_i in the second. Then the curve C defined by $f = 0$ is the union of pieces B_i, of multiplicities $min(k_i, n_i)$, and such that each branch of B_i has Puiseux series of the form*

$$y = ax^{n_i/k_i} \text{ plus higher terms, with } a \neq 0,$$

and correspondingly $f = \prod f_i$ where the Newton polygon of f_i is a single edge from $(0, k_i)$ to $(n_i, 0)$.

Proof Consider any branch B of C, given say by $x = t^n$, $y = \sum a_r t^r$, and with equation $g(x, y) = 0$. If k is the least value of r such that $a_r \neq 0$ then g has Newton polygon consisting of a single edge, and if h is the highest common factor (k, n) and we set $k = hk'$, $n = hn'$ then the corresponding terms in the equation are $(y^{n'} - a_k^{n'} x^{k'})^h$. Thus the Newton polygon of the branch is the segment from $(k, 0)$ to $(0, n)$.

Now suppose C has branches B_j whose Newton polygons are the segments from $(k_j, 0)$ to $(0, n_j)$; we order the branches so that the fractions $\frac{n_j}{k_j}$ are non-decreasing. Then we see by inspection that the Newton polygon of C starts at $(\sum_j k_j, 0)$ and proceeds in turn through the points $(\sum_{j \leq r} k_j, \sum_{j > r} n_j)$.

Hence conversely, given the Newton polygon of the original curve, we obtain the desired information about the branches by taking the edges separately, and for each edge factorising the polynomial corresponding to the terms of the equation on that edge. □

2.5 Notes

Section 2.1 The algorithm is due to Newton. Newton's contributions were outlined in letters [142]: see also [143]. Two letters, including the crucial one of 24/10/1676 to Oldenburg, are reproduced in [23]pp 495–498. This emphasises the use of the Newton polygon to obtain the initial term of a series solution and shows that Newton was well able to continue to find further terms. There is no discussion of convergence. Puiseux' own paper [152] is dated 1850.

Section 2.2 The development follows standard texts in complex variable theory: see e.g. [81]. Both the statements and proofs of Theorem 2.2.2, Theorem 2.2.3 and Corollary 2.2.4 remain valid if x is interpreted as denoting a sequence of complex variables $(x_1, \dots, x_k)$ rather than a single such variable. Further references are given in [190]. An alternative proof is given in [79]. The original reference for the Weierstrass Preparation Theorem is [198]. Although it seems an immediate consequence, the Division Theorem (Corollary 2.2.4) is due (independently) to Stickelberger [173] and Späth [170].

Section 2.3 The method for going from a parametrisation to an equation is due to [65]pp 56–58.

Section 2.4 The use of the preparation theorem to establish algebraic properties of $\mathcal{O}_n$ was pioneered by Rückert [157].

It follows by the same proof as for Theorem 2.2.5, together with induction on n, that, for any n, each of the rings $\mathbb{C}[[x_1, \dots, x_n]]$, $\mathbb{C}\{x_1, \dots, x_n\}$ is a unique factorisation domain. Similar arguments show that these rings possess other properties standard for polynomial rings – noetherian, of global (homological) dimension n. We refer the interested reader to [81]pp 74–76; see also [79]Section 2.2.

Lemma 2.4.4 is a special case of Hensel's Lemma, which can be stated as follows.

Theorem 2.5.1 *Let R be either $\mathbb{C}[[x]]$ or $\mathbb{C}\{x\}$; write $\mathfrak{m}$ for the ideal of series with zero constant term. Let $f(y) \in R[y]$ be a polynomial, and suppose $a \in R$ such that $f(a)$ belongs to the ideal $\langle f'(a)^2\rangle\mathfrak{m}$. Then there is a unique $b \in R$ such that $f(b) = 0$ and $b - a \in \langle f'(a)\rangle\mathfrak{m}$.*

See [64]Theorem 7.3 for a fuller statement and discussion. A simpler version is given in [79]p 44. Some authors deduce these results from Artin's approximation theorem.

Finite characteristic We now discuss how the results in the text need to be modified in characteristic p. As noted above, Puiseux' theorem is false in all the forms stated. The example given: $y = R(x) = \sum_{k=0}^{\infty} x^{2-2^{-k}}$ is algebraic, as it satisfies $y^2 + x^2y + x^2 = 0$. However, this curve does admit the rational parametrisation $(x, y) = (\frac{t^2}{1+t}, \frac{t^2}{1+t^2})$, easily developed as power series in t, e.g. $x = t^2 + t^3 + t^4 + \cdots$. The trouble arises when trying to invert this to express t as fractional power series in x: we obtain $t = \sum_{k=1}^{\infty} x^{1-2^{-k}} + x$. Note also that projection on the x-axis is a double cover, with covering transformation τ, where $\tau(t) = \frac{1}{1+t}$, giving $\tau(y) = y + x^2$ and $\tau(x) = x$.

The general theory is developed in [27]. Campillo's viewpoint is the formal one: study the ring $\mathcal{O}$, or rather its completion $\hat{\mathcal{O}}$ in the m-adic sense. Then [27]1.3.1 if Γ has just one branch at O the integral closure $\tilde{\mathcal{O}}$ of $\hat{\mathcal{O}}$ is a complete, rank 1 discrete valuation ring, and hence is isomorphic to the power series ring $k[[t]]$. Thus choosing an isomorphism gives a local parametrisation and the composite map

$$k[x, y] \to k[\Gamma] \to \mathcal{O} \to \tilde{\mathcal{O}} \cong k[[t]]$$

gives expressions of x and y as power series in t. The most effective replacement for Puiseux' theorem is the Hamburger-Noether expansion: see [27] for an exposition.

In fact there are always fractional power series expansions $y = \sum a_\alpha x^{r_\alpha}$. The denominators of the exponents r_α are no longer bounded, but there exists an integer N such that all the Nr_α have denominator a power of p. Convergence cannot be defined, but the set of exponents r_α is well-ordered, which makes manipulations possible. Indeed, the series satisfying these conditions form an algebraically closed field [172]. A precise characterisation of the algebraic closure of $k[[x]]$ was given by Kedlaya [98].

One aspect of what goes wrong in characteristic p is that the infinite cyclic group $\pi_1(D^*_\eta)$ is replaced by the Galois group over $k[[t]]$ of its (separable) algebraic closure. Although in any example, only a finite quotient of this is relevant, this need not even be abelian. Here are two examples in characteristic 2. In each case we give an polynomial g_i defining a curve $\Gamma_i : g_i = 0$, and consider the projection π_i of Γ_i on the x-axis.

$g_2 = x+y^4+y^5$. Here π_2 is not a normal covering; its Galois group has order 12 or 24, since if we put the equation for y in normal form using Weierstrass preparation (multiply by a unit $1 + y + y^2 + y^3 + \cdots$) we get $y^4 + xy^3 + xy^2 + xy + x$ plus higher terms, which has cubic resolvent $\lambda^3 + (x + x^2)\lambda^2 + x^2\lambda + (x^2)$ plus higher terms, or $\lambda^3 + x^2$ plus higher terms, which clearly has no root in $k[[x]]$.

$g_3 = x + y^4 + y^7$. We have a Galois covering π_3 with group the four group. Covering transformations are $y \mapsto y + \alpha y^2 + \alpha^2 y^3 + \alpha^3 y^4 + \alpha y^5 + \sum a_n y^n$, where $\alpha^4 = \alpha$, so α is any element of the Galois field $\mathbb{F}_4$, and a_n is found inductively by equating the coefficients of y^{n+6} in the equation expressing invariance of x under the transformation.

2.6 Exercises

Exercise 2.6.1 Show that $(t^3, t^2 + t^4)$ is a good parametrisation of a curve B, and find an equation for B.

Exercise 2.6.2 For the curve given by $y^3 - 9x^3y - x^4 = 0$, find a parametrisation of the form $x = t^3$, $y = \psi(t)$, and obtain the first four non-vanishing coefficients in ψ.

Exercise 2.6.3 Suppose we are given a *convergent* power series $f(t) = \sum_0^\infty a_r t^r$. Show that the power series $\phi_s(x) = \sum_0^\infty a_{nq+s}(x)x^q$ are all convergent.

Exercise 2.6.4 Show that the curve given by $x^5 - x^2y^2 + y^5 = 0$ has 2 branches, and find the first 2 terms of the Puiseux series for y in terms of x for each of them.

Exercise 2.6.5 Let $f \in \mathbb{C}[[x, y]]$ have zero constant term. Assume that, in the notation of the proof of Theorem 2.1.1, all the m_r are equal to m. Show that f has a root of multiplicity m. (Hint: by Theorem 2.4.1 you may assume that f is a Weierstrass polynomial. Show that the algorithm yields a power series solution $y = \sum_1^\infty C_r x^r$, and that the order of $f - \left(y - \sum_1^N C_r x^r\right)^m$ tends to infinity with N.)

Exercise 2.6.6 Let $f \in \mathbb{C}\{x, y\}$ have zero constant term and no repeated factor. Prove the convergence of the formal power series constructed in Theorem 2.1.1 as follows. Show that, for some r, the algorithm leads to an equation $f_r(x_r, y_r) = 0$ such that $f_r \in \mathbb{C}\{x, y\}$, the constant term vanishes, and the coefficient of y_r is non-zero. Deduce from the Implicit Function Theorem that y_r can be expressed as a convergent power series in x_r. Deduce that the corresponding power series for y is also convergent.

Exercise 2.6.7 Find the Weierstrass decomposition of $y^2 - xy^2 - xy$ in the form $U(x, y)(y^2 + ya_1(x) + a_2(x))$.

Exercise 2.6.8 Show (following the proof of Theorem 2.2.7) that if the reduced curve C consists of a curve C' of multiplicity 2 and a smooth branch B transverse to it, then in suitable coordinates it admits an equation $y(x^2 + y^n) = 0$ for some $n \geq 2$.

Exercise 2.6.9 Show that $y^3 + xy^2 + x^4$ is reducible in $\mathbb{C}\{x, y\}$.

Exercise 2.6.10 Show that $y^2 - x^2 - x^3$ is irreducible in $\mathbb{C}[x, y]$, but not in $\mathbb{C}\{x, y\}$.

Exercise 2.6.11 Find an equation for the curve parametrised by $x = t^6$, $y = t^8 + t^{13}$.

Exercise 2.6.12 Find the Newton polygons of the branches of $x^4 + x^3y + y^5 = 0$.

Exercise 2.6.13 How many branches at O have the following curves?

(a) $x^2 + 2xy + 2y^2 = 0$,
(b) $x^6 - x^2y^3 - y^5 = 0$,
(c) $x - y + (x + y)^2 = 0$.

Exercise 2.6.14 Calculate the intersection number of the branches B : $x = t^4$, $y = t^6 + t^7$ and B' : $x = t^6$, $y = t^9 + t^{10}$ by substituting the parametrisation for B' in the equation for B.

3

Resolutions

The central result of this chapter is that a curve singularity may be resolved by successive blowings up. We define blowing up in Section 3.2 and prove the result in Section 3.3. The result is related to Puiseux' Theorem, so we begin the chapter by defining the Puiseux characteristic, which will be a dominant concept throughout the discussion of invariants of a single branch, and here provides a convenient vehicle for the inductive proof.

In the second half of the chapter we begin to develop some numerical invariants of a single branch using the resolution. We introduce the classical notion of 'infinitely near points', and also the dual graph of the configuration of exceptional curves arising in the resolution, and obtain a number of interrelations between the different concepts we have defined.

3.1 Puiseux characteristics

Suppose B a branch of the germ at O of a holomorphic curve in $\mathbb{C}^2$. If $x = 0$ is not tangent to B, we have a good parametrisation $x = t^m$, $y = \sum_{r=m}^{\infty} a_r t^r$. Recall that 'good' means that a point on the curve corresponds to only one value of t: here this amounts to saying that the values of r with $a_r \neq 0$, together with m, have highest common factor 1. The notation does not exclude the possibility that $a_m = 0$, which holds if $y = 0$ is the tangent to B at O.

Define β_1 to be the exponent of the first term in the power series which is not a power of t^m, and e_1 to be the highest common factor of m and β_1:

$$\beta_1 = min\{k | a_k \neq 0, m \not| k\}, \quad e_1 = hcf(m, \beta_1)$$

and, inductively,

$$\beta_{i+1} = min\{k|a_k \neq 0, e_i \not| k\}, \quad e_{i+1} = hcf(e_i, \beta_{i+1}),$$

continuing till we reach g with $e_g = 1$, which exists since the parametrisation was supposed good. Thus β_i is the least exponent appearing in the series which does *not* belong to the additive group generated by m and the preceding β_j.

We shall call the sequence of numbers

$$(m; \beta_1, \ldots, \beta_g)$$

the *Puiseux characteristic* of B; the β_i are sometimes called the (Puiseux) characteristic exponents. We will keep the notation e_i for the numbers defined above; we also define $\beta_0 := e_0 := m$.

In Example 2.1.1, we had the parametrisation $(x, y) = (t^4, t^6 + t^7)$. The Puiseux characteristic is thus $(4; 6, 7)$.

We will see that the characteristic is not only independent of the choice of coordinates, but also contains deep information about the chosen branch. This independence can be proved by manipulating coordinates, but we prefer first to introduce a new geometrical construction. We will obtain enough intrinsic information from the resolution to determine the Puiseux characteristic; a formal summary will be given in Proposition 4.3.8.

3.2 Blowing up

Although blowing up may be defined in much more general situations, in this book we will only use the simplest form of it. The basic idea is as follows. We start from a point P on a smooth algebraic surface S, and will construct a new surface T and a map $\phi : T \to S$ such that $\phi^{-1}(P)$ is a curve E; ϕ gives a bijection, indeed an isomorphism, from $T - E$ to $S - \{P\}$, and the points on E correspond to the different directions in S at P. The map ϕ is called the *blowing up* of S with *centre* P.

Let us carry this through for the case $S = \mathbb{C}^2$, with coordinates (x, y) on S, and $P = O$. Introduce the projective line $P^1(\mathbb{C})$ with coordinates $(\xi : \eta)$. Then define $T = T_1$ to be the subspace of points in the product $\mathbb{C}^2 \times P^1(\mathbb{C})$ satisfying the equation $x\eta = y\xi$. The projection of the product to $\mathbb{C}^2$ defines a map $\pi : T \to \mathbb{C}^2$. Any point $(x, y) \neq O$ determines a unique $(\xi : \eta) = (x : y)$, hence a unique point $\pi^{-1}(x, y)$, while corresponding to the point O we have the entire projective line $P^1(\mathbb{C})$, so that $\pi^{-1}O$ is a curve E isomorphic to $P^1(\mathbb{C})$, and called the *exceptional curve* of the blow up.

For a general surface S we may introduce local coordinates (x, y) in a neighbourhood of a smooth point P. These define a biholomorphic equivalence between a neighbourhood U of P in S and a neighbourhood V of O in $\mathbb{C}^2$. The required blowing up of S is now obtained by piecing together $S - \{P\}$ and $\pi^{-1}(V)$ using the equivalence of $U - \{P\}$ and $\pi^{-1}(V - \{O\})$ via the equivalence of each with $V - \{O\}$.

The reason for introducing the general surface S above and not restricting entirely to the case $S = \mathbb{C}^2$ is that we will need to blow up with centre a point on the resulting surface $T = T_1$, giving a surface T_2, then blow up with centre a point in T_2, and so on. However, all our calculations will be performed by taking local coordinates in all these surfaces, which are constructed as follows.

Recall that $P^1(\mathbb{C})$ is the union of two (affine) coordinate charts: U_0, where $\xi \neq 0$ and we can take η/ξ as coordinate, and U_1, where $\eta \neq 0$ and we can take ξ/η as coordinate. On the part of T where $\xi \neq 0$ we write Y for η/ξ, and the equation $x\eta = y\xi$ then simplifies to $y = xY$, showing that this part of T can be identified with $\mathbb{C}^2$ by taking the coordinates (x, Y). Similarly, on the part of T where $\eta \neq 0$ we write X for ξ/η; the equation $x\eta = y\xi$ simplifies to $x = Xy$, and we identify this part of T with $\mathbb{C}^2$ using the coordinates (X, y).

The existence of these local coordinates exhibits the fact that the blow up of $P^2(\mathbb{C})$ is another non-singular surface.

Note in particular that the preimage E of O is isomorphic to $P^1(\mathbb{C})$, and is given in the first chart by $x = 0$ (with coordinate Y) and in the second by $y = 0$ (with coordinate X).

From the viewpoint of calculations, we can simply introduce $Y = y/x$ (or $X = x/y$) as a new coordinate. However the geometric description shows us more. The first important observation is that the construction of the blow up, while conveniently expressed in terms of local coordinates, is not dependent on them.

Lemma 3.2.1 *The result of blowing up a smooth surface S with centre a point P is intrinsically well-defined.*

Proof Let (x, y) and (x', y') be two systems of local coordinates at P on S. Let $\pi : T \to S$, $\pi' : T' \to S$ be the blowings up defined using the two systems of local coordinates. We can locally express x', y' as functions $\phi(x, y)$, $\psi(x, y)$ of x and y. The derivatives at the origin give the linear terms of the power series expansions

$$\phi(x,y) = ax + by + \dots, \qquad \psi(x,y) = cx + dy + \dots.$$

Since (x', y') is also a local coordinate system, the determinant $ad - bc \neq 0$. On the preimage of the origin we can write $X' = \frac{x'}{y'} = \frac{ax+by}{cx+dy} = \frac{aX+b}{cX+d}$. This is a typical change of coordinates in a projective space $P^1(\mathbb{C})$.

We wish to show that there is (at least in a neighbourhood of the preimage of O) a unique map $\theta : T \to T'$ with $\pi' \circ \theta = \pi$; it will follow that there is a unique map in the reverse direction providing an inverse, and hence showing that T and T' are equivalent. Consider a point with coordinates (X, y) in T: such that $cX + d \neq 0$. Then we obtain $x' = \phi(Xy, y)$ and $y' = \psi(Xy, y)$ directly, so it will suffice to show that X' is well defined and smooth. Since each of $\phi(Xy, y)$ and $\psi(Xy, y)$ vanish along $y = 0$, by Lemma 1.4.3 there are holomorphic functions such that $\phi(Xy, y) = y\alpha(X, y)$ and $\psi(Xy, y) = y\beta(X, y)$. Setting $y = 0$, we have $\alpha(X, 0) = aX + b$ and $\beta(X, 0) = cX + d$. Thus in a neighbourhood of a point of T in the preimage of O at which $cX + d \neq 0$, $\beta(X, y)$ is invertible, so $X' = \frac{x'}{y'} = \frac{\alpha(X,y)}{\beta(X,y)}$ is indeed smooth. □

The second observation is that we can explore the geometry of the curve E which appears on blowing up, and its relation to other curves which appear when we repeat the process.

Let $\phi : T \to S$ be the blowing up with centre the point $P \in S$, and $E = \phi^{-1}(P)$ the exceptional curve of the blow up. If C is a curve in S not passing through P, it corresponds to the unique curve $\phi^{-1}(C)$ in T. If C is a curve through P, $\phi^{-1}(C)$ is called its *total transform*. This contains the exceptional curve E; the closure of $\phi^{-1}(C) - E$ is called the *strict transform* of C.

For the blowing up of $\mathbb{C}^2$ with centre O, consider a branch B parametrised as above. Then we obtain a parametrisation of the strict transform by setting $Y = \sum_{r=k}^{\infty} a_r t^{r-n}$. This intersects the exceptional curve at the point given by $t = 0$, where $Y = a_n$. Observe that this intersection is a single point, and that this is the point on E corresponding to the tangent to the branch B. A neighbourhood of this point is contained in the chart with coordinates (x, Y).

3.3 Resolution of singularities

We first consider the case of a single branch (the general case will be discussed in the next section). Write C for a branch at $O_0 = O$ of a holomorphic plane curve in $T_0 = \mathbb{C}^2$. Blowing up with centre O_0 produces a smooth surface T_1, an exceptional curve E_0 in it, and a strict transform $C^{(1)}$ meeting E_0 at a unique point O_1. Now blow up T_1 with centre O_1.

Inductively, suppose we have constructed a surface T_i containing curves E_j for $0 \leq j \leq i-1$ and a curve $C^{(i)}$ meeting E_{i-1} at a unique point O_i. Then blowing up T_i with centre O_i gives a new smooth surface T_{i+1} and a map $\pi_i : T_{i+1} \to T_i$. We write E_j again for the strict transform of E_j if $j < i$, E_i for the exceptional curve of π_i, $C^{(i+1)}$ for the strict transform of the curve (branch) $C^{(i)}$: it follows as before that this meets E_i in a unique point O_{i+1}.

We now show that this process eventually yields a smooth curve. This process is known as *resolving* the singularity by blowing up. If $C^{(N)}$ is smooth, the projection $\pi : T_N \to T_0$ is a *resolution* of C.

Recall that for any $x \in \mathbb{R}$ the integer part, which we will denote $\lfloor x \rfloor$, is the integer M such that $M \leq x < M+1$. Similarly we write $\lceil x \rceil$ for the integer N with $N < x \leq M+1$: thus $\lceil x \rceil = -\lfloor -x \rfloor$.

Theorem 3.3.1 *In the situation described above, there exists an integer N such that $C^{(N)}$ is smooth (and hence $C^{(n)}$ is smooth for $n > N$).*

Proof We induct on the multiplicity m; we wish to make a subsidiary induction on the Puiseux invariant β_1 of Section 3.1; as we have not yet proved its invariance under coordinate change, we must proceed carefully.

For the curve C we have the parametrisation

$$x = t^m, \quad y = \sum_{r=1}^{\infty} a_r t^r,$$

which we rewrite as $y = b_1 t^m + b_2 t^{2m} + \ldots + b_q t^{qm} + ct^{\beta_1} + \ldots$, where we write $q := \lfloor \frac{\beta_1}{m} \rfloor$.

Blowing up once, we obtain $C^{(1)}$, parametrised by

$$x = t^m, \quad Y = b_1 + b_2 t^m + \ldots + b_q t^{(q-1)m} + ct^{\beta_1 - m} + \ldots.$$

If $q \geq 2$, we shift the origin, writing $y_1 := Y - b_1$, again have multiplicity m, and continue. After q blowings up we find (after shifting the origin as necessary) that the expansion of y_q starts with $ct^{\beta_1 - qm}$, so $C^{(q)}$ has multiplicity $\beta_1 - qm < m$.

Since, if $m > 1$, we can blow up to reduce the multiplicity m, the result follows by induction on m. □

Once the invariance of β_1 is established, we will be able to give the proof by ordering the set of pairs (m, β_1) lexicographically, then observe that this is a well-ordering, and note that in either case above, the pair (m, β_1) is decreased on blowing up.

Example 3.3.1 Let C be the curve $y^8 = x^{11}$. For the first blow up we set $(x, y) = (x_1, x_1y_1)$ (note that here, and in other examples where we blow up repeatedly, it is convenient to distinguish coordinates at different stages of the blowing up by a suffix). Substituting in f gives $x_1^8(y_1^8 - x_1^3) = 0$, which is the equation of the total transform. The first factor represents the exceptional curve E_0, counted 8 times (corresponding to the fact that C has multiplicity 8 at O); the second factor f_1 is the equation of the strict transform $C^{(1)}$. Note that the singular point of $C^{(1)}$ does not lie in the coordinate chart given by the substitution $(x, y) = (x_1'y_1', y_1')$.

For the second blow up we make the substitution $(x_1, y_1) = (x_2y_2, y_2)$. This produces a total transform of $C^{(1)}$ consisting of the exceptional curve E_1 given by $y_2 = 0$, counted thrice, and the strict transform $C^{(2)}$ given by $y_2^5 - x_2^3 = 0$. The strict transform of E_0 is given by $x_2 = 0$.

For the third blow up we set $(x_2, y_2) = (x_3y_3, y_3)$. In this chart, the strict transform of E_0 is given by $x_3 = 0$; the strict transform of E_1 does not meet the domain of this chart; the exceptional curve E_2 is given by $y_3 = 0$; and the strict transform $C^{(3)}$ of C is given by $y_3^2 - x_3^3 = 0$.

For the fourth blow up we set $(x_3, y_3) = (x_4, x_4y_4)$; in this chart, E_0 and E_1 do not appear; E_2 is given by $y_4 = 0$ and E_3 by $x_4 = 0$; the strict transform $C^{(4)}$ of C is the smooth curve $y_4^2 = x_4$. The four blowings up are illustrated in Figure 3.1.

A close examination of the proof reveals that the induction adopted here is essentially the same as that used in the previous chapter in the proof of Theorem 2.1.1. Indeed, these two results are closely related. For if B is resolved by blowing up, then we need only quote the inverse function theorem to obtain a parametrisation of the blow up of the form $y = g(x)$ (compare Exercise 2.6.6).

It is worth emphasising that the blowing up procedure, even when iterated, is easy to do explicitly on reasonable examples, and leads to algorithms which can be effectively implemented. As a simple example, consider the effect of blowing up on the Newton polygon – see Figure 3.2.

Let the polygon have vertices $(r_i, s_i)\ :\ 0 \leq i \leq k$ with $0 = r_0 < r_1 < \ldots < r_k$ and $s_0 > s_1 > \ldots > s_k = 0$ and (by convexity) the

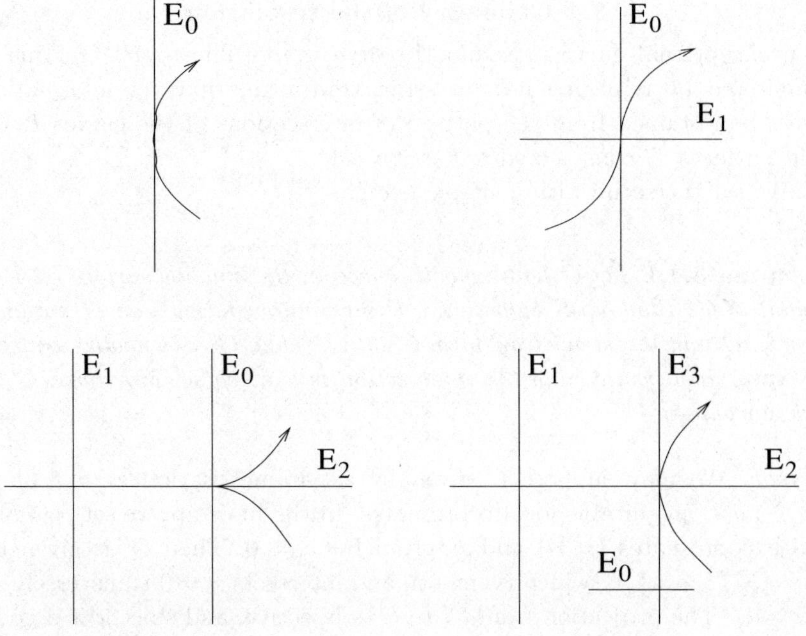

Fig. 3.1. Successive stages in a resolution

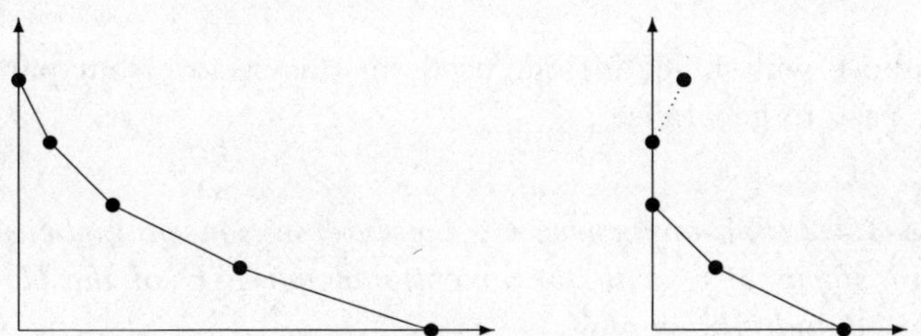

Fig. 3.2. Blowing up a Newton polygon

negatives $\sigma_i = \frac{s_{i-1}-s_i}{r_i-r_{i-1}}$ of the slopes strictly decreasing with i. Suppose that $\sigma_p \geq 1 > \sigma_{p+1}$. Then f has order $m = r_p + s_p$, and in the chart where we substitute $(x, y) = (x, xY)$ and divide by $x^{r_p+s_p}$, the term Y^{s_p} occurs with non-zero coefficient, so no terms with higher powers of Y contribute to the Newton polygon. The new Newton polygon thus has vertices $(r_i + s_i - m, s_i)$ arising from the vertices (r_i, s_i) with $i \geq p$.

3.4 Geometry of the resolution

An exceptional curve is a smooth curve isomorphic to $P^1(\mathbb{C})$: this is unaltered by replacing it by a strict transform. However information may be obtained from the pattern of intersections of the curves E_j in the surfaces T_i constructed in Section 3.3.

We set the scene with a simple lemma.

Lemma 3.4.1 *Let C be a smooth curve in the smooth surface S, P a point of C. Blow up S with centre P giving a surface T, an exceptional curve E and the strict transform C' of C. Then C' is smooth, it meets E in a single point, and the intersection is transverse. Moreover, C' is isomorphic to C.*

Proof We may suppose C given by a parametrisation $x = t$, $y = \sum_{r=1}^{\infty} a_r t^r$ in suitable local coordinates. In the blow up, we set $y = xY$, with coordinates (x, Y) and E given by $x = 0$. Then C' is given by $Y = \sum_{r=1}^{\infty} a_r x^{r-1}$, which is smooth and intersects $x = 0$ transversely at $Y = a_1$. The projection from C' to C is bijective, and since x is a good parameter for both, the projection is an isomorphism. □

In particular we may take $S = T_1$ and $C = E_0$. The lemma then tells us that E_0 and E_1 are smooth and intersect transversely in a single point in T_2.

The above will suffice for our needs in this section, but part of the result is easy to generalise.

Lemma 3.4.2 *For any curve C, the intersection multiplicity of the strict transform of C with the exceptional curve E of the blow up is equal to the multiplicity of C.*

Proof It is sufficient to consider the case of a single branch, since both expressions are additive for a union of branches. Choosing suitable coordinates, we may take a good parametrisation $x = t^m$, $y = \sum_{r=m+1}^{\infty} a_r t^r$. Blowing up as above, we find $C^{(1)}$ given by $x = t^m$, $Y = \sum_{r=m+1}^{\infty} a_r t^{r-m}$. Substituting this parametrisation in the equation $x = 0$ of the exceptional curve gives t^m, so the intersection number is equal to the multiplicity m of C. □

We return to the study of the sequence of blowings up.

Proposition 3.4.3 *The exceptional curve E_i in T_{i+1} intersects E_{i-1} and at most one curve E_j with $j < i-1$. These intersections are transverse, and no three of the curves E_i pass through a common point.*

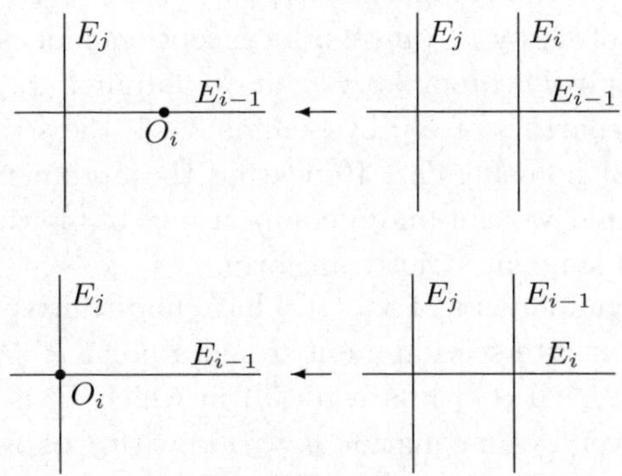

Fig. 3.3. Exceptional curves produced by blowing up

Proof We proceed by induction on i: we have already established the result for $i = 1$. Inductively assume that the result holds for $E_i \subset T_{i+1}$. The surface T_{i+2} is obtained by blowing up with centre a point O_{i+1} on E_i.

If O_{i+1} lies on E_j ($j < i-1$) or on E_{i-1} as well as on E_i, then by Lemma 3.4.1 the two curves meet transversely there, so have different directions. These directions correspond to distinct points on the exceptional curve E_{i+1}. By Lemma 3.4.1 again, E_{i+1} meets each of E_j and E_i transversely, at distinct points.

Since E_{i+1} meets E_j in T_{i+2} only if $O_{i+1} \in E_j$ in T_{i+1}, the result follows. □

A collection of curves in a smooth surface is said to have *normal crossings* if each curve is smooth, no three meet in a point, and any intersection of two of them is transverse. We have just seen that the configuration of exceptional curves always has normal crossings.

In general, given a singular point P of a curve C in a smooth surface S, a *good resolution* is a map $\pi : T \to S$ such that, if $E = \pi^{-1}(P)$, then π gives an isomorphism $(T - E) \to (S - P)$; and the collection $\pi^{-1}(C)$ of curves has normal crossings. The curves in this collection are the exceptional curves of π and the components of the strict transform of C.

Theorem 3.4.4 *Any plane curve singularity has a good resolution.*

Proof Let the curve C have branches B_j $(1 \leq j \leq k)$. By Theorem 3.3.1, we can find a sequence of blowings up such that the strict transform of B_1 by the composite $\pi_{(1)} : T^1 \to T^0 = \mathbb{C}^2$ is smooth. The strict transform of B_2 by $\pi_{(1)}$ meets the exceptional locus of $\pi_{(1)}$ in a single point. We apply Theorem 3.3.1 again to obtain $\pi_{(2)} : T^2 \to T^1$ which resolves the singularities of B_2; by Lemma 3.4.1, the strict transform by $\pi_{(1)} \circ \pi_{(2)}$ is still non-singular. Repeating the argument, it follows by induction on k that we can find a composite π' of blowings up in which each B_j has non-singular strict transform.

Suppose two components of $\pi'^{-1}(O)$ have non-transverse intersection: say B and B' have intersection number s at a point P. We then blow up P. We may take local coordinates (x, y) in which B' is given by $y = 0$ and B is then given by an equation $y = f(x)$ with f of order s. The blow up $(x, y) = (x_1, x_1 y_1)$ produces a similar situation with $f(x)$ replaced by $f(x_1)/x_1$, of order $s - 1$. Since no intersection numbers are increased by the blow up, and the new exceptional curve is (by Lemma 3.4.1) transverse to all other components, we may iterate this procedure to reduce all intersection numbers to 1.

Finally, if there is still a point where three or more of the curves in question meet, then since any two of them are transverse, blowing up this point will separate them all; they will meet the new exceptional curve transversely in distinct points. Thus blowing up each such point will yield a good resolution of C. □

Example 3.4.1 We return to the curve $y^8 = x^{11}$. Recall that in the chart after the fourth blow up, E_0 and E_1 do not appear; E_2 is given by $y_4 = 0$ and E_3 by $x_4 = 0$; the strict transform $C^{(4)}$ of C is the smooth curve $y_4^2 = x_4$. This curve touches E_3, so we must continue blowing up.

After the fifth blow up $(x_4, y_4) = (x_5 y_5, y_5)$ we have the exceptional curves E_3 and E_4 given by $x_5 = 0$ and $y_5 = 0$ and the transform $C^{(5)}$ given by $x_5 = y_5$. Any two of these are transverse, but all three go through a single point, so we must blow up once more. This produces a final exceptional curve E_5 which meets each of E_3, E_4 and $C^{(6)}$ transversely, all at different points, so that at last we have a good resolution. The final two blowings up are illustrated in Figure 3.4.

The procedure we have described is very simple: whenever there is a singular point or one where the normal crossing condition fails, choose

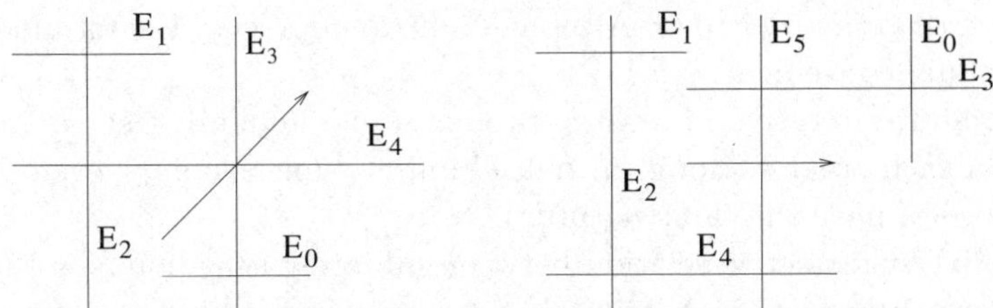

Fig. 3.4. A good resolution

one such point, and blow it up. A resolution obtained by following this procedure is called *minimal* (as opposed to one where additional unnecessary blowings up are performed). The order in which we blow points up does not affect the result, since if two points in a surface T are both to be blown up, blowing up one of the points does not change what happens in a neighbourhood of the other.

For any resolution, the strict transform of C is a smooth curve $\widetilde{C}$, and we have a projection $\pi : \widetilde{C} \to C$, which is (up to isomorphism) independent of the resolution. This projection, or the curve $\widetilde{C}$, is called the *normalisation* of C. This statement is not very interesting when we are studying germs of curves: in the case of a single branch, π can be regarded as a good parametrisation; if C has r branches, there are r corresponding smooth points on $\widetilde{C}$ and we have a parametrisation for each branch of C. When C is a curve in projective space, as we will study in Chapter 7, the normalisation $\widetilde{C}$ itself carries important information.

The above argument establishes rather more than we have stated, and several different systems of terminology exist for utilizing this information. In the next section, we introduce the somewhat old-fashioned terminology of ‘infinitely near points’; in the following one we introduce the dual graph of the resolution.

3.5 Infinitely near points

We return to the case of a single branch B, and the resolution obtained as in Theorem 3.3.1 by repeatedly blowing up, though we now prefer to continue blowing up till a good resolution is obtained.

A point of the curve $E_0 \in T_1$ is said to be an *infinitely near point* of the first order to $O \in \mathbb{C}^2$; a point of $E_{r-1} \subset T_r$ is an infinitely near point of the rth order to O. Although a little imprecise, the terminology is certainly convenient.

The above iterative procedure produces one infinitely near point O_r on B of each order r. However, not all infinitely near points, even of the same order, have the same nature.

The basic geometric relation between infinitely near points is that of proximity. Suppose given infinitely near points O_i and O_j of respective orders i and j, with $j < i$, so that O_i determines a sequence of surfaces T_s with $s \leq i$ and maps $\pi_s : T_{s+1} \to T_s$ for $s < i$. Then O_i is said to be *proximate* to O_j if the curve E_j has a strict transform (also denoted E_j) in T_i, and O_i lies on the curve E_j, so that E_i intersects E_j in T_{i+1}. We note that O_i is always proximate to O_{i-1}.

Along with proximity we consider the multiplicity sequence $m_i(B)$, where $m_i(B)$ is the multiplicity at O_i of the strict transform $B^{(i)}$ of B in T_i. The following are the basic properties of the proximity relation.

Proposition 3.5.1

(i) *For each i there is at most one value $j < i - 1$ such that O_i is proximate to O_j.*

(ii) *If O_i is proximate to O_j, and $j < k < i$, then O_k is proximate to O_j.*

(iii) *The multiplicity $m_j(B)$ is equal to the sum of all $m_i(B)$ such that O_i is proximate to O_j.*

Proof Assertion (i) is an immediate consequence of Proposition 3.4.3. Since $\pi_{i-1}(O_i) = O_{i-1} \in T_{i-1}$, it follows inductively that O_i projects to $O_k \in T_k$. Since $O_i \in E_j$, it follows that O_k lies on the projection $E_j \subset T_k$ of $E_j \subset T_i$.

As to (iii), it suffices to consider the case $j = 0$. As in the proof of Theorem 3.3.1, we parametrise B by $x = t^m$, $y = b_1 t^m + b_2 t^{2m} + \ldots + b_q t^{qm} + ct^{\beta_1} + \ldots$, where $q = \lfloor \frac{\beta_1}{m} \rfloor$. Blowing up once, we obtain $B^{(1)}$, parametrised by

$$x_1 = t^m, \quad y_1 = b_1 + b_2 t^m + \ldots + b_q t^{(q-1)m} + ct^{\beta_1 - m} + \ldots,$$

in coordinates given by $(x_0, y_0) = (x_1, x_1 y_1)$, so E_0 is given by $x_1 = 0$. The point O_1 is $(0, b_1) \in E_0$ and we shift the origin, writing $y_1' := y_1 - b_1$.

There are two cases. If $\beta_1 > 2m$ the multiplicity is still $m_1(B) = m$. A further blowing up is given by $(x_1, y_1') = (x_2, x_2 y_2)$ in a chart where $x_1 \neq 0$, hence disjoint from E_0. The blown up curve $B^{(2)}$ is given by $x_2 = t^m$,

$y_2 = b_2 + b_3t^m + \ldots + b_qt^{(q-2)m} + ct^{\beta_1-2m} + \ldots$, so $O_2 = (0, b_2) \notin E_0$. Thus only O_1 is proximate to O_0 and the result holds in this case.

If, however, $\beta_1 < 2m$, the first term in the expansion of y_1 is that in t^{β_1-m}. Thus $m_1(B) = \beta_1 - m$: let us write m_1 for short. Divide m by m_1, obtaining $m = Qm_1 + R$ with $0 \le R < m_1$. Re-parametrise $B^{(1)}$ as $y_1' = u^{m_1}$, $x_1 = du^m + \ldots$, with $d \neq 0$. Now blow up a further Q times.

For $1 \le i \le Q$ we can write the coordinates in T_{i+1} as $x_{i+1} = y_1^{-i}x_1$ and $y_{i+1} = y_1$. The curve E_i is given by $y_{i+1} = 0$; the chart is defined where $y_i \neq 0$, so is disjoint from E_{i-1}; but the strict transform of E_0 is $x_{i+1} = 0$. The strict transform $B^{(i+1)}$ of B is given by $y_{i+1} = u^{m_1}$, $x_{i+1} = du^{m-im_1} + \ldots$.

If $i < Q$, the point $O_{i+1} = (0,0)$ lies on the strict transforms of E_0 and E_i, and the multiplicity $m_{i+1}(B) = m_1$. If $i = Q$ and $R = 0$, $O_{i+1} = (d, 0)$ is no longer on the strict transform E_0 of O_0. Thus only the points O_i for $1 \le i \le Q$ are proximate to O_0; each of these multiplicities $m_i(B) = m_1$, and their sum is indeed $Qm_i = m$.

If $R > 0$, $O_{Q+1} = (0,0)$ lies on the strict transforms of E_0 and E_Q, but the multiplicity $m_{Q+1}(B) = R$, since x_{Q+1} has lower order R than y_{Q+1}. Thus $B^{(q+1)}$ is tangent to $E_Q : y_{Q+1} = 0$ and not to $E_0 : x_{Q+1} = 0$. So after the next blow up we will have $O_{Q+2} \in E_q$, $O_{Q+2} \notin E_0$, so O_{Q+2} is proximate to O_Q but not to O_0. Thus the points proximate to O_0 are $O_1, \ldots, O_{Q+1}$, with multiplicities m_1 (Q times) and R, adding up to $Qm_1 + R = m$. □

We can think of the first assertion as follows. Since the curves E_j in T_i have normal crossings, O_i lies on E_{i-1} and at most one other E_j. In the first case there is a degree of freedom in the choice of the point O_i; in the second, O_i is the unique point of intersection $E_{i-1} \cap E_j$. In the latter case, O_i is sometimes called a *satellite point*.

Any sequence of proximity relations satisfying conditions (i) and (ii) of the Proposition corresponds to some branch: see Exercise 4.7.19.

It follows from the proposition that O_{i+2} is proximate to O_i if and only if $m_i(B) > m_{i+1}(B)$: the deduction of proximity relations from the sequence of multiplicities is now very simple.

Corollary 3.5.2 *For two irreducible curve-germs, with sequences $\{O_i\}$, $\{O_i'\}$ of infinitely near points, the following conditions are equivalent:*

(i) *we have the same proximity relations: O_i is proximate to O_j if and only if O_i' is proximate to O_j',*

(ii) *the sequence of multiplicities is the same for both.*

Proof First suppose the proximity relations are the same. Note that all multiplicities $m_i(B) = 1$ for i sufficiently large. If the multiplicities coincide for $i > j$, it follows from our hypothesis and Proposition 3.5.1(iii) that they also coincide for $i = j$; so by induction they coincide for all i.

Conversely if the multiplicities are the same, then for each j choose i such that $m_j(B) = \sum_{k=j+1}^{i} m_k(B)$. Then O_k is proximate to O_j if and only if $j < k \leq i$ and correspondingly for the O'_i. □

A convenient way to present the data of proximity relations is as follows. We define the *proximity matrix* $P(B)$ of the branch B to have entries $p_{i,j}$ given by

$$p_{i,j} := \begin{cases} 1 & \text{if } i = j, \\ -1 & \text{if } O_j \text{ is proximate to } O_i, \text{ so } i < j \\ 0 & \text{otherwise.} \end{cases} \tag{3.1}$$

Here $0 \leq i, j \leq N$, where N must be taken large enough to correspond to a good resolution. Since this matrix is upper unitriangular, it has determinant 1 and an inverse $Q(B) := P(B)^{-1}$ (also unitriangular) with integer entries.

Proximity relations may be graphically represented by writing the points O_i in sequence, and connecting O_i by an arc to O_j whenever O_j is proximate to O_i. Rather than write the symbols O_i it is more informative to write the multiplicities m_i.

Example 3.5.1 We return again to the curve $y^8 = x^{11}$ of Example 3.3.1. We had the sequence of multiplicities 8,3,3,2,1,1, ... ; each of O_1, O_2, O_3 is proximate to O_0; only O_2 is proximate to O_1; O_3, O_4 are proximate to O_2; O_4, O_5 are proximate to O_3; and only O_{i+1} is proximate to O_i if $i > 3$. We represent this by the diagram

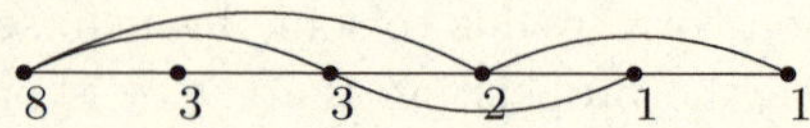

The proximity matrix and its inverse are given by

$$P(B) = \begin{pmatrix} 1 & -1 & -1 & -1 & 0 & 0 \\ 0 & 1 & -1 & 0 & 0 & 0 \\ 0 & 0 & 1 & -1 & -1 & 0 \\ 0 & 0 & 0 & 1 & -1 & -1 \\ 0 & 0 & 0 & 0 & 1 & -1 \\ 0 & 0 & 0 & 0 & 0 & 1 \end{pmatrix},$$

$$P(B)^{-1} = \begin{pmatrix} 1 & 1 & 2 & 3 & 5 & 8 \\ 0 & 1 & 1 & 1 & 2 & 3 \\ 0 & 0 & 1 & 1 & 2 & 3 \\ 0 & 0 & 0 & 1 & 1 & 2 \\ 0 & 0 & 0 & 0 & 1 & 1 \\ 0 & 0 & 0 & 0 & 0 & 1 \end{pmatrix}.$$

Lemma 3.5.3 *The multiplicity* $m_r(B)$ *defined above is the* (r, N) *entry in the inverse matrix* $Q(B)$.

Proof We can reformulate Proposition 3.5.1 (iii) as giving $\sum_i p_{j,i} m_i(B) = 0$. This equation holds for all $j < N$: the exception at $j = N$ occurs since we have not included points proximate to O_N. Thus for all i we have $\sum_i p_{j,i} m_i(B) = \delta_{j,N}$ (here we use Kronecker's delta notation: $\delta_{i,j}$ is equal to 1 if $i = j$ and to 0 if $i \neq j$). Now multiply by $q_{r,j}$ and sum over j. Since Q is the inverse matrix to P, we obtain $m_r(B) = q_{r,N}$. □

We can also interpret the other entries in this matrix. For any infinitely near point O_k we may choose a smooth curve $\tilde{C}_k$ crossing E_k transversely (in T_{k+1}) at a point not lying on any E_i with $i < k$. It projects to a curve in the original surface $T_0 = \mathbb{C}^2$ giving a branch ϵ_k which is in general singular. Such a branch is called a *curvette* at O_k. The terminology 'curvette' is relatively new: in older language such a branch was said to be 'primitive at O_k'.

The choice of a curvette allows a lot of freedom. We will see that many of the invariants of a curvette are determined by the definition. One example where it is easy to write down equations is for a monomial curve B, given by $x^r = y^s$, say. Then the charts in a resolution are all of the form $(x, y) = (x_k^a y_k^b, x_k^c y_k^d)$ with $ad - bc = 1$, so $(x_k, y_k) = (x^d y^{-b}, x^{-c} y^a)$; we can take $\tilde{C}_k$ to be a straight line with either x_k or y_k constant, so ϵ_k is a monomial curve.

Resolving the singularity of ϵ_k by blowing up recovers the sequence of points O_i for $i \leq k$. The same argument as above shows that the sequence of multiplicities $m_r(\epsilon_k)$ is given by the entries in the kth column of $Q(B)$.

To relate the sequence of multiplicities $m_r(B)$ to the Puiseux characteristic, we need to know the effect of blowing up on the Puiseux characteristic. We will assume invariance of the Puiseux characteristic under coordinate change (shortly to be proved in Corollary 4.1.4), since the direct verification is tedious.

We have already seen in Lemma 2.2.1 that any power series $f(t)$ of order n can be expressed as $g(t)^n$ and in Theorem 1.4.2 that if $h(t)$ has order 1, we can invert the substitution $u = h(t)$ to express t in the form $k(u)$; moreover, g is convergent if and only if f is, and k is convergent if and only if h is. We need information relating the set of exponents of t with non-vanishing coefficients in these various series.

Let $S \subset \mathbb{Z}$ be a set of non-negative integers including 0 and closed under addition, hence a semigroup. Let S_0 be a subset such that if $s = s' + s''$ with $s \in S_0$ and $s', s'' \in S$ then either $s' = 0$ or $s'' = 0$. Write $\mathcal{O}_S$ for the set of (convergent) power series $\sum_r a_r t^r$ such that $a_r = 0$ for all $r \notin S$; and $\mathcal{O}_S^*$ for the subset satisfying the further condition $a_r \neq 0$ for all $r \in S_0$.

Lemma 3.5.4

(i) Let $(t\alpha(t))^m = t^m\gamma(t)$ with $\alpha(0) \neq 0$. Then $\alpha \in \mathcal{O}_S$ if and only if $\gamma \in \mathcal{O}_S$, and $\alpha \in \mathcal{O}_S^*$ if and only if $\gamma \in \mathcal{O}_S^*$.

(ii) Let α be a power series with $\alpha(0) \neq 0$, and let β be such that $t = u\beta(u)$ solves $u = t\alpha(t)$. Then $\alpha \in \mathcal{O}_S$ if and only if $\beta \in \mathcal{O}_S$, and $\alpha \in \mathcal{O}_S^*$ if and only if $\beta \in \mathcal{O}_S^*$.

Proof Write $\alpha(t) = \sum_0^\infty \alpha_r t^r$ (so that $\alpha_0 \neq 0$), and similarly for γ. Then

$$\sum_0^\infty \gamma_r t^r = \left(\sum_0^\infty \alpha_r t^r\right)^m = \sum_{r_1,\dots,r_m} (\alpha_{r_1} \dots \alpha_{r_m}) t^{r_1+\dots+r_m}.$$

If $\alpha \in \mathcal{O}_S$, $\alpha_r = 0$ for all $r \notin S$, so the only terms on the right hand side with non-zero coefficient are those with each $r_i \in S$, hence also $\sum r_i \in S$. Conversely, suppose $\gamma \in \mathcal{O}_S$ and that for each $r < k$ with $r \notin S$ we have $\alpha_r = 0$. If $k \notin S$, the coefficient of t^k on the right hand side is $m\alpha_0^{m-1}\alpha_k$. Thus $\alpha_k = 0$, and it follows by induction on k that $\alpha \in \mathcal{O}_S$.

If further $p \in S_0$, to find the coefficient of t^p on the right hand side we must again have $r_i = 0$ for all but one value of i, so $\gamma_p = m\alpha_0^{m-1}\alpha_p$. Thus indeed $\gamma_p \neq 0$ if and only if $\alpha_p \neq 0$.

Here substituting one power series in the other gives the identity

$$u = \sum_s \sum_{r_0,\dots,r_s} \alpha_s \beta_{r_0} \dots \beta_{r_s} u^{s+1+r_0+\dots+r_s}.$$

If $k \notin S$ and α_r and β_r vanish for all $r < k$ with $r \notin S$, the coefficient of u^{k+1} on the right hand side is given by terms with one of $s, r_0, \ldots, r_s$ equal to k and the rest zero, hence is $\alpha_0\beta_k + \alpha_k\beta_0^{k+1}$. Since $\alpha_0 \neq 0$ and $\beta_0 \neq 0$ we have $\alpha_k = 0$ if and only if $\beta_k = 0$. Hence by induction, $\alpha \in \mathcal{O}_S$ if and only if $\beta \in \mathcal{O}_S$.

Similarly, if these conditions hold and $p \in S_0$, the coefficient of u^{p+1} on the right hand side is $\alpha_0\beta_p + \alpha_p\beta_0^{p+1}$. Thus $\alpha_p = 0$ if and only if $\beta_p = 0$. □

Theorem 3.5.5 *Suppose given an irreducible curve whose Puiseux characteristic is* $(m; \beta_1, \ldots, \beta_g)$. *Then the Puiseux characteristic of the curve obtained by blowing up is given by*

$\beta_1 > 2m$	$(m\,; \beta_1 - m, \ldots, \beta_g - m)$
$\beta_1 < 2m, (\beta_1 - m) \nmid m$	$(\beta_1 - m\,; m, \beta_2 - \beta_1 + m, \ldots, \beta_g - \beta_1 + m)$
$(\beta_1 - m) \mid m$	$(\beta_1 - m\,; \beta_2 - \beta_1 + m, \ldots, \beta_g - \beta_1 + m)$.

Proof Write a parametrisation for the curve as $x = t^m$, $y = \sum_1^\infty a_r t^r$. Making the substitution $y' = y - \sum_{ms<\beta_1} a_{ms}x^s$, we may suppose that $a_r = 0$ for $r < \beta_1$. Then the blown up curve is parametrised by $x_1 = t^m$, $y_1 = \sum_{\beta_1}^\infty a_r t^{r-m}$. If $\beta_1 > 2m$, this is in standard form, so the Puiseux characteristic is as given.

If, however, $\beta_1 < 2m$, the blown up curve has multiplicity $\beta_1 - m$, and we take a new parametrisation with $y = u^{\beta_1 - m}$. Now apply Lemma 3.5.4, taking

$$S := \{r \in \mathbb{Z} \mid \text{for some } q \geq 1,\ r \geq \beta_q - \beta_1 \text{ and } e_q|r\}.$$

and $S_0 := \{\beta_q - \beta_1 \mid q \geq 1\}$.

Our hypothesis gives $y_1 = t^{\beta_1 - m}\alpha(t)$ with $\alpha \in \mathcal{O}_S^*$. Hence $y_1 = (t\beta(t))^{\beta_1 - m}$, with $\beta \in \mathcal{O}_S^*$. Set $u = t\beta(t)$, so that $y_1 = u^{\beta_1 - m}$. By the lemma again, $t = u\gamma(u)$ with $\gamma \in \mathcal{O}_S^*$. Thus $x = t^m = (u\gamma(u))^m = u^m\delta(u)$ with $\delta(u) \in \mathcal{O}_S^*$.

Provided m is not divisible by the multiplicity $\beta_1 - m$, m itself is the first characteristic exponent in the series $u^m\delta(u)$; the highest common factor of $\beta_1 - m$ and m is e_1, and the Puiseux exponents can be read off from the definition of $\mathcal{O}_S^*$.

If however $(\beta_1 - m) \mid m$, then $\beta_1 - m = e_1$, and the first exponent not divisible by e_1 is $\beta_2 - \beta_1 + m$. The remaining assertions are immediate. □

Theorem 3.5.6 *The Puiseux characteristic of a branch B determines the sequence of multiplicities $m_i(B)$, and conversely.*

Proof It follows at once from Theorem 3.5.5 that from the Puiseux characteristic of the branch B we can calculate those of the branches obtained by successive blowings up, and hence their multiplicities $m_i(B)$. To see the converse, first observe that the three cases of the theorem are distinguished by the multiplicities $m_0(B)$ and $m_1(B)$: in the first case, $m_1(B) = m_0(B)$; in the second case, $m_1(B)$ does not divide $m_0(B)$, and in the third, $m_1(B)$ is a proper divisor of $m_0(B)$.

We induct on the number of values of i for which $m_i(B) > 1$. If there are none, the curve is smooth. If there is just one, we must have $\beta_1 - m = 1$, so the Puiseux characteristic is $(m; m+1)$, with $m = m_0(B)$. In general we may suppose that the sequence of multiplicities determines the Puiseux characteristic of the blown up curve, which is $(m_1; \beta_1, \ldots, \beta_g)$. It follows that:

if $m_1(B) = m_0(B)$, the Puiseux characteristic of the original curve B is $(m_1; \beta_1 + m_1, \ldots, \beta_g + m_1)$;

if $m_1(B)$ does not divide $m_0(B)$, the Puiseux characteristic of B is $(m_0; \beta_1 + m_1, \ldots, \beta_g + m_1)$;

if $m_1(B)$ is a proper divisor of $m_0(B)$, the Puiseux characteristic of B is $(m_0; m_0 + m_1, \beta_1 + m_1, \ldots, \beta_g + m_1)$. □

This result appears to establish invariance of the Puiseux characteristic under coordinate change, but we have already assumed this in the foregoing arguments.

The cases in Theorem 3.5.5 also correspond to the different cases for proximity relations. In the first case $m_1(B) = m_0(B)$ and so only O_1 is proximate to O_0. Otherwise $m_1(B) = m' < m_0(B) = m$ and we distinguish cases according as the remainder r given by division $m = qm' + r$ vanishes (the third case of the theorem) or not: only if $r \neq 0$ is O_{q+1} (of multiplicity $< m'$) proximate to O_0.

The calculus of infinitely near points permits also an efficient description of singularities of curves with more than one branch. Indeed, the points O_n were described abstractly and the property that (the strict transform of) C passes through O_n is intrinsic. However, the definition of an infinitely near point O_i as a point in a particular blow up T is inadequate. Perhaps the best formal definition of an infinitely near point is in terms of ideals of functions: see Chapter 11.

Given a curve C with several branches, we may take a good resolution of it. If the resolution is minimal, each exceptional curve arises in the resolution of at least one component of C, and its relation to its predecessors is thus the same as before. However, instead of all the points being arranged in a single sequence, the sequences for the different components overlap, so the result takes the form of a tree. To each point of this tree is attached a multiplicity, which is the sum of those arising from the different branches of C. It thus follows that Proposition 3.5.1 (iii) extends to the case of several branches. We will return to this in Chapter 8.

3.6 The dual graph

A geometric way to present resolution data of a curve C is by the dual graph. The term 'graph' has several meanings in mathematics, but we will need it in its combinatorial sense. So for us a *graph* Γ consists of

a set $\mathcal{V}(\Gamma)$ of vertices V_i,

and

a set $\mathcal{E}(\Gamma)$ of edges, each one incident to a pair of vertices. We picture the edge as joining the vertices.

We now give some basic terminology about graphs. Our graphs will have the additional properties that no edge joins a vertex to itself, and there is at most one edge joining any given pair of vertices. Graphs with these properties are said to be *simple*. For a simple graph there is no need for a special notation for the edges: we may simply refer to 'the edge V_iV_j'.

The *valence* of a vertex of Γ is the number of edges incident to it. Thus the sum of all the valences of the vertices is twice the total number of edges. We will say that a vertex of valence at least 3 is a *rupture point* of Γ.

A sequence of edges $V_{i_1}V_{i_2}, V_{i_2}V_{i_3}, \ldots V_{i_{N-1}}V_{i_N}$ is called a *path* from V_{i_1} to V_{i_N} in Γ. A path is *simple* if all the vertices in the sequence are mutually distinct. The path is *closed* if $V_{i_1} = V_{i_N}$; a closed path is a *cycle* if all the edges are distinct. A graph containing no cycles is called a *forest.*

One can define an equivalence relation on $\mathcal{V}(\Gamma)$ by letting $V_i \sim V_j$ if there is a path in Γ from V_i to V_j. The equivalence classes are the *components* of Γ, which is *connected* if there is just one component.

A connected forest is called a *tree*. The graphs of importance in this book will all be trees.

A tree can be built up from a single point by repeatedly adding an edge joining some vertex of the graph to a new vertex. Thus the number of vertices exceeds the number of edges by 1. Conversely, a connected graph such that the number of vertices exceeds the number of edges by 1 is a tree.

For any two vertices V_i, V_j of a tree there is a unique simple path joining them. It is called the *geodesic* from V_i to V_j.

In questions of graph theory there is often some additional structure attached to the graph. This may consist in having orientations defined for some or all of the edges. For the graphs that arise in this book we will often have numerical functions on the set of vertices of the graph.

Let C have branches B_j, and let $\pi : T_N \to T_0$ be a good resolution (we nearly always take the minimal good resolution). Consider the exceptional curves E_i $(0 \leq i < N)$ and the strict transforms $B_j^{(N)}$ in T_N. The *dual graph* $\Gamma_R(C)$ is defined to be the abstract graph with vertices V_i corresponding to the curves E_i, and with an edge joining V_i to V_k if and only if the curves E_i and E_k intersect.

We can build this up one step at a time. Suppose we already have a graph for the curves in T_{i-1}. If O_i is proximate only to O_{i-1} we adjoin a new vertex V_i and a new edge $V_{i-1}V_i$. If O_i is proximate to both O_{i-1} and some O_j with $j < i-1$, then E_{i-1} and E_j intersect in T_i, so there is already an edge $V_{i-1}V_j$. We replace this by two edges $V_{i-1}V_i$ and V_iV_j – one may think of this as subdividing the edge $V_{i-1}V_j$ at a new vertex V_i. These two cases are those pictured in Figure 3.3. We see inductively that the dual graph is in all cases a tree.

The reader should construct the graphs corresponding to all possible cases for proximity relations between O_0, O_1, O_2 and O_3.

We define the *augmented dual tree* $\Gamma_R^+(C)$ in the same way, but including also vertices W_j corresponding to the curves $B_j^{(N)}$. It is convenient to distinguish these pictorially by attaching an arrow along each new edge pointing towards its vertex W_j, so they are sometimes called 'arrowhead vertices'. Note that each W_j has valence 1 in this graph: this holds at each stage of the inductive process of building up the tree.

Example 3.6.1 We return to the curve of Example 2.1.1, with parametrisation $x = t^4$, $y = t^6 + t^7$, and hence Puiseux characteristic $(4; 6, 7)$. Blowing up to get a good resolution gives successive multiplicities

4,2,2,1,1. Thus O_2 is proximate to O_0 and O_4 to O_2, as well as O_{i+1} to O_i for each i. Proximities are thus given by the diagram

and the proximity matrix is

$$P = \begin{pmatrix} 1 & 0 & 0 & 0 & 0 \\ -1 & 1 & 0 & 0 & 0 \\ -1 & -1 & 1 & 0 & 0 \\ 0 & 0 & -1 & 1 & 0 \\ 0 & 0 & -1 & -1 & 1 \end{pmatrix}.$$

The dual graph $\Gamma_R^+(C)$ is shown in the following picture.

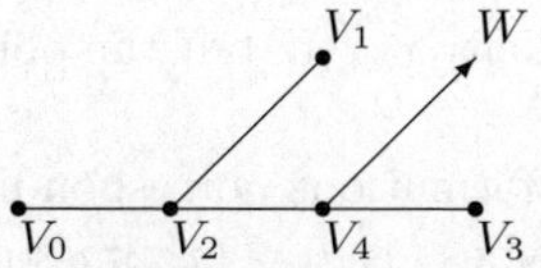

Example 3.6.2 Let C be the union of $B_1 : y^2 = x^3$ and $B_2 : y^3 = x^4$. The first blow up $y = x_1y_1$ gives $y_1^2 = x_1$ and $y_1^3 = x_1$; the second blow up $x_1 = x_2y_2$ gives $y_2 = x_2$ and $y_2^2 = x_2$, which both go through the origin but are no longer tangent to each other; thus B_1 and B_2 have no further infinitely near points in common. The diagram of infinitely near points is thus

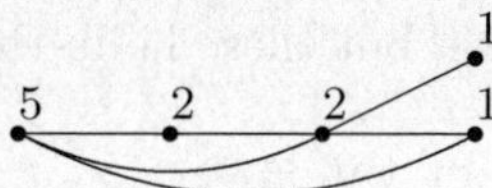

and the dual graph is

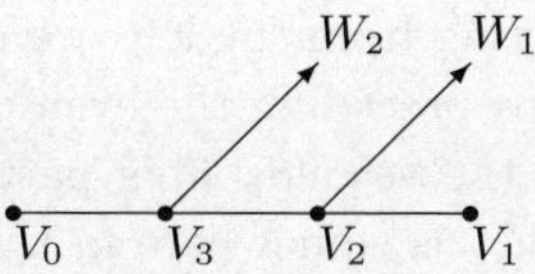

For a single branch B we can analyse all our invariants in terms of the Puiseux characteristic, and we now investigate how $\Gamma_R^+(B)$ is built up. We begin with the Newton polygon. By Lemma 2.4.4, the Newton polygon for a single branch consists of a single edge, say from $(ad, 0)$ to

$(0, ad + bd)$ with a and b coprime. The first group of blowings up corresponds to the steps in the Euclidean algorithm for finding the highest common factor, d, of ad and bd.

Suppose the steps in this algorithm as follows:

$$\begin{aligned} a &= bq_1 + r_1 \quad (0 < r_1 < b) \\ b &= r_1 q_2 + r_2 \quad (0 < r_2 < r_1) \\ &\dots \\ r_{f-1} &= r_f q_{f+1}; \end{aligned} \tag{3.2}$$

(so $r_f = d$); we will write $s_k = \sum_{i=1}^{k} q_i$. Then O_0 has multiplicity ad; the next q_1 points have multiplicity bd and are proximate to O_0, as is the next, with multiplicity $r_1 d$. The next $q_2 - 1$ points also have multiplicity $r_1 d$, and are proximate to O_{q_1}. In general, O_{s_k} has multiplicity $r_{k-1}d$; the next q_{k+1} points are proximate to it, each with multiplicity $r_k d$, as is $O_{s_{k+1}+1}$, with multiplicity $r_{k+1}d$; but the point following is proximate only to $O_{s_{k+1}}$.

The coordinate transformations corresponding to these blowings up are all of the standard types $(x, y) = (x', x'y')$ or $(x'y', y')$, with the new centre at the origin in each case except for O_{s_f+1}. As in the proof of Proposition 3.5.1, we have a sequence of transformations of one type, followed by a sequence of those of the other type; the change-over corresponding to a proximity relation. We observe that the composite of all these coordinate changes is still a monomial transformation – i.e. it is of the form $(x, y) = (X^p Y^b, X^q Y^a)$, with $ap - bq = 1$.

It is easy to follow what happens to the dual graph, but the result is a little unexpected. As far as V_{s_f}, the graph may be considered as a sequence of points on a line, but these lie in the order

$$\{0, s_1 + 1, \dots, s_1 + q_2 = s_2, \cdots, \cdots s_2 + q_3 = s_3, \dots, \\ s_2 + 1, q_1 = s_1, \dots, 2, 1\}$$

We emphasise that, starting from the left, we have the first group, then the third; the odd groups preceding the even ones which conclude with the fourth group, then the second. This pattern will be explained in Section 8.5. The point V_{s_f} is somewhere in the middle; V_{s_f+1} is joined to it by a segment leaving the above line; then we have a sequence of segments end-to-end until we again start obtaining proximate points.

We can construct the whole of $\Gamma_R^+(B)$ inductively, using the effect (Theorem 3.5.5) of blowing up on the Puiseux characteristic. The general shape of the result is as follows: see Figure 3.5 for an illustration.

Lemma 3.6.1 *If B is an irreducible curve with Puiseux characteristic $(m; \beta_1, \dots, \beta_g)$, then $\Gamma_R^+(B)$ consists of a single chain of edges from the initial vertex V_0 to the vertex W, with g side branches, each a single chain, attached at distinct vertices of the original chain.*

Proof We have described above the sequence of blowings up required to reduce the length of the Puiseux characteristic. We also found that the dual graph up to this point consists of a chain from V_0 to V_{s_f}, a side chain attached at V_{s_f}, and that any later vertices of the graph will be attached at V_{s_f}. We can now repeat the procedure; the result follows by induction on g. □

We will call the chain of vertices and edges from V_0 to W the *core* of $\Gamma_R^+(B)$.

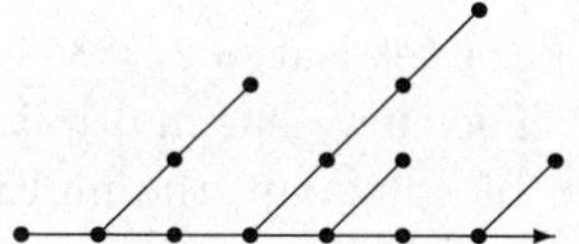

Fig. 3.5. Typical dual graph of branch

For a curve C with several branches, $\Gamma_R^+(C)$ is in some sense the union of the trees $\Gamma_R^+(B_j)$ corresponding to the different branches, overlapping as appropriate. We will give a more precise statement in Chapter 9.

Although the sequence of proximity relations (or of multiplicities) determines $\Gamma_R^+(C)$, the abstract structure of the graph does not suffice to allow us to reconstruct this data, as is clear from the above discussion, where many different sequences yield straight line graphs. The graph will thus always be used in conjunction with additional data.

We can simply label each vertex V_r by the integer r corresponding to its position in the sequence of blowings up of infinitely near points O_r. This already suffices to determine the proximity relations and hence, by Corollary 3.5.2, the sequence of multiplicities.

Lemma 3.6.2 *For a single branch B the graph $\Gamma_R^+(B)$, labelled by the order in which vertices are generated, determines the proximity relations.*

Proof Consider the last vertex V_n added to the graph. We saw in the second paragraph of this section, as a corollary of Proposition 3.5.1, that there are edges joining V_n to V_{n-1} and at most one further vertex V_j with $j < (n-1)$. Correspondingly, O_n is proximate to O_{n-1} and – if there is such a j – O_j.

In the same paragraph we described how $\Gamma_R^+(B)$ is built up inductively. Thus the preceding stage is obtained by, if there is no j, simply deleting the edge $V_{n-1}V_n$; and if there is, replacing the two edges V_jV_n and $V_{n-1}V_n$ by the single edge V_jV_{n-1}. This construction uses only the labelling, so the result follows by induction. □

We may also label each vertex V_i with the corresponding multiplicity $m_i(B)$. In the case of a single branch, since we may rearrange these in decreasing order, it follows from Corollary 3.5.2 that this determines the pattern of proximity relations for B.

Example 3.6.3 Consider a curve $y^2 = x^{k+1}$ with an A_k singularity. If $k = 2r$ is even, there is a parametrisation $x = t^2$, $y = t^{2r+1}$, with Puiseux characteristic $(2; 2r+1)$. The curve has multiplicity 2 and its blow up is $y_1^2 = x_1^{2r-1}$. Thus after r blowings up, the multiplicity becomes 1. The point O_{r+1} as well as O_r is proximate to O_{r-1}.

O_0 O_1 O_{r-1} O_r O_{r+1}
2 2 2 1 1

If $k = 2r-1$ is odd, the curve has two smooth branches: $y = x^r$ and $y = -x^r$. Again, blowing up yields a sequence $O_0, \ldots, O_{r-1}$ of points of multiplicity 2. One further blow up separates the branches, and we have distinct points O_r, O'_r, each proximate only to O_{r-1}. Since substituting for y from one equation in the other gives $2x^r$, the branches have intersection number r.

Since by Theorem 2.2.7 every singularity with multiplicity 2 has type A_k for some k, these descriptions show how k is determined by the properties of successive blowings up.

3.7 Notes

Section 3.1 We have adhered to a single notation for the Puiseux characteristic to avoid confusing the reader, but other conventions are also in use in the literature. The 'characteristic Puiseux pairs' (m_i, n_i) $1 \leq i \leq g$ are defined by $m_i = \beta_i/e_i$, $n_i = e_{i-1}/e_i$, so that we can recover e_i as $m/(n_1 n_2 \ldots n_i)$ and β_i as $e_i m_i$. The 'Zariski characteristic pairs' (p_i, q_i) $1 \leq i \leq g$ are defined by $q_i = n_i$, $p_1 = m_1$ and $p_i = m_i - n_i m_{i-1}$ for $1 < i \leq g$. These arise by direct application of the algorithm of Section 2.1, and occur in the normal forms

$$y = x^{q_1/p_1} \left(a_1 + x^{q_2/p_1 p_2} \left(a_2 + \ldots \right) \right),$$

as in (2.1), for a parametrisation and

$$\ldots((y^{p_1} - A_1 x^{q_1})^{p_2} - A_2 x^{q_2}) \ldots = 0$$

for an equation.

Section 3.2 The direct generalisation of blowing up to higher dimensions applies when we have a complex manifold M and a closed submanifold V. For a local model, suppose M a vector space and V a vector subspace. Then the blow up $\tilde{M}$ is a subspace of $M \times P(M/V)$, where $P(M/V)$ denotes the projective space of the vector space M/V. The projection $\tilde{M} \to M$ is an isomorphism over $M \setminus V$, but the preimage of each point of V is isomorphic to $P(M/V)$. The preimage of V is $V \times P(M/V)$: it has codimension 1, so is a divisor, called the exceptional divisor of the blow up. The definition can be given globally, also for algebraic varieties. A yet more general definition begins with an algebraic variety M and a sheaf $\mathcal{I}$ of ideals over the structure sheaf $\mathcal{O}_M$, and produces a variety $\tilde{M}$ such that the pull back of $\mathcal{I}$ becomes a (sheaf of) principal ideal.

Sections 3.3,3.4 Resolution of singularities of curves in the projective plane was first proved by Max Nœther [145]. In early work on resolution, instead of blowing up, 'standard quadratic transformations' were used. In convenient coordinates, this is just the correspondence $(x : y : z) \mapsto (yz : zx : xy)$ from $P^2(\mathbb{C})$ to itself: it is not well-defined at the 'base points' (1,0,0), (0,1,0) and (0,0,1). While this has the advantage that all the arguments can take place in the fixed space $P^2(\mathbb{C})$, it has the inconvenience of being a mixture of blowing up and the opposite blowing down so that while some singularities are being simplified, others are being introduced. Thus the end result was more complicated, allowing singularities consisting of a number of smooth branches meeting mutually transversely. See [165] for a traditional textbook treatment. The move to the version described here began with Zariski [203].

A different approach uses 'toric' blowings up, which extend maps $\mathbb{C}^2 \setminus (0,0) \to \mathbb{C}^2 \setminus (0,0)$ of the form $(x,y) \mapsto (x^a y^b, x^c y^d)$. An irreducible curve singularity with g Puiseux pairs can be resolved by g toric blowings up. A treatment is given in [146]. Yet another version is to give an à priori construction of an object, and prove it to be a non-singular curve which resolves the singularities of the given curve C: such an account is given in [99].

The problem of resolution of singularities of algebraic varieties of higher dimensions was resolved in dimension 2 by the Italian school, in dimension 3 by Zariski, and in arbitrary dimension in the landmark paper of Hironaka [88]. Although there is a recent literature giving alternative proofs, the problem of extending the result to varieties (of dimension greater than 2) over fields of finite characteristic remains open.

The preferred procedure for resolution is to consider a variety X embedded in a smooth variety M and blow M up along a smooth submanifold V: this produces a strict transform $\tilde{X} \subset \tilde{M}$ and an exceptional divisor E. For a good resolution we require that the collection of strict transforms of X and all the exceptional divisors arising has normal crossings, in that at any point of the final blow up of M there are local coordinates such that each of these transforms is defined locally by the vanishing of one or more of these coordinates.

A proof of resolution of curves in finite characteristic is given in [27]1.5.10. Campillo starts with the local ring $\mathcal{O}$ and shows that the quotient $\tilde{\mathcal{O}}/\hat{\mathcal{O}}$ has finite length, and that blowing up the origin replaces $\hat{\mathcal{O}}$ by a strictly larger subring of $\tilde{\mathcal{O}}$ unless the two are already equal and so Γ smooth.

The existence of good resolutions for curves with several branches now follows by the same arguments as over $\mathbb{C}$. The entire apparatus of infinitely near points, proximity relations, and dual graphs now develops as before. Campillo also observes that any graph satisfying the proximity relations corresponds to a curve. We will show in Section 5.3 that we can always find an example where the parametrisations all have integer coefficients.

Section 3.5 An alternative notation for describing proximity relations for infinitely near points is due to Enriques [66]. We refer to Casas' book [35] – which makes infinitely near points the central tool of the treatment – for a precise description. For the resolution of a branch, one draws a curve in the plane containing the O_i in the order of their suffices such that

(i) If O_{i+1} is not a satellite point, the edge O_iO_{i+1} is smooth but curved; if $i > 0$ it has the same tangent at O_i as the edge $O_{i-1}O_i$.

(ii) The edges connecting a maximal sequence of vertices proximate to O_i form a straight line segment, orthogonal to the tangent at O_{i+1} to O_iO_{i+1}.

Classical Italian geometers such as Enriques [66] thought of infinitely near points as actual 'hidden' points on the curve which could be treated in the same way as other points; and indeed we will obtain (in Lemma 4.4.2 and Theorem 6.5.9) some formulae where this philosophy does apply, though care is always necessary.

The terminology 'curvette' is due to P. Deligne.

The proximity matrix $P(B)$ was introduced by du Val [60].

Section 3.6 The dual graph of the minimal good resolution of a branch is described by Brieskorn [23] as a *mobile*: he thinks of the chain of points corresponding to an application of the Euclidean algorithm to $(a : b)$ as a horizontal bar, suspended at the point V_{s_f} from the next point; then we have a vertical chain until the next bar. The book [23] gives an alternative introduction to a number of the topics in this book.

We will study the dual graph in much greater detail in Chapter 8 and Chapter 9.

3.8 Exercises

Exercise 3.8.1 Determine the Puiseux characteristic of each of the following curves:

(a) $(t^2, t^3 + t^4)$ (b) $(t^2, t^4 + t^6 + t^7)$

(c) $y = x^{5/2} + x^{13/4}$ (d) $y = x^{7/6} + x^{37/24} + x^{20/9}$.

Exercise 3.8.2 Classify irreducible singularities of multiplicity 4 up to equisingularity by listing all possibilities for Puiseux characteristics. In each case, determine the sequence of multiplicities.

Exercise 3.8.3 Draw pictures illustrating the successive stages of blowing up for obtaining good resolutions of (i) $y^3 = x^5$, (ii) $y^3 = yx^3$.

Exercise 3.8.4 Give explicit examples of curvettes for the curve B_1 given by $x^3 = y^5$.

Exercise 3.8.5 Let B_1 be a branch with Puiseux characteristic $(2;5)$. List the sequence of multiplicities $m_i(B_1)$ occurring in a minimal good resolution of B_1. Hence determine the proximity relations arising. Write down the proximity matrix P. Draw the augmented dual graph of the resolution. Sketch the pattern of strict transforms of B_1 and the exceptional curves at each stage of the resolution process.

Exercise 3.8.6 Give an alternative proof of Proposition 3.5.1 (iii) by arguing in terms of the Newton polygons of B and the relevant blown up curves.

Exercise 3.8.7 For each of the following cases, list the sequence of multiplicities occurring in a good resolution of a curve with the given Puiseux characteristic, find the proximity relations, write down the proximity matrix, and describe the augmented dual graph Γ^+_R: (a) $(3;8)$, (b) $(4;10,13)$, (c) $(6;15,22)$.

Exercise 3.8.8 Let $f = x^5 + x^2y^2 + y^5$. Show, by blowing up, that the curve C defined by $f = 0$ has two branches at O, each with Puiseux characteristic (2;3), and with distinct tangents.

Exercise 3.8.9 Let C_1, C_2 be the curves defined by the respective equations $(y^3 - x^4)^2 = x^7y^2$, $(y^2 - x^3)^3 = x^{11}$. By repeated blowing up, and noting the sequence of multiplicities obtained, show that each C_i has a single branch, determine their Puiseux characteristics, and their resolution graphs.

By checking the two series of blowings up, determine how many infinitely near points C_1 and C_2 have in common. Hence obtain the resolution graph for $C_1 \cup C_2$.

Exercise 3.8.10 Deduce from Theorem 3.5.5 that the length $g + 1$ of the Puiseux characteristic first becomes shorter after s_{f+1} blowings up, and that it then has the form $(e_1; \beta_2 - \alpha, \ldots, \beta_g - \alpha)$. Determine the value of α. Determine also the Puiseux characteristic of the blown up curve corresponding to the first non-satellite infinitely near point.

4

Contact of two branches

In order to give a full discussion of the case of curves with several branches at the singular point it is necessary to discuss the type of contact that two such branches can have. This also allows us to fill in several details in our discussion of the geometry of a single branch.

We will express our result in terms of the exponent of contact of two branches. In fact, we obtain a more flexible concept by introducing the notion of pro-branch and exponents of contact of pro-branches.

The most important result is a formula relating exponent of contact to intersection multiplicity. This is the key to numerous later developments. The basic formula relates to the case when each curve has just one branch. We then develop a notation to express the type of contact of curves with several branches. It takes the form of a tree with numerical information attached, which seems best suited to describe the numerical invariants of curves with several branches.

We use our main formula to give a complete description of the semi-group of a branch. The intersection multiplicity can also be expressed in terms of the calculus of infinitely near points, and establish the essential equivalence of these two approaches, which is formalised by the notion of equisingularity.

A further section gives an application of these techniques to give a proof of a recent theorem on the decomposition of polar curves. As this is somewhat outside the main line of development of the first half of this book, it may be omitted on a first reading.

4.1 Exponents of contact and intersection numbers

In this section we define pro-branches and their exponents of contact. We express the intersection number of two branches as a sum of exponents of contact of their pro-branches, and deduce an important formula for it in terms of the exponent of contact of the two branches. This is applied, among others, to the contact of a curve with a transverse polar curve.

Let C be a single branch, with equation $f(x, y) = 0$, and assume that the y axis is not tangent to C at O. By Puiseux' theorem, we may write y as a fractional power series in x

$$y = \sum_{r \geq m} a_r x_m^r, \qquad \text{with } x_m^m = x. \tag{4.1}$$

The m roots of the equation for y are obtained by taking successively for x_m the mth roots of x.

We want to be able to distinguish a single such root. In view of the monodromy, to do this it is necessary to restrict x to lie in a sector $|arg\, x - \alpha| < \epsilon$ of the plane of complex numbers. Fixing a root gives what we will call a *pro-branch* defined over this sector. For example, if the sector contains the positive real axis, we have a uniquely defined x_m, for any m, which takes positive real values on the positive real axis, so we can write fractional powers of x without ambiguity. Over any sector it is possible to make systematic choices similarly. Although the terminology 'pro-branch' is new, making a choice of x_m (in fixed coordinates) is standard procedure.

Suppose γ, γ' are pro-branches of branches B, B', defined over the same sector. We may write their equations in the form

$$y = \sum_s c_s x^s, \qquad y = \sum_s c'_s x^s,$$

where the exponents s satisfy $s \geq 1$ and $m(B)s \in \mathbb{Z}$, $m(B')s \in \mathbb{Z}$ respectively. We define the *exponent of contact* to be $\mathcal{O}(\gamma, \gamma') := \min\{s \mid c_s \neq c'_s\}$. This notion appears to depend on coordinates, but we see at once that it may also be expressed geometrically in more invariant terms.

Lemma 4.1.1 *Suppose given pro-branches γ, γ' with exponent of contact κ. Suppose for $P \in \gamma'$ that the distance of P from γ is $d(P, \gamma)$ and its distance from O is $d(P, O)$: then as $P \to O$ we have $d(P, \gamma)/d(P, O)^\kappa \to C$ for some constant $C \neq 0$.*

Proof If the two branches have distinct tangents, the exponent of contact is 1 and the ratio $d(P, \gamma)/d(P, O)$ converges to a value determined by the angle between the branches (the precise value depends on the choice of metric).

Otherwise we may take the common tangent as $y = 0$. Then for points close enough to O all the tangent lines make small angles with $y = 0$ and the line joining P to the closest point on γ is nearly perpendicular to this. Thus the distance of P from the point of γ with the same value of x differs from $d(P, \gamma)$ by a factor that tends to 1 as $P \to O$. The result follows. □

It will be convenient to make the convention that $\mathcal{O}(\gamma, \gamma) = \infty$. If we have three pro-branches γ, γ' and γ'' then in all cases

$$\mathcal{O}(\gamma'', \gamma) \geq \min(\mathcal{O}(\gamma, \gamma'), \mathcal{O}(\gamma', \gamma'')), \tag{4.2}$$

so the two smaller of the three orders coincide.

In practice, we fix the sector (which we may think of as including the positive real axis in x) and then consider the set of all pro-branches defined in the given sector. The following observation, which shows that we obtain an invariant of branches, not just pro-branches, is crucial.

Lemma 4.1.2 *Let γ be a pro-branch of B. Then the set of exponents of contact of γ with the different pro-branches of B' does not depend on the choice of the pro-branch γ.*

Proof Clearly the exponents of contact are not affected if we shrink the sector. Each exponent of contact is also not changed if we rotate the sector round the origin in $\mathbb{C}$, using analytic continuation; hence the set as a whole is unaltered. But a full rotation round O permutes the pro-branches of B transitively, so the set is also independent of the choice of γ. □

This result allows us to use pro-branches without reference to choices. We may fix a sector, and write pro$\,(B)$ for the set of all pro-branches of B (over that sector), so the set defined above may be written $\{\mathcal{O}(\gamma', \gamma) \mid \gamma \in \text{pro}\,(B)\}$. This will be the basis for arguments to be developed below. We emphasise that although the set is unordered, we do count its elements with multiplicities.

The exponent of contact of the branches is defined as

$$\mathcal{O}(B, B') = \min\{\mathcal{O}(\gamma, \gamma') \mid \gamma \in \text{pro}\,(B), \gamma' \in \text{pro}\,(B')\}.$$

Here we may fix γ or γ' (but not, of course, both). It follows from (4.2) that given three branches B, B' and B'', we have

$$\mathcal{O}(B, B'') \geq \min(\mathcal{O}(B, B'), \mathcal{O}(B', B'')) \tag{4.3}$$

so the two smaller of the three numbers $\mathcal{O}(B, B')$, $\mathcal{O}(B, B'')$ and $\mathcal{O}(B', B'')$ coincide. We have $\mathcal{O}(B, B') = \infty$ if and only if $B = B'$.

For further analysis, it is convenient to change slightly the notation of the preceding chapter, and write $\alpha_q = \frac{\beta_q}{m}$: we may call these the *Puiseux exponents*: they are the most important exponents appearing in the Puiseux series $y = \sum a_q x^q$. Notice that if two branches have Puiseux series which agree up to some exponent $\kappa > \alpha_q$, then they have the same Puiseux exponents $\alpha_1, \ldots, \alpha_q$, though the later exponents will in general differ.

Proposition 4.1.3

(i) *Let* $\gamma' \in \mathrm{pro}\,(B')$, *let* $\mathcal{O}(B, B') = \kappa$, *and let* $\alpha_q < \kappa \leq \alpha_{q+1}$. *Then* $\{\mathcal{O}(\gamma', \gamma) : \gamma \in \mathrm{pro}\,(B)\}$ *consists of:*

α_i, *occurring* $(e_{i-1} - e_i)$ *times for* $1 \leq i \leq q$;

κ, *occurring* e_q *times.*

(ii) *Let* $\gamma' \in \mathrm{pro}\,(B)$. *Then* $\{\mathcal{O}(\gamma', \gamma) : \gamma \in \mathrm{pro}\,(B), \gamma \neq \gamma'\}$ *consists of:*

α_i, *occurring* $(e_{i-1} - e_i)$ *times for* $1 \leq i \leq g$.

Proof Suppose the pro-branch γ' has exponent of contact κ with the pro-branch γ given by (4.1). The remaining pro-branches γ_s of B in the chosen sector are obtained by multiplying x_m formally by $e^{2\pi i s/m}$ for $0 < s < m$.

First suppose $r > 1$. Then unless s is divisible by $\frac{m}{e_1}$, this substitution will change the coefficient of $x^{\alpha_1} = x_m^{\beta_1}$, for $s\beta_1/m$ will not be an integer. Thus for $m - e_1 = e_0 - e_1$ values of s, the exponent of contact will be α_1.

Similarly for any $i \leq q$, the exponent of contact will be α_i if $s\alpha_{i-1}$ is an integer, but $s\alpha_i$ is not, i.e. if s is divisible by $\frac{m}{e_{i-1}}$ but not by $\frac{m}{e_i}$, which occurs for just $(e_{i-1} - e_i)$ values of s.

Finally if s is divisible by $\frac{m}{e_q}$, all the terms up to x^κ are the same for γ_s as for γ, so the exponent of contact with γ' is again κ (by hypothesis, it cannot be higher).

In view of our conventions, assertion (ii) is a special case of (i). □

Corollary 4.1.4 *The Puiseux characteristic of a branch is invariant under holomorphic change of coordinates.*

Proof By Lemmas 4.1.1 and 4.1.2, the list $\{\mathcal{O}(\gamma', \gamma) : \gamma \in \mathrm{pro}\,(B), \gamma \neq \gamma'\}$ is independent of all choices. By the Proposition, the number of entries in this list is $e_0 - e_q = m - 1$, so m is invariant; the distinct entries are the α_i, hence also $\beta_i = m\alpha_i$ is invariant. □

We now apply Proposition 4.1.3 (i) to determine the intersection number. We check that the total number of entries in the list (i) is $e_0 = m$, the number of pro-branches of B.

Proposition 4.1.5 *The intersection number $B.B'$ is equal to the sum*

$$\sum_{\gamma \in \mathrm{pro}\,(B), \gamma' \in \mathrm{pro}\,(B')} \mathcal{O}(\gamma, \gamma').$$

Proof We saw in Section 1.2 that intersection numbers may be calculated as follows. Choose a (reduced) equation $f'(x, y) = 0$ for B', a parametrisation $(x, y) = (a(t), b(t))$ for B, substitute, and obtain the order of the result $f(a(t), b(t))$ as a function of t.

Choose a coordinate system with the y-axis not tangent to either branch. Then by Theorem 2.2.2 we can write the equation of B in the form of a monic polynomial (of degree m) in y, and by Theorem 2.2.6 we can factorise this as

$$f(x, y) = \prod_{1}^{m} (y - a_r(x)),$$

where the a_r are the Puiseux series corresponding to the pro-branches, so if we substitute $x = t^m$ we have a power series $\tilde{a}(t)$, and can take $a_r(x) = \tilde{a}(e^{\frac{2\pi i r}{m}} t)$. The same goes for B' with an equation $f'(x, y) = \prod_1^{m'} (y - a'_r(x))$ and a parametrisation $x = u^{m'}$, $y = \tilde{a}'(u)$, say.

Substituting thus gives

$$f'(t^m, \tilde{a}(t)) = \prod_{r=1}^{m'} (\tilde{a}(t) - a'_r(t^m)),$$

whose order in t is

$$\sum_{r=1}^{m'} \mathrm{ord}_t\,(\tilde{a}(t) - a'_r(t^m)).$$

Since $x = t^m$, $\mathrm{ord}_t = m\,\mathrm{ord}_x$. Also, the order in x of the difference is precisely the exponent of contact of the pro-branches. We thus obtain, for a chosen pro-branch γ of B, $m \sum_{\gamma' \in \mathrm{pro}\,(B')} \mathcal{O}(\gamma, \gamma')$ which, in view of Lemma 4.1.2, yields the result. □

This yields the following important formula.

Theorem 4.1.6

(i) Let $\mathcal{O}(B, B') = \kappa$ and $\alpha_q < \kappa \leq \alpha_{q+1}$. Then

$$B.B' = \frac{m(B')}{m(B)}\{(e_0 - e_1)\beta_1 + \ldots + (e_{q-1} - e_q)\beta_q + e_q m(B)\kappa\}.$$

(ii) Given three branches B, B' and B'', $\mathcal{O}(B, B') = \mathcal{O}(B, B'')$ if and only if $B.B'/m(B') = B.B''/m(B'')$. □

Formula (i) is a classical result which goes back to M. Noether; (ii) follows at once from (i).

Observe that, if we write $\Phi(t) = \prod_{r=1}^{m'}(\tilde{a}(t) - \tilde{a}'(e^{\frac{2\pi ir}{m'}} t^{\frac{m}{m'}}))$, the resultant $R_y(f, f')$ of the equations for y defining B and B' is just the product $\prod_{s=1}^{m} \Phi(e^{\frac{2\pi is}{m}} t)$. Hence

$$\mathrm{ord}_x R_y(f, f') = m^{-1}\mathrm{ord}_t R(f, f') = \mathrm{ord}_t \Phi(t) = B.B'.$$

We record this in

Lemma 4.1.7 *If C, C' are curves at O, with no branch of either curve tangent to the y-axis, and with equations $f(x, y) = 0$, $f'(x, y) = 0$ where $f, f' \in \mathbb{C}\{x\}[y]$ are monic polynomials, then $(C.C')_O = \mathrm{ord}_x R_y(f, f')$.*

Instead of looking at the resultant of the equations defining two branches, we may consider the discriminant of the equation for a single branch. This leads to the following.

Lemma 4.1.8 *If C is a curve at O, with no branch tangent to the y-axis, and with equation $f(x, y) = 0$ where $f \in \mathbb{C}\{x\}[y]$ is a monic polynomial, and P is the curve defined by $\partial f/\partial y = 0$, then*

(i) $(C.P)_O = \mathrm{ord}_x D_y(f)$.

(ii) $(C.P)_O$ *is the sum over all ordered pairs of distinct pro-branches γ_r, γ_s, of C of their mutual exponent of contact.*

Proof Formula (i) follows from Lemma 4.1.7 together with the fact that (up to sign) $D_y(f) = R_y(f, \partial f/\partial y)$.

As to (ii), we factorise $f = \prod(y - a_i(x))$, with one factor for each pro-branch (or we may first substitute $x = t^N$ for suitable N), and then use Lemma 1.3.2 (i). □

We can sharpen (ii) as follows.

Lemma 4.1.9 *Let C be a curve, with equation $f(x,y) = 0$, such that the y-axis is tangent to no branch of C at O, let γ be any pro-branch of (any branch of) C, and let P be given by $\partial f/\partial y = 0$. Then the set of exponents of contact of γ with the remaining pro-branches of C coincides with the set of exponents of contact of γ with the pro-branches of P.*

Proof As above, using Theorem 2.2.2 and Theorem 2.2.6, we can factorise f as a product over pro-branches: $f(x,y) = U(x,y)\prod(y - a_i(x))$, where U is a unit and the a_i are the corresponding Puiseux series. As usual, we may make a substitution $x = t^N$ to turn these into power series, but it is more convenient not to do so.

First suppose, for convenience, that $U = 1$. Substitute $z = y - a_1(x)$ and write $f(x, z + a_1(x)) = zg(x,z)$. Then the exponents of contact of the pro-branch $y = a_1(x)$ with the other branches of C are just the exponents of contact of $z = 0$ with the pro-branches of $g(x,z) = 0$. According to Lemma 2.4.4, these are determined straightforwardly by the Newton polygon of g.

On the other hand, the same substitution carries $\partial f/\partial y$ to $\partial(zg)/\partial z$. So the exponents of contact of γ with the pro-branches of P are determined in the same way by the Newton polygon of $\partial(zg)/\partial z$. It remains only to observe that this is the same as the Newton polygon of g, since the coefficient in $\partial(zg)/\partial z$ of a monomial $x^r z^s$ is $s+1$ times the coefficient of the same monomial in g, so the two contain the same monomials.

We check that the presence of U does not affect the argument. Suppose the terms giving the corners of the Newton polygon for $\prod_{i>1}(z + a_1(x) - a_i(x))$ are the $a_j x^{\rho_j} z^{s_j}$: then each other term is obtained from one of these by increasing the exponent of x or that of z or both. Multiplying by a unit does not affect this conclusion, but multiplies each a_j by $U(0,0)$. Now multiply further by z, and then differentiate with respect to z. This produces the terms $U(0,0)s_j a_j x^{\rho_j} z^{s_j}$ and others which are above these (in the same sense). The Newton polygon is thus unaltered. □

This result will be important in Section 4.5. We see that in the case when C is a single branch B, all the exponents of contact in (ii) are given by (iii), so that

$$(B.P)_O = m\sum_{q=1}^{g}(e_{q-1} - e_q)\frac{\beta_q}{m} = \sum_{q=1}^{g}(e_{q-1} - e_q)\beta_q.$$

Theorem 4.1.6 leads us to introduce the following important function. Let B be a branch with Puiseux characteristic $\{m; \beta_1, \ldots, \beta_g\}$. Then the *Herbrand function* H of B is defined (for $t \geq 0$) as follows. We have $H(t) = t$ if $t \leq \beta_1$, and if $\beta_i \leq t \leq \beta_{i+1}$,

$$H(t) = \frac{1}{m}\{m\beta_1 + e_1(\beta_2 - \beta_1) + \cdots + e_{i-1}(\beta_i - \beta_{i-1}) + e_i(t - \beta_i)\}, \tag{4.4}$$

where we may interpret β_{g+1} as ∞.

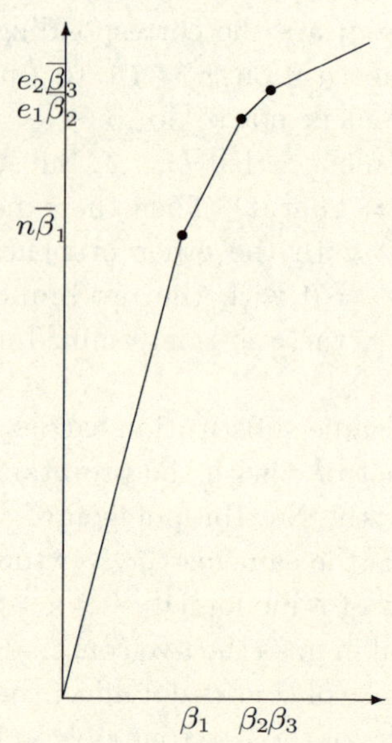

Fig. 4.1. Graph of a Herbrand function

Since the two values given for $H(\beta_i)$ agree, H is well-defined, continuous and strictly increasing. It gives a bijection of $\mathbb{R}^+$ to itself. Its general appearance is illustrated in Figure 4.1. Using the Herbrand function, we can give a more compact statement of Theorem 4.1.6.

Corollary 4.1.10 *Let B and B' be branches at O with multiplicities m, m' and exponent of contact κ. Let H be the Herbrand function for B. Then the intersection number of B and B' is equal to $m'H(m\kappa)$.*

Since H defines a bijection, it follows that (for given B), each of the exponent c of contact and the intersection number $(B.B')_0$ determines the other.

We observed above that the two values given for $H(\beta_i)$ agree. This number will play an important role in the sequel, and we set

$$\overline{\beta}_i = \frac{m}{e_{i-1}} H(\beta_i) \qquad (1 \leq i \leq g). \tag{4.5}$$

We will also write $\overline{\beta}_0 = m$. Since the slope of $H(t)$ in the interval $\beta_i < t < \beta_{i+1}$ is equal to $\frac{e_i}{m}$, it follows that

$$\frac{e_i \overline{\beta}_{i+1}}{m} - \frac{e_{i-1} \overline{\beta}_i}{m} = \frac{e_i}{m}(\beta_{i+1} - \beta_i),$$

so we obtain the formula, to which we will frequently refer,

$$\overline{\beta}_{i+1} - \frac{e_{i-1}}{e_i} \overline{\beta}_i = \beta_{i+1} - \beta_i. \tag{4.6}$$

Since $\overline{\beta}_0$ is an integer, and $\frac{e_{i-1}}{e_i} \in \mathbb{Z}$ for all i, it follows inductively that each $\overline{\beta}_i$ is an integer.

Example 4.1.1 For the curve $y^8 = x^{11}$ we have $m = 8$ and $\beta_1 = 11$. Thus

t	$t \leq 11$	$t \geq 11$
$H(t)$	t	$11 + \frac{1}{8}(t - 11)$

;

$\overline{\beta}_0 = 8$ and $\overline{\beta}_1 = 11$. This is typical of branches with $g = 1$.

For a curve with Puiseux characteristic $(4; 6, 7)$ we have $m = e_0 = 4$, $e_1 = 2$, $e_2 = 1$ and $\beta_1 = 6$, $\beta_2 = 7$. Thus

t	$t \leq 6$	$6 \leq t \leq 7$	$t \geq 7$
$H(t)$	t	$6 + \frac{1}{2}(t - 6) = 3 + \frac{1}{2}(t)$	$6\frac{1}{2} + \frac{1}{4}(t - 7) = \frac{1}{4}(19 + t)$

;

and $\overline{\beta}_0 = 4$, $\overline{\beta}_1 = 6$ and $\overline{\beta}_2 = 13$. Thus if B is given by $x = t^4$, $y = t^6 + t^7$ and B' is the branch given by $x = t^4$, $y = t^6 + t^7 + t^8$, we have $t = 8$ and so $B.B' = m(B').\frac{1}{4}(19 + 8) = 27$.

4.2 The Eggers tree

In this section we introduce a convenient notation for exponents of contact of a curve with several branches.

First consider a single branch B, with multiplicity m. The tree $\Gamma_E(B)$ is a copy of the compactified positive real axis, which we denote $[0, \infty]$,[1] and write

$$v_B : \Gamma_E(B) \to [0, \infty], \quad \pi_B : [0, \infty] \to \Gamma_E(B)$$

[1] We denote closed intervals on the line by $[a, b]$, half open ones $[a, b)$ etc.

for inverse isomorphisms. We define $\Gamma_E(B)$ to have vertices $A_0 := \pi_B(0)$, $A_q := \pi_B(\alpha_q)$ for $1 \le q \le g$, and $B^\infty := \pi_B(\infty)$.

Define the function ν_B on $\Gamma_E(B)$ to take the value $\frac{m}{e_{q-1}}$ on the interval $A_{q-1}A_q$, including the left hand end point. Thus in particular on A_0A_1 we have the value $\frac{m}{e_0} = 1$ and on A_gB^∞ we have $\frac{m}{e_g} = m$.

The Herbrand function defines a function h_B on $\Gamma_E(B)$ by $h(\pi_B(\kappa)) := m^{-1}H(m\kappa)$. It follows from (4.4) that $\frac{\partial h_B}{\partial v_B} = \frac{1}{\nu_B}$. Hence any two of the functions v_B, ν_B, h_B on $\Gamma_E(B)$ determine the third; and these three also determine the interior vertices $A_1, \dots, A_g$ as the set of points of discontinuity of ν_B.

Now suppose we have a curve C with two branches B, B', with $\mathcal{O}(B, B') = \kappa$. We define $\Gamma_E(C)$ to be obtained from $\Gamma_E(B) \dot\cup \Gamma_E(B')$ by identifying the images of $[0, \kappa]$ under π_B and $\pi_{B'}$; hence the functions v_B and $v_{B'}$ combine to define a function v_C on $\Gamma_E(C)$. Let us check that the identification is compatible with the structures we have defined. By hypothesis, there are Puiseux series of B and B' which agree up to (but not including) the term in x^κ. It follows from the definition of the Puiseux characteristic that, for all i with $\alpha_i < \kappa$ (say this holds for $i < q$) $\alpha_i = \frac{\beta_i}{m} = \frac{\beta'_i}{m'}$, and $\frac{e_i}{m} = \frac{e'_i}{m'}$. Hence the points $A_1, \dots, A_{q-1}$ for the two branches correspond, and the functions ν_B and $\nu_{B'}$ agree on the common interval $[0, \kappa]$, as do h_B and $h_{B'}$, so they combine to define functions ν_C and h_C on the whole of $\Gamma_E(C)$. We introduce the branch point $I := \pi_B(\kappa)$ as a vertex whether or not it coincides with the point A_q for one or both of the branches B and B'.

In general, let the branches of C be $\{B_j \,|\, j \in J\}$, with $\kappa_{j,k} := \mathcal{O}(B_j, B_k)$, and define $\Gamma_E(C)$ to be the quotient of the disjoint union of the $\Gamma_E(B_j)$ by the identification, for any pair j, k, of the images of $[0, \kappa_{j,k}]$ under π_{B_j} and π_{B_k}. More precisely, we proceed as follows. Define a relation on $[0, \infty] \times J$ by

$$(a, j) \sim (a, k) \;:\; \text{if } a \le \kappa_{j,k}.$$

This is clearly reflexive and symmetric; it is transitive since if $(a, j) \sim (a, k)$ and $(a, k) \sim (a, l)$ then $a \le \min(\kappa_{j,k}, \kappa_{k,l}) \le \kappa_{j,l}$, by (4.3), so $(a, j) \sim (a, l)$. We define the *Eggers tree* $\Gamma_E(C)$ to be the quotient of $[0, \infty] \times J$ by this equivalence relation: this is a tree. The inclusion of $[0, \infty] \times j$ induces an injection $\pi_{B_j} : [0, \infty] \to \Gamma_E(C)$, whose image we may identify with $\Gamma_E(B_j)$. The first projection induces a map $v : \Gamma_E(C) \to [0, \infty]$, whose restriction to $\Gamma_E(B_j)$ is v_{B_j}.

We have already verified that the functions ν_{B_j} on $\Gamma_E(B_j)$ agree on the regions of overlap, so they combine to define a function ν_C on $\Gamma_E(C)$.

Similarly we obtain a function h_C defined on the whole tree. We say that two Eggers trees are isomorphic if there is a homeomorphism of the underlying trees which respects each of the functions v_C, ν_C and h_C; though as before, any two of these determine the third.

We regard $\Gamma_E(C)$ as a graph whose vertices are the images of those defined for the separate branches, which are the same as the end points and the points A_q of discontinuity of ν_C, together with all branch points $I_{j,k} := \pi_{B_j}(\kappa_{j,k})$ (each of which may or may not be included among the A_q). Let us re-phrase this. We have the end points, of valence 1 in $\Gamma_E(C)$, and the interior vertices, which consist of the points of discontinuity of ν_C, and the branch, or rupture points, of valence ≥ 3 in $\Gamma_E(C)$; a rupture point may or may not be a point of discontinuity of ν_C.

We orient each edge of $\Gamma_E(C)$ in the sense in which v increases: thus all edges point away from the vertex A_0. This gives a partial ordering on the set of vertices of $\Gamma_E(C)$. Any two points X, Y in the graph have a unique infimum, i.e. a point $\inf(X, Y)$ such that

$$Z \leq \inf(X, Y) \iff Z \leq X \text{ and } Z \leq Y.$$

For if X, Y belong to a common branch $\Gamma_E(B_j)$, then as this is isomorphic to $\mathbb{R}$, it is totally ordered, so $X \leq Y$ or $Y \leq X$; and if not, but $X \in \Gamma_E(B_j)$ and $Y \in \Gamma_E(B_k)$, then $\inf(X, Y)$ exists and is equal to $I_{j,k}$. In particular, $I_{j,k} = \inf(B_j^\infty, B_k^\infty)$.

Lemma 4.2.1 *For any branch B' at O there is a unique point $X_{B'} \in \Gamma_E(C)$ such that for each $j \in J$, $\mathcal{O}(B', B_j) = v(\inf(X_{B'}, B_j^\infty))$.*

Proof Choose j to maximise $\mathcal{O}(B', B_j)$ – the value is $\mathcal{O}(B', C) = \kappa$, say – and define $X_{B'} := \pi_{B_j}(\kappa)$. Then by the triangle inequality (4.3), for any $k \neq j$ we either have $\mathcal{O}(B', B_k) = \kappa \leq \mathcal{O}(B_j, B_k)$ or $\mathcal{O}(B', B_k) = \mathcal{O}(B_j, B_k) \leq \kappa$. In the former case, $X_{B'} \leq I_{j,k} < B_k^\infty$, so $v(\inf(X_{B'}, B_j^\infty)) = v(X_{B'}) = \kappa = \mathcal{O}(B', B_k)$. In the latter case, $X_{B'} \cap B_k^\infty = I_{j,k}$, so $v(\inf(X_{B'}, B_j^\infty)) = v(I_{j,k}) = \mathcal{O}(B_j, B_k) = \mathcal{O}(B', B_k)$. Uniqueness is immediate. □

We measure the contact of B' with C by the 0-chain $[B'] := m(B')X_{B'}$. The point $X_{B'}$ need not be a vertex of $\Gamma_E(C)$ as we have just defined it, so it may sometimes be necessary to subdivide the graph. The above definition leads to $[B_j] = m(B_j)B_j^\infty$. For a curve germ C' with several branches B_i', the contact with C is measured by the 0-chain $[C'] := \sum_i [B_i']$. In particular, $[C] = \sum_{j \in J} m(B_j)B_j^\infty$.

For each j, the exponent of contact of B' with B_j is $v(\inf(X_{B'}, B_j^\infty))$. Hence $B'.B_j = m(B')m(B_j)h(\inf(X_{B'}, B_j^\infty))$, so

$$B'.C = m(B')\sum_j m(B_j)h(\inf(X_{B'}, B_j^\infty)).$$

We can write this in the form $m(B')\overline{H}(X_{B'})$, where $\overline{H}$ is the function defined on $\Gamma_E(C)$ by $\overline{H}(Z) := \sum_j m(B_j)h(\inf(Z, B_j^\infty))$.

Example 4.2.1 For a curve B with Puiseux characteristic $(4; 6, 7)$, the Eggers tree is a copy of $[0, \infty]$, with vertices A_0 at 0, A_1 at $\frac{3}{2}$, A_2 at $\frac{7}{4}$ and B^∞ at ∞.

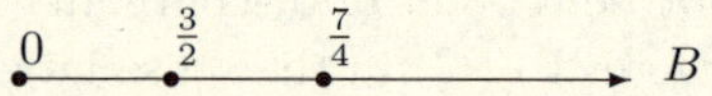

Range of t	$[0, \frac{3}{2}]$	$[\frac{3}{2}, \frac{7}{4}]$	$[\frac{7}{4}, \infty)$
$\nu_B(t)$	1	2	4
$h_B(t)$	t	$\frac{1}{2}t + \frac{3}{4}$	$\frac{t}{4} + \frac{19}{16}$

For Example 3.6.2, C is the union of $B_1 : y^2 = x^3$ and $B_2 : y^3 = x^4$. The exponent of contact is $\frac{4}{3}$. We thus identify the intervals $[0, \frac{4}{3}]$ on two copies of $[0, \infty]$; the vertices are A_0, the point I at $\frac{4}{3}$, the ends B_1^∞ and B_2^∞, and the point A_1 at $\frac{3}{2}$ between I and B_1^∞.

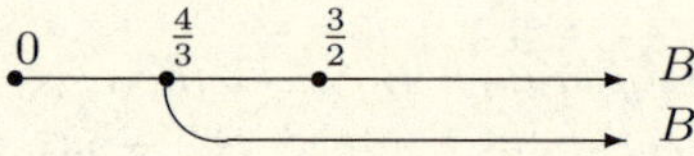

Range of t	$[A_0I]$	$[IA_1]$	$[A_1B_1^\infty)$	$[IB_2^\infty)$
$\nu_C(t)$	1	1	2	3
$h_C(t)$	t	t	$\frac{1}{2}t + \frac{3}{4}$	$\frac{t}{3} + \frac{8}{9}$

4.3 The semigroup of a branch

The set of intersection numbers of B with other curves in the plane forms a semigroup $S(B)$. We begin by determining the structure of this semigroup in terms of the Puiseux characteristic of the branch, and establishing an important duality property. We then establish the equivalence of several invariants of a branch: for example, the Puiseux characteristic, the semigroup and the sequence of multiplicities: two branches with the same invariants are called equisingular. We obtain a corresponding result for curves with several branches.

If B is a branch with good parametrisation (or equivalently, normalisation) $\gamma : \mathbb{C} \to \mathbb{C}^2$, composition with γ defines a ring homomorphism $\gamma^* : \mathcal{O} = \mathbb{C}\{x, y\} \to \mathbb{C}\{t\}$. The kernel of γ^* is the ideal of functions vanishing on B. It follows from the formulary of Section 2.3 that this is the ideal $\langle f \rangle$ in $\mathbb{C}\{x, y\}$ generated by f, where $f(x, y) = 0$ is the defining equation of B.

For any curve C, with defining equation $f(x, y) = 0$, the quotient $\mathbb{C}\{x, y\}/\langle f \rangle$ is defined to be the *local ring* $\mathcal{O}_C$ of C. The local ring $\mathcal{O}_C$ is of considerable importance; we will discuss it in more detail in Chapter 11. For a single branch, we may use a parametrisation γ to identify $\mathcal{O}_B$ with the image of γ^*, a subring of $\mathbb{C}\{t\}$. In general we can use a normalisation
$\pi : \widetilde{C} \to C$ similarly.

We define the *semigroup of the branch* $S(B)$ by

$$S(B) := \{\text{ord}\,\phi \,|\, \phi \in \mathcal{O}_B\}.$$

We also define the *double point number* $\delta(B)$ to be the cardinality of the set of gaps: $\delta(B) := \sharp\{r \geq 0 \,|\, r \notin S(B)\}$. It is immediate from the definition that $S(B)$, and hence also $\delta(B)$, is unaltered by a change of holomorphic local coordinates at $O \in \mathbb{C}^2$. It is also unaffected by a change of good parametrisation for B. For if u is an alternative good parameter, we can express each of t and u in terms of the other by power series with zero constant term and non-zero linear term. Substituting such an expression $t = \phi(u)$ in a power series $a(t)$ yields a power series $a(\phi(u))$ which is easily seen to have the same order as a.

Clearly $S(B)$ is a semigroup since $\text{ord}\,\phi_1\phi_2 = \text{ord}\,\phi_1 + \text{ord}\,\phi_2$. We will investigate these invariants in terms of the Puiseux characteristic of B.

Lemma 4.3.1

(i) *The set $S(B)$ consists of the intersection numbers $(B.C)_0$ of B with germs at $(0, 0)$ not having B as a component.*

(ii) *For each $q \geq 0$, $\overline{\beta}_q \in S(B)$.*

(iii) *For any branch B, $\delta(B)$ is finite.*

Proof follows since we know that intersection numbers are obtained by inserting the equation of C in the parametrisation of B, and this equation can be an arbitrary element of $\mathbb{C}\{x, y\}$ that does not vanish on B.

Suppose B has parametrisation

$$x = t^m, \quad y = \sum_{r=1}^{\infty} a_r t^r,$$

with notation as in Section 3.1. Define curves B_q^- by the parametrisations

$$x = t^m, \quad y = \sum_{1 \le r < \beta_q} a_r t^r, \tag{4.7}$$

Then all terms are powers of $t^{e_{q-1}}$, so B_q^- has multiplicity m/e_{q-1} (the parametrisation is not good). By construction, the exponent of contact is β_q/m. Hence by Theorem 4.1.6 the intersection number $(B.B_q^-)_0 = \frac{m}{e_{q-1}} H(\beta_q)$, which by definition is $\overline{\beta}_q$.

We claim that the highest common factors of $(\beta_0, \dots, \beta_g)$ and $(\overline{\beta}_0, \dots, \overline{\beta}_g)$ are the same. For it follows from (4.6) that $(\beta_i, \beta_{i+1}) = (\overline{\beta}_i, \overline{\beta}_{i+1})$, and the claim follows by induction. Since the former highest common factor is 1, so is the latter.

Hence we can choose integers a_i with $1 = \sum_0^g a_i \overline{\beta}_i$. We can change our choice by adding $m = \overline{\beta}_0$ to a_i and subtracting $\overline{\beta}_i$ from a_0. Doing this sufficiently often, we may suppose $a_i \ge 0$ for all $i > 0$. Thus if $c := \sum_{i=1}^g a_i \overline{\beta}_i$ we have $c \in S(B)$ and $c \equiv 1 \pmod m$. But for any $n \in \mathbb{N}$, we can write $n = mq + r$ with $0 \le r < m$ and so if $n \ge c(m-1)$ we have $n - rc \ge c(m-1) - c(m-1) = 0$ and $n - rc$ is divisible by m. Thus n belongs to the semigroup generated by m and c, and hence to $S(B)$. □

We now give a basic structural result for $\mathcal{O}_B$.

Proposition 4.3.2

(i) *For N sufficiently large, any element of $\mathbb{C}\{t\}$ of order $> N$ belongs to $\mathcal{O}_B$.*

(ii) *The powers t^r with $r \notin S(B)$ form a basis of $\mathbb{C}\{t\}/\mathcal{O}_B$. Hence $\dim_{\mathbb{C}}(\mathbb{C}\{t\}/\mathcal{O}_B) = \delta(B)$.*

Proof Choose a parametrisation $x = t^m$, $y = a(t)$ of B. We have inclusion maps of rings $\mathbb{C}\{x\} \hookrightarrow \mathcal{O}_B \hookrightarrow \mathbb{C}\{t\}$.

In the proof of (iii) of Lemma 4.3.1 we constructed an element c of $S(B)$ with $c \equiv 1 \pmod{m}$, so $c = 1 + qm$ for some $q \in \mathbb{N}$ and there exists $A \in \mathbb{C}\{x, y\}$ with $\gamma^* A$ of order c. For each j with $0 \leq j < m$, write $\gamma^*(A^j) = f_j(t) = \sum_{r=0}^{\infty} a_{r,j} t^r$. By Exercise 2.6.3, for each k with $0 \leq k < m$, the series $\alpha_{j,k}(x) := \sum_{q=0}^{\infty} a_{mq+k,j} x^q$ is convergent. We have

$$\gamma^*(A^j) = \sum_{k=0}^{m-1} t^k \alpha_{j,k}(x). \tag{4.8}$$

Since the left hand side has order exactly $jc = j + jqm$ in t, $\alpha_{j,j}$ has order exactly jq while for $j \neq k$, the order of $\alpha_{j,k}$ is strictly greater than $jq + \frac{j-k}{m}$.

We regard (4.8) as a system of linear equations over the ring $\mathbb{C}\{x\}$ for the powers t^k. In view of the estimates of the orders of the $\alpha_{j,k}$ we see that the determinant $D(x) := \det(\alpha_{j,k})$ has order exactly $\sum_{j=0}^{m-1} jq = \frac{1}{2}m(m-1)q$. Cramer's rule gives an expression for $D(x)t^i$ as a linear combination of the functions $\gamma^*(A^j)$. Hence $D(x)t^k \in \mathcal{O}_B$ for $0 \leq k < m$.

If $\phi \in \mathbb{C}\{t\}$ has order $> \frac{1}{2}m^2(m-1)q$, ϕ is divisible by $D(t^m)$, and as above we can express the quotient $\psi(t) := \phi(t)/D(t^m)$ in the form $\psi(t) = \sum_{k=0}^{m-1} t^k \psi_k(x)$ with each $\psi_k(t) \in \mathcal{O}_B$. Hence $\phi(t) = \sum_{k=0}^{m-1} D(x) t^k \psi_k(x) \in \mathcal{O}_B$.

Choose N such that any element of $\mathbb{C}\{t\}$ of order $> N$ belongs to $\mathcal{O}_B$. Then $\mathbb{C}\{t\}/\mathcal{O}_B$ is a quotient of $\mathbb{C}\{t\}/t^{N+1}\mathbb{C}\{t\}$, which has basis $\{t^i \,|\, 0 \leq i \leq N\}$. For each element $s \leq N$ of $S(B)$, choose $\epsilon_s \in \mathcal{O}_B$ of order s. Then the images of the elements ϵ_s and the powers t^r with $r \notin S(B)$ also form a basis of $\mathbb{C}\{t\}/t^{N+1}\mathbb{C}\{t\}$, since their matrix in terms of the above basis is triangular, hence non-singular.

It follows that the powers t^r with $r \notin S(B)$ span $\mathbb{C}\{t\}/\mathcal{O}_B$. That they map to independent elements follows since if $\sum c_r t^r \in \mathcal{O}_B$, it would follow that the least r with $c_r \neq 0$ was in $S(B)$. □

We will write $N(S(B))$ for the greatest integer not in $S(B)$.

Corollary 4.3.3 *Any element of $\mathbb{C}\{t\}$ of order $> N(S(B))$ belongs to $\mathcal{O}_B$.*

This follows from the argument at the end of the proof. We observe that the corresponding results for the power series ring $\mathbb{C}[[t]]$ hold, and are easier to prove, since we can omit discussions of convergence.

Before proceeding with the detailed analysis of $S(B)$, we make some general remarks about semigroups, which in this section will always be subsets S of $\mathbb{Z}$ closed under addition and containing 0.

If S contains positive and negative elements, it is a subgroup of $\mathbb{Z}$, hence is of the form $d\mathbb{Z}$ for some $d \in \mathbb{N}$. For it is enough to see that $r \in S$ implies $-r \in S$. We may suppose that $r > 0$ and that S contains a negative element $-s$. Then $-r = r(-s) + (s-1)r \in S$.

From now on we suppose that $S \subseteq \mathbb{N}$, and that $S \neq \{0\}$. Write $d(S)$ for the highest common factor of all elements of S.

S contains all but finitely many positive multiples of $d(S)$. It is enough to explain this in the case when $d(S) = 1$. Choose $0 \neq r \in S$. Then S projects to a subsemigroup of the quotient group $\mathbb{Z}/r\mathbb{Z}$. This contains inverses, since the sum of x with itself $r-1$ times is the inverse of X. So it is a subgroup; since $d(S) = 1$, it is the whole of $\mathbb{Z}/r\mathbb{Z}$. Let $s \in S$ project to 1. Then if n is any integer $> rs$, write $n = ar + b$ with $0 \leq b < r$: it follows that $n - bs$ is positive and divisible by r. Hence n is a nonnegative linear combination of $r, s \in S$, so belongs to S.

Write $N(S)$ for the largest multiple of $d(S)$ not belonging to S. The simplest example is $S = d\mathbb{N}$, with $N(S) = -d$. We call S a *dual semigroup* if, for r divisible by $d(S)$, $r \in S \Leftrightarrow (N(S) - r) \notin S$. This condition clearly holds for $S = d\mathbb{N}$.

Now consider the semigroup $S_1 = S + \mathbb{N}b$ generated by S and a further element b. The divisor $d(S_1)$ is equal to the highest common factor $hcf(b, d(S))$. Set $q = \frac{d(S)}{d(S_1)}$: this is the least integer such that qb is divisible by $d(S)$.

Lemma 4.3.4 *If S is a dual semigroup and $qb > N(S)$, then S_1 is a dual semigroup, with $N(S_1) = N(S) + b(q-1)$.*

Proof Since $qb > N(S)$, we have $qb \in S$. Hence any element of S_1 can be written in the form $s + ub$ with $s \in S$ and $0 \leq u < q$. Moreover, this expression is unique.

Any multiple of $d(S_1) = hcf(b, d(S))$ can be expressed in the form $x = rd(S) + ub$, with $r, u \in \mathbb{Z}$ and this expression also becomes unique if we insist that $0 \leq u < q$. With u so restricted, we claim that $x \in S_1$ if and only if $rd(S) \in S$. For this is clearly sufficient; conversely, if $x = s + nb$ with $s \in S$ and $n \geq 0$, we have $n \equiv u \pmod{q}$ and hence $n \geq u$ so that $rd(S) - s$ is a nonnegative multiple of b and of $d(S)$ and hence (if not zero) is $\geq qb > N(S)$, so belongs to S.

Set $M = N(S) + b(q-1)$. We claim that for x divisible by $d(S_1)$, $x \in S_1$ if and only if $M - x \notin S_1$. Since 0 is the least element of S_1 it follows that M is the greatest multiple of $d(S_1)$ not in S_1, so $N(S_1) = M$ and S_1 is a dual semigroup.

Any x divisible by $d(S_1)$ may be uniquely expressed in the form $x = y + ub$, with y divisible by $d(S)$ and $0 \leq u < q$. By the above, $x \in S_1$ if and only if $y \in S$. We have $(M-x) = (N(S)-y)+(q-1-u)b$ expressed in the same form. Hence $(M-x) \in S_1$ if and only if $(N(S)-y) \in S$. But since S is a dual semigroup, this is equivalent to $y \notin S$. □

For example, consider a 2-generator semigroup $\mathbb{N}a + \mathbb{N}b$: set $d = hcf(a,b)$. Take $S = \mathbb{N}a$, which is dual with $d(S) = a$, $N(S) = -a$. We have $q = a/d$. The hypothesis of the lemma is immediate since $N(S) < 0$. Hence S_1 is a dual semigroup with

$$N(S_1) = -a + b\left(\tfrac{a}{d} - 1\right) = \tfrac{ab}{d} - a - b.$$

In particular, if a and b are coprime, the largest integer not expressible as a nonnegative linear combination of a and b is $ab - a - b$.

This concludes our digression. We return to the consideration of the semigroup $S(B)$ belonging to a plane curve branch B, which we suppose to have Puiseux characteristic $\{m; \beta_1, \dots, \beta_g\}$.

Theorem 4.3.5

(i) *$S(B)$ is generated by $\overline{\beta}_0, \overline{\beta}_1, \dots, \overline{\beta}_g$.*

(ii) *This is a minimal set of generators: $\overline{\beta}_q$ is the least element of $S(B)$ not divisible by e_{q-1}.*

(iii) *$S(B)$ is a dual semigroup.*

Write S_q for the semigroup generated by $\overline{\beta}_0, \dots, \overline{\beta}_q$. The proof depends on the following lemma, which is proved by induction on q.

Lemma 4.3.6

(i) $d(S_q) = e_q$

(ii) $N(S_q) = -m - \beta_q + \frac{e_{q-1}}{e_q}\overline{\beta}_q = -\overline{\beta}_0 + \overline{\beta}_{q+1} - \beta_{q+1}$.

(iii) *The hypothesis of Lemma 4.3.4 holds for adjoining the generator $\overline{\beta}_q$ to the semigroup S_{q-1}. In fact, $\frac{e_{q-1}}{e_q}\overline{\beta}_q > N(S_q) > N(S_{q-1})$.*

(iv) *S_q is a dual semigroup.*

Proof We first recall that, by definition, $\overline{\beta}_0 = m$, $\overline{\beta}_1 = \beta_1$, and for $q > 1$

$$\tfrac{e_q}{m}\overline{\beta}_{q+1} - \tfrac{e_{q-1}}{m}\overline{\beta}_q = H(\beta_{q+1}) - H(\beta_q) = \tfrac{e_q}{m}(\beta_{q+1} - \beta_q),$$

so the two formulae in (ii) are equivalent.

The assertions all hold for $q = 1$, since S_1 has the two generators m and β_1, and the case of 2-generator semigroups was analysed above. We suppose them true for q and seek to proceed to $q + 1$.

It follows from the definition that $\overline{\beta}_{q+1} - \beta_{q+1}$ is an integer linear combination of $\overline{\beta}_q$ and β_q, and hence is divisible by e_q. Hence $hcf(\overline{\beta}_{q+1}, e_q) = hcf(\beta_{q+1}, e_q) = e_{q+1}$, and (i) follows.

Define A_q by $A_q = -m - \beta_q + \frac{e_{q-1}}{e_q}\overline{\beta}_q$. We have

$$N(S_q) + \left(\tfrac{e_q}{e_{q+1}} - 1\right)\overline{\beta}_{q+1} = -\overline{\beta}_0 + \overline{\beta}_{q+1} - \beta_{q+1} + \left(\tfrac{e_q}{e_{q+1}} - 1\right)\overline{\beta}_{q+1}$$

$$= -\overline{\beta}_0 - \beta_{q+1} + \tfrac{e_q}{e_{q+1}}\overline{\beta}_{q+1} = A_{q+1}.$$

In particular, $\frac{e_q}{e_{q+1}}\overline{\beta}_{q+1} \geq A_{q+1} \geq A_q$.

This proves that Lemma 4.3.4 applies to adjoining the generator $\overline{\beta}_{q+1}$ to S_q, so (iv) holds, and the above calculation shows that $N(S_{q+1}) = A_{q+1}$ and hence that (ii) and (iii) hold also. □

Proof [of Theorem 4.3.5] In view of the lemma, (ii) and (iii) will follow from (i). We have already seen that each $\overline{\beta}_q \in S(B)$.

Since any curve-germ at $(0, 0)$ is made up of branches, and its intersection multiplicity with B is the sum of theirs, it suffices to consider a branch B' and to show that its intersection number with B belongs to the above semigroup S_g. Suppose in fact that the exponent κ of contact satisfies $\beta_q < m\kappa \leq \beta_{q+1}$: then we will show that $(B.B')_0 \in S_q$.

Since the exponent of contact is κ, the Puiseux series agree up to the x^κ term, so the power series a, a' are the same up to this point up to replacing t, t' by powers of a variable t''. Hence the vector $(m = \beta_0, \beta_1, \ldots, \beta_q)$ for B is proportional to the corresponding vector for B'. Thus the same assertion holds for the vector

$$(m = \beta_0, \beta_1, \ldots, \beta_q, e_1, \ldots, e_q, \overline{\beta}_1, \ldots, \overline{\beta}_q, m\kappa).$$

Hence

$$(B.B')_0 = m'H(m\kappa) = m'\left\{H(\beta_q) + (m\kappa - \beta_q).\tfrac{e_q}{m}\right\}$$

$$= \tfrac{m'}{m}\left\{e_{q-1}\overline{\beta}_q + e_q(m\kappa - \beta_q)\right\} = e'_{q-1}\overline{\beta}_q + e_q(m'\kappa - \beta'_q)$$

is a nonnegative linear combination of $\overline{\beta}_q$ and e_q, hence is divisible by e_q. To show that it belongs to S_q it will suffice to show that it is greater than $N(S_q)$. But since $(m'\kappa - \beta'_q) > 0$ and $e'_{q-1} \geq \frac{e'_{q-1}}{e'_q} = \frac{e_{q-1}}{e_q}$, we have $(B.B')_0 \geq \frac{e_{q-1}}{e_q}\overline{\beta}_q > N(S_q)$. □

Corollary 4.3.7 *We have $2\delta(B) = 1 + N(S(B))$.*

Proof The assertion follows since, as $S(B)$ is a dual semigroup, exactly half of the integers $\{0, 1, 2, \dots, N\}$ belong to $S(B)$, while all larger integers do. □

Example 4.3.1 For a curve with Puiseux characteristic $(4; 6, 7)$, the semigroup is generated by 4, 6 and 13; the first elements of the semigroup are 0, 4, 6, 8, 10, 12, 13, 14, 16, 17, 18; and $r \in S \Leftrightarrow 15 - r \notin S$.

Now consider the number of gaps in $S(B)$. We recall from Lemma 4.3.4 that $N(S_1) = N(S) + b(q-1)$, with $q = \frac{d(S)}{d(S_1)}$ and that by (iii) of Lemma 4.3.6 this situation holds for each of the inclusions $S_0 \subset S_1 \subset \dots S_g = S(B)$. Since $N(S_0) = -m$ and $d(S_q) = e_q$, we obtain

$$N(S(B)) = -m + \sum_{q=1}^{g} \overline{\beta}_q\left(\tfrac{e_{q-1}}{e_q} - 1\right).$$

We can rewrite this in several ways.

$$\begin{aligned} N(S(B)) &= -m + \textstyle\sum_{q=1}^{g-1}\{\overline{\beta}_{q+1} - \beta_{q+1} + \beta_q - \overline{\beta}_q\} + \overline{\beta}_g(e_{g-1} - 1) \\ &= -m + \overline{\beta}_g e_{g-1} - \beta_g \\ &= \textstyle\sum_{q=1}^{g}(e_{q-1} - e_q)(\beta_q - 1) - 1. \end{aligned}$$

This number will play an important role in the sequel.

We conclude this section with a first summary of the equivalence of several relations between branches.

Proposition 4.3.8 *Any one of the following sets of data determines the others:*

The Puiseux characteristic $\{m; \beta_1, \dots, \beta_q\}$,
The Herbrand function $H(t)$,
The sequence $\{\overline{\beta}_0, \dots, \overline{\beta}_q\}$,
The Eggers tree,
The semigroup $S(B)$,
The sequence of multiplicities $m_i(B)$,
The proximity relations between infinitely near points,
The proximity matrix.

In particular, all the above are independent of any choices of coordinates or parametrisation.

Proof The Herbrand function $H(t)$ is continuous, piecewise-linear, in fact linear in the intervals defined by the β_q, has $H(0) = 0$, and $H'(t)$ is equal to 1 for $0 < t < \beta_1$ and to e_q/m for $\beta_q < t < \beta_{q+1}$ (in particular to $e_g/m = 1/m$ for $t > \beta_g$). Thus the Puiseux characteristic determines H.

Since $H(\beta_q) = \overline{\beta}_q$, the inverse function H^{-1} is the unique continuous function on $[0, \infty)$, linear in the intervals $[0, \overline{\beta}_0]$, $[\overline{\beta}_q, \overline{\beta}_{q+1}]$ for $0 \leq q \leq g-1$ and $[\overline{\beta}_g, \infty)$, and having derivatives 1, m/e_q, m in these intervals. Since $m = \overline{\beta}_0$, the Herbrand function determines the $\overline{\beta}_q$.

The $\overline{\beta}_q$ determine the $e_q = hcf(\overline{\beta}_0, \ldots, \overline{\beta}_q)$, also $\beta_0 = \overline{\beta}_0$, and by induction, using (4.6), the remaining β_q and hence the Puiseux characteristic.

The Eggers tree was explicitly defined in terms of the Puiseux characteristic. Conversely, the tree determines the sequence m/e_q of values of ν_B, in particular the final value $m = m/e_g$. It also determines the values α_q of v_B at the points of discontinuity of ν_B, and hence the remaining terms $\beta_q = m\alpha_q$ in the Puiseux characteristic.

The first part of Theorem 4.3.5 states that $S(B)$ is generated by, hence determined by, $\overline{\beta}_0 \ldots, \overline{\beta}_g$. The second shows that given $S(B)$ we can recover these numbers: $\overline{\beta}_0 = e_0$ is the least non-zero element of $S(B)$, and inductively $\overline{\beta}_q$ is the least element of $S(B)$ not divisible by e_{q-1} and e_q is the highest common factor of e_{q-1} and $\overline{\beta}_q$.

Since $S(B)$ is independent of any choices, so are all the others discussed so far; thu we have a second proof of invariance of the Puiseux characteristic.

We may now apply Theorem 3.5.6, which states that the Puiseux characteristic of a branch determines the sequence of multiplicities $m_i(B)$, and conversely. By Corollary 3.5.2, giving the proximity relations between infinitely near points is equivalent to giving the sequence of multiplicities $m_i(B)$. By definition, the proximity matrix encodes the set of proximity relations. □

Two branches with the same Puiseux characteristic, and hence sharing the other properties listed, are said to be *equisingular*. This does not imply that they are equivalent up to change of coordinates, and an example is given in Exercise 4.7.22 where this is not the case, but most of the properties studied in this book are shared by any two equisingular curves.

For the case of curves with several branches, there is a corresponding notion. It is convenient to formulate the result taking account of the case of a single branch.

Proposition 4.3.9 *For plane curve singularities C and C', the following are equivalent.*

(a) *There is a bijection $B_i \leftrightarrow B'_i$ between branches of C and C' such that, for each i, B_i and B'_i are equisingular and*

  (i) *for all i, j, $B_i, B_j = B'_i.B'_j$, or equivalently,*

  (ii) *for all i, j, B'_i and B'_j have the same exponent of contact as B_i and B_j.*

(b) *There is an isomorphism of Eggers trees $\Gamma_E(C) \to \Gamma_E(C')$.*

(c) *There is an isomorphism between the trees of infinitely near points in good resolutions of C and C' preserving proximity relations.*

Proof The equivalence between (i) and (ii) follows from Theorem 4.1.6.

If (a) holds, there are isomorphisms $\Gamma_E(B_i) \to \Gamma_E(B'_i)$, and since the exponents of contact are the same, there is an isomorphism $\Gamma_E(C) \to \Gamma_E(C')$. Conversely, an isomorphism of Eggers trees implies isomorphisms of Eggers trees, and hence equisingularity, of the branches; and also equality of exponents of contact, and hence (a(ii)).

The equivalence of (a) and (c) will follow from Lemma 4.4.2. □

We say that the curves C and C' are *equisingular* if the equivalent conditions of Proposition 4.3.9 hold.

Results stating that two curves are equisingular if and only if the dual trees of minimal good resolutions are isomorphic, may also be formulated, but some care is needed in the formulations to ensure that sufficient information is attached to the tree. See Exercise 4.7.14 for an example showing that the tree of infinitely near points, with multiplicities attached, does not suffice. See Theorem 8.1.7 and Proposition 8.3.1 for correct versions.

The fact that an isomorphism of Eggers trees needs to preserve all the structure is illustrated in Example 4.4.2 (b).

Example 4.3.2 A singular point equisingular to one defined by the equation $xy^2 = x^{k-1}$ is said to be of *type* D_k (for $k \geq 4$). We also give names to the following:

$$E_6: \; y^3 = x^4, \quad E_7: \; y^3 = x^3 y, \quad E_8: \; y^3 = x^5.$$

Singularities of types A_k, $k \geq 1$, D_k, $k \geq 4$, E_k, $k = 6, 7, 8$ are said to be *simple singularities*. It can be shown that any curve with singular point of one of these types can be reduced to the given equation by holomorphic change of coordinates.

4.4 Intersections and infinitely near points

The intersection number of two branches can also be expressed in the terminology of infinitely near points. The key to this is the following (generalising Lemma 2.3.3).

Lemma 4.4.1 *Let the branches B and B' respectively have multiplicities m and m'; suppose that blowing up the origin gives branches B_1 and B'_1. Then if B and B' have distinct tangents at O, their intersection number is mm'. If they have a common tangent, B_1 and B'_1 are both centred at the point P corresponding to it, and $(B.B')_O = mm' + (B_1.B'_1)_P$.*

Proof We recall from Section 4.1 that the intersection multiplicity is the order of the power series

$$\prod_1^m (a'(t') - a(e^{\frac{2\pi ir}{m}} t'^{\frac{m'}{m}})),$$

where B is given by $x = t^m$, $y = a(t)$, and B' by $x = t'^{m'}$, $y = a'(t')$. The effect of blowing up is to replace a by $b(t) = t^{-m}a(t)$, and similarly for a'. Substituting in the above expression for a and a', we find that it reduces to

$$\prod_1^m (t'^{m'} b'(t') - (e^{\frac{2\pi ir}{m}} t'^{\frac{m'}{m}})^m b(e^{\frac{2\pi ir}{m}} t'^{\frac{m'}{m}})),$$

and hence to

$$t'^{mm'} \prod_1^m (b'(t') - b(e^{\frac{2\pi ir}{m}} t'^{\frac{m'}{m}})),$$

from which the lemma is immediate. □

One might try to argue as follows. By Lemma 4.1.7, $B.B'$ is the sum of the exponents of contact of the m pro-branches γ of B and the m' pro-branches γ' of B'. The exponent of contact of γ and γ' decreases by 1 on blowing up and the result follows. However, more care is necessary since the multiplicity may decrease on blowing up, and then the pro-branches of the blow up do not correspond in a simple manner with those of the original.

Given two branches B and B', blowing up yields branches B_1 and B'_1, say. If B and B' have the same tangent, B_1 and B'_1 have the same centre, and we can blow up again, obtaining B_2 and B'_2. We obtain a sequence of branches B_i and B'_i, with multiplicities m_i and m'_i, say. Applying the above inductively yields

$$(B.B')_O = \sum_{i=0}^{q-1} m_i m'_i + (B_q.B'_q).$$

The next result follows at once.

Lemma 4.4.2 *If B and B' are distinct branches, then after finitely many blowings up, they will have different centres. Their intersection number is given by the sum*

$$(B.B')_O = \sum_i m_i m'_i$$

of products of multiplicities over those infinitely near points that they have in common.

There is a corresponding result where we use the discriminant in place of the resultant. Recall that in Lemma 4.1.8 we considered the invariant $B.P$, where B is a branch at O not tangent to the y-axis, and with equation $f(x, y) = 0$ where $f \in \mathbb{C}\{x\}[y]$ is a monic polynomial, and P is the polar curve defined by $\partial f/\partial y = 0$. Let B_1 be defined by blowing B up once, and P_1 be related to B_1 as P is to B; and similarly for higher blowings up.

Lemma 4.4.3 $(B.P)_O = m(m-1) + (B_1.P_1)_O$.

Proof The effect of the blow up $(x, y) = (x_1, x_1 y_1)$ is to replace f by $f_1(x_1, y_1) = x_1^{-m} f(x_1, x_1 y_1)$. Apply (i) of Lemma 1.3.2: since, if the a_i are the roots of f, the roots of f_1 will be $x^{-1}a_i$, we have $D(f_1) = x^{-m(m-1)}D(f)$. The result now follows from (i) of Lemma 4.1.8. □

Since B_1 may be tangent to the y-axis (if a proximity relation holds) we cannot conclude that $(B.P)_O = \sum_i m_i(m_i - 1)$. A correct result will be obtained in Theorem 6.5.9.

We can obtain the formula by using (ii) of Lemma 4.1.8, the fact that the exponent of contact of two pro-branches decreases by 1 on blow up, and that there are $m(m-1)$ pairs of distinct pro-branches. But again this fails to give a proof since we cannot directly compare the pro-branches of B and its blow up.

Suppose that B, B' are two branches with B' smooth. Then all multiplicities for B' are equal to 1, and no proximity relations hold. Thus the set of infinitely near points common to B and B' is a sequence where there are no proximity relations, so the sequence of multiplicities must be constant except, perhaps, for the last term. We recall from Proposition 3.5.1 that if we divide β_1 by m to obtain $\beta_1 = mq + r$, the first multiplicities in the sequence are m q times, followed by r; the next point is proximate to the last point of multiplicity m. Our intersection number is thus either of the form im with $1 \leq i \leq q$ or equal to $rm + q = \beta_1$. In the latter case, B' is said to have *maximal contact* with B. Thus the Puiseux series for y begins with the term $x^{\beta_1/m}$ if and only if we choose the x-axis to have maximal contact with B.

Example 4.4.1 Let B have type A_{2k}. Then a smooth branch B' may have the points $O_0, \ldots, O_r$ in common with B for any $r \leq k$, but not O_{k+1} since this is proximate to O_{k-1} and a smooth curve does not pass through any satellite points. The intersection number is $2(r+1)$ if $0 \leq r < k$ and $2k+1$ if $r = k$.

If B is given by $y^2 = x^{2k+1}$ then, for example, the curve $y = x^r$ has intersection number $2r$ with B; the line $y = 0$ has intersection number $2k+1$.

We can apply Lemma 4.4.2 to obtain a further interpretation of the proximity matrix. We recall that the sequence of multiplicities $m_r(\epsilon_k)$ is given by the entries in the kth column of $Q(B)$. It follows that

Lemma 4.4.4 *The intersection number $\epsilon_k.\epsilon_l$ is the (k, l) entry of the matrix*

$$Q(B)^t.Q(B) = (P(B).P(B)^t)^{-1}.$$

We now have two ways to calculate intersection numbers: Theorem 4.1.6, by exponents of contact, and Lemma 4.4.2, by infinitely near points. It is natural to seek a direct relation between the exponent of contact of two branches and the set of infinitely near points they have in common: the 'Enriques coefficient of contact' may be defined as the number of infinitely near points in common. Although more precise information will be given in Section 8.3, we now give simple examples to show that not only does neither determine the other, but they do not even increase together.

Example 4.4.2

(a) Let B be a simple branch $y = 0$, and B' the monomial curve $y = x^\kappa$, for some $\kappa > 0 \in \mathbb{Q}$. These clearly have exponent of contact κ. We see by repeated blowing up that the number of infinitely near points in common is $\lceil \kappa \rceil$, the next integer $\geq \kappa$, which does not, of course, determine κ.

(b) Both B ($y = 0$) and B'' ($y^3 = 2x^5$) have exponent of contact $\frac{5}{3}$ with B' ($y^3 = x^5$). However, B and B' have just the infinitely near points O_0 and O_1 in common, while B' and B'' share also O_2 and O_3. The intersection number $B.B' = 5 = 1.3 + 1.2$, while the intersection number $B''.B' = 15 = 3.3 + 2.2 + 1.1 + 1.1$ (the latter expression is in the form $\sum m_i m_i'$). Observe also that the Eggers trees of $B \cup B'$ and $B' \cup B''$ are isomorphic if we ignore h and ν: in each case there is just one interior vertex, at $\frac{5}{3}$.

(c) Let B_1 be the curve $y^5 = x^{12}$; B_2 the curve $y^3 = x^7$ and B_3 the x-axis $y = 0$. Then $\kappa(B_1, B_2) = \frac{7}{3} < \frac{12}{5} = \kappa(B_1.B_3)$; while B_1 and B_2 have 5 infinitely near points in common; B_1 and B_3 have only 3. The intersection number $B_1.B_2 = 15+15+2+2+1 = 35$; $B_1.B_3 = 5 + 5 + 2 = 12$, as we can also calculate from parametrisations or from the Herbrand function.

4.5 Decomposition of transverse polar curves

In this section, we present an important decomposition theorem for transverse polar curves of curves with arbitrary numbers of branches. The decomposition obtained depends only on the equisingularity type of the curve.

This is not necessarily the complete decomposition of a transverse polar into irreducible components, even for curves with a single branch. Indeed, as Exercise 4.7.21 shows, the number of branches can vary for different transverse polars of the same curve, and P_i need not even be reduced. To obtain a full analysis, it is necessary to impose further conditions.

It is convenient to associate with the Eggers tree combinatorial information in the form of several cycles and cocycles. The function ν_C is a locally constant function on the tree, with discontinuities (at most) at the vertices. We may identify this with the 1-chain on $\Gamma_E(C)$, regarded as a simplicial complex, such that the coefficient of any edge (i.e. 1-simplex) is the constant value taken by ν_C on the interior of that edge.

Recall that, by Lemma 4.2.1, for any branch B' at O there is a unique point $X_{B'} \in \Gamma_E(C)$ such that for each $j \in J$, $\mathcal{O}(B', B_j) = v(\inf(X_{B'}, B_j^\infty))$. We measure the contact with C of a single branch B' by the 0-chain $[B'] := m(B')X_{B'}$; for a curve C' with branches B'_i, we use the 0-chain $[C'] := \sum_i [B'_i]$.

Since $\Gamma_E(C)$ is a tree, its chain groups (with $\mathbb{Z}$ coefficients) form a short exact sequence $0 \to C_1 \xrightarrow{\partial} C_0 \xrightarrow{\epsilon} \mathbb{Z} \to 0$, which is split by the map from $\mathbb{Z}$ to C_0 taking 1 to A_0. This induces a splitting map $s : C_0 \to C_1$. The value of s on a vertex X is the sum of the edges forming the path from A_0 to X.

For any vertex X, write $\widehat{X}$ for the 1-cochain dual to the edge immediately below X (and $\widehat{A_0} := 0$). If C' is, as above, a germ with branches B'_i, we claim that $\langle \widehat{X}, s[C'] \rangle$ is the sum of the multiplicities of the branches B' of C' passing through X (i.e. with $X \leq X_{B'}$). It suffices to check for a single branch B'. But then $s[B']$ is the sum of the edges below $X_{B'}$, each with coefficient $m(B')$, so has non-trivial product with $\widehat{X}$ if and only if $X \leq X_{B'}$.

Let us calculate the list of exponents of contact with C of the pro-branches of a general branch B'. First consider the case when C has a single branch B.

By Proposition 4.1.3 (1), if γ' is a pro-branch of B', $\mathcal{O}(B, B') = \kappa$, and $\alpha_q < \kappa \leq \alpha_{q+1}$, then $\{\mathcal{O}(\gamma', \gamma) \mid \gamma \in \text{pro}\,(B)\}$ consists of α_i, occurring $(e_{i-1} - e_i)$ times $(1 \leq i \leq q)$; and κ, occurring e_q times. We may represent this list by the 0-chain $\sum_1^q (e_{i-1} - e_i)A_i + e_q X_{B'}$ on $\Gamma_E(B)$. Applying the splitting map s to this 0-chain gives the 1-chain

$$\zeta_{B,B'} := \sum_1^q e_{r-1}(A_{r-1}A_r) + e_r(A_q X_{B'}),$$

and the 0-chain is then $\partial \zeta_{B,B'} + m(B)A_0$.

If C has branches B_j, we define $\zeta_{C,B'} := \sum_j \zeta_{B_j,B'}$. This is determined by the list of exponents of contact of a pro-branch of B' with the pro-branches of the several branches of C, so depends only on the point $X_{B'}$. We define $\eta_C(X_{B'}) := \zeta_{C,B'}$, and extend to an additive homomorphism $\eta_C : C_0 \to C_1$. Thus if C' has branches B'_i, we have $\eta_C[C'] := \sum_i m(B'_i)\zeta_{C,B'_i}$.

For each point $M \neq A_0$, the highest point occurring with non-zero coefficient in $\partial\eta_C(M)$ is M itself. Thus the matrix of η_C is triangular, and η_C has kernel $\mathbb{Z}A_0$. Hence to recover the characteristic 0-cycle $[C']$ of the curve-germ C' it is sufficient to know the information $\eta_C[C']$ determined by the exponents of contact. This remark applies not only to the tree $\Gamma_E(C)$ but also to any tree obtained by subdividing at a finite number of points: the transitions η_C and its inverse do not require additional subdivision.

The explicit inversion of η_C is given by the following.

Lemma 4.5.1 *With the above notation, we have*

$$\langle\widehat{M},\nu\rangle\langle\widehat{M},\eta_C[C']\rangle = \langle\widehat{M},s[C']\rangle\langle\widehat{M},s[C]\rangle. \tag{4.9}$$

Proof Since both sides are additive in $[C']$, we may assume $C' = B'$ irreducible. Now calculate explicitly. For any point M, we have

$$\langle\widehat{M},\eta_C[B']\rangle = \sum_j m(B')\langle\widehat{M},\zeta_{B_j,B'}\rangle = \sum_j m(B')m(B_j)c_j(M), \tag{4.10}$$

where $c_j(M)$ is zero unless $M \le X_{B'}$ and $M \in \Gamma_E(B_j)$, while if $M \le X_{B'}$ and $M \in \pi_{B_j}(A_{r-1}, A_r]$, we have $c_j(M) = \frac{e_{r-1}}{m(B_j)} = \frac{1}{\langle\widehat{M},\nu\rangle}$: though the definition of e_{r-1} depends on the choice of branch, the quotient $\frac{e_{r-1}}{m}$ is determined by the point of $\Gamma_E(C)$.

Hence the right hand side of (4.10) is 0 unless $M \le X_{B'}$, when it becomes

$$m(B') \sum_{M \le B_j^\infty} m(B_j)/\langle\widehat{M},\nu\rangle = m(B')\langle\widehat{M},s[C]\rangle/\langle\widehat{M},\nu\rangle.$$

Thus in all cases it is equal to $\langle\widehat{M},s[B']\rangle\langle\widehat{M},s[C]\rangle/\langle\widehat{M},\nu\rangle$. □

Theorem 4.5.2 *Let C be a reduced curve germ, P a transverse polar curve of C. Then the contact of P with C is measured by the 0-chain $[P] = [C] - \partial\nu - A_0$. In particular, the points occurring with non-zero coefficients in $[P]$ are the vertices of $\Gamma_E(C)$ with values in $(0,\infty)$.*

Proof By Lemma 4.1.9, for any pro-branch γ of C,

$$\{\mathcal{O}(\gamma,\delta) \mid \delta \in \text{pro}\,(P)\} = \{\mathcal{O}(\gamma,\gamma') \mid \gamma \ne \gamma' \in \text{pro}\,(C)\}.$$

Thus if $\gamma \in \text{pro}\,(B_j)$, the 0-chains corresponding to

$$\{\mathcal{O}(\gamma,\delta) \mid \delta \in \text{pro}\,(B')\}$$

for B' equal to P or C differ by B_j^∞. Hence the 1-chains differ by $s(B_j^\infty)$, and

$$\eta_C[P] = \sum_j m(B_j)(\eta_C[B_j^\infty] - s[B_j^\infty]) = \eta_C[C] - s[C]. \tag{4.11}$$

Applying (4.9) to $B' = P$, for any point M of the tree, gives

$$\langle \widehat{M}, s[P]\rangle\langle \widehat{M}, s[C]\rangle = \langle \widehat{M}, \nu\rangle\langle \widehat{M}, \eta_C[P]\rangle = \langle \widehat{M}, \nu\rangle\langle \widehat{M}, (\eta_C[C] - s[C])\rangle, \tag{4.12}$$

while, applying (4.9) to $B' = C$ itself, we have

$$\langle \widehat{M}, s[C]\rangle\langle \widehat{M}, s[C]\rangle = \langle \widehat{M}, \nu\rangle\langle \widehat{M}, \eta_C[C]\rangle. \tag{4.13}$$

Substituting (4.13) in the right hand side of (4.12), and cancelling the factor $\langle \widehat{M}, s[C]\rangle$, gives

$$\langle \widehat{M}, s[P]\rangle = \langle \widehat{M}, s[C]\rangle - \langle \widehat{M}, \nu\rangle,$$

and since the 1-cochains $\widehat{M}$ span C^1 we infer $s[P] = s[C] - \nu$.

Applying ∂ we see that $[P]$ differs from $[C] - \partial\nu$ by a multiple of A_0. We determine this multiple by applying ϵ: $\epsilon[P] = m(P) = m(C) - 1$, $\epsilon[C] = m(C)$, and $\epsilon \circ \partial = 0$, so $[P] = [C] - \partial\nu - A_0$, as claimed.

By inspection, the coefficients of A_0 and the B_j^∞ in this expression vanish (as they must), and no point other than the original vertices of $\Gamma_E(C)$ can appear. To see that each of these has non-zero coefficient, observe that for a single branch, the values of ν on the edges increase strictly, so each A_i occurs with strictly positive coefficient in $-\partial\nu$; while for a branch point $I_{j,k}$ the values of ν on each of the edges immediately above the point are at least equal to the value on the edge immediately below, so again we have a strictly positive coefficient. □

4.6 Notes

Section 4.1 Orders and exponents of contact have belonged to the theory from early days. Much of the development in this chapter (including the notation $\overline{\beta}_i$) is due to Zariski: see particularly [205], which includes the main properties of the semigroup. Although the term 'pro-branch' is new, essentially the same concept is used by other authors, sometimes called 'Puiseux root' (with respect to a fixed coordinate system and choice of root $x^{1/m}$).

We have borrowed the term 'Herbrand function' from number theory, following IV.3 of Serre [166]. The situation there is that of a Galois extension L/K of fields, with Galois group G, and a discrete valuation v of L. The *ramification group* G_i is defined as the set of $\sigma \in G$ such that $v(a) \geq 0$ implies $v(\sigma(a) - a) \geq i + 1$: this definition also makes sense for $i \notin \mathbb{Z}$, but depends only on the least integer $\geq i$. The Herbrand function is then defined by

$$\phi(u) = \int_0^u \frac{dt}{|G_0 : G_t|} = \frac{1}{g_0}(g_1 + \cdots + g_k + (u-k)g_{k+1}), \tag{4.14}$$

where g_k denotes the order $|G_k|$ and $k \le u \le k+1$.

Since this definition can be rephrased in terms of the rings of integers $\mathcal{O}_K \subset \mathcal{O}_L$, it is very closely related to our own situation, where the group G consists of the substitutions $t \mapsto e^{2ai\pi/n}t$ and $v(\sigma(y) - y)$, where v is interpreted as the order quâ Puiseux series in x, gives the exponent of contact of two pro-branches. See Exercise 4.7.5 for a parallel definition of the Herbrand function of the text.

It is striking that, though the use of the Herbrand function here is quite different from its importance in the number-theoretic situation, it is the same construction that is required. Formula (4.14) is also related to the important definition of the Swan conductor in characteristic p.

Section 4.2 The introduction of Eggers trees is due to Harald Eggers [62]. Eggers calls the vertices B^∞ white and the rest black, and defines certain edges to be dotted. An essentially equivalent, though in the author's opinion less convenient, 'tree-model' is due to Kuo and Lu [105]. The Eggers tree itself can be identified with the Hasse diagram of the poset of the vertices, but this fails to describe the functions v_C, h_C and ν_C.

The version given here was developed to facilitate the exposition of the Decomposition Theorem 4.5.2, and refined when the relation with the resolution tree was clarified for Section 9.4. Our definition of isomorphism is essentially that of García Barroso and González Pérez [77].

Section 4.3 It is also possible to define a semigroup in the case of a curve C with several branches B_i $(1 \le i \le k)$. Here $S(C)$ is the subsemigroup of $\mathbb{N}^k$ consisting of the vectors $(\mathrm{ord}_{B_1}\phi, \ldots, \mathrm{ord}_{B_k}\phi)$ for ϕ in $\mathbb{C}[[x,y]]$ or $\mathbb{C}\{x,y\}$ not vanishing identically on any B_i; or equivalently, consisting of the $(C'.B_1, \ldots, C'.B_k)$ where C' is a curve containing none of the B_i and all intersection numbers are taken at $(0,0)$. We can also define this via a normalisation $\pi : (\widetilde{C}, \{x_1, \ldots, x_k\}) \to (C, O)$, taking the induced embedding $\mathcal{O}_{(C,O)} \to \prod_i \mathcal{O}_{(\widetilde{C},x_i)} \cong \prod_i \mathbb{C}\{t_i\}$, and taking the orders.

This semigroup is, however, less convenient than in the case of a single branch. In the simplest case of a curve with 2 smooth branches, transverse to each other, e.g. that given by $xy = 0$, it consists of the set of pairs (r,s) with either $r = s = 0$ or $r \ge 1, s \ge 1$, for each of $r = 0$, $s = 0$ is equivalent to $\phi(0,0) = 0$; that all other pairs can occur is shown by taking C to be parametrised by $t \mapsto (t^s, t^r)$. Already in this example, $\mathbb{N}^2 - S(C)$ is an infinite set, and $S(C)$ is not finitely generated. We will thus not develop the theory of this semigroup: the reader may refer to [46]. A more sophisticated 'extended semigroup' was defined in [29]. The result of Exercises 4.7.10 and 5.7.7 are due to these authors, who also obtain an extension to several variables for curves with several components.

The notion of equisingularity was developed by Zariski in a series of papers [205], [206], [207]: the case of equisingularity of plane curves provided the springboard for his general theory.

Section 4.4 Parts of the proof of the Resolution Theorem 3.4.4 could have been expressed using maximal contact: see for example [23].

It follows from Lemma 4.4.4 that the final column of ${}^tQ(B).Q(B)$ gives the intersection numbers $\epsilon_k.B$. We will see in Lemma 8.5.1 that the curves B_q^- of Lemma 4.3.1 are curvettes, and we saw in the proof of that lemma that $B_q^-.B = \overline{\beta}_q$. Thus the entries in the final column of ${}^tQ(B).Q(B)$ generate $S(B)$. This gives a direct way to see that the proximity matrix determines the semigroup.

Section 4.5 It had long been known that the germ P of a polar curve is usually reducible. Indeed, the first result concerning the contact between branches of a curve and of its generic polar was due to Henry Smith [168] in 1875.

In 1976 a general statement about the decomposition of the generic polar of a curve with a single branch was obtained by Michel Merle [125]: this result yields a (local) decomposition $P = \bigcup_{1 \leq i \leq g} P_i$, where both the multiplicity of P_i, and the exponent of contact with C of each branch Q of P_i are given.

This raised the problem of finding a corresponding statement for polars of curves with several branches. The result was extended to the case of curves C with two branches by Delgado [48], who also showed that the method of proof, which uses intersection numbers and estimates on multiplicities, does not work for curves with more branches.

The problem was essentially solved in 1983 by Eggers [62]. A definitive result was given in the 1996 thesis of García Barroso [75], [76]. Another proof was found by Assi [14]. Our argument, like those of [62], [75] and [14], hinges on Lemma 4.1.9, which is due to Kuo and Lu [105]. We have followed Eggers in using exponents of contact, but the combinatorial arguments are greatly simplified by the effective use of cycles and cocyles on the Eggers tree. A different approach appears in Lê, Michel and Weber [111]. García Barroso gives a detailed comparison of her results with those of [125], [48], [62], [111] and Casas. She also relates the invariants to the geometry of curvature of the Milnor fibre $f^{-1}(t)$ in the limit as $t \to 0$.

Easy examples show that the equisingularity type of a transverse polar curve of C is not determined by the equisingularity type of C: see e.g. Exercise 4.7.22. A more precise result for a 'general' curve C and a general polar P is given by Casas [31]. In [32] and [34], Casas determines the equisingularity type of a generic polar for the case when C is a general member of its equisingularity class; in particular, he shows that in that case, the pieces determined by the decomposition Theorem 4.5.2 are irreducible, so no further decomposition is possible. We refer to these papers and the book [35] (especially Chapter 6) for further discussion and numerous examples.

One generalisation of polar curves is the Jacobian $\partial(f,g)/\partial(x,y) = 0$ of two given curves $f = 0$ and $g = 0$. There are some results on decomposition of such curves. See e.g. [47] and [2].

Curves over an arbitrary field The formula for intersection numbers in terms of infinitely near points remains valid in characteristic p. From this one can develop the semigroup, which behaves essentially as in characteristic 0. The equivalent conditions which characterise equisingularity remain equivalent. However a different approach is necessary as exponents of contact are not immediately available as before.

In characteristic p there is a clear notion of equisingularity which can be defined by the resolution graph, by the tree of infinitely near points, or equivalently by the semigroup.

Over a field k which is not algebraically closed the results about infinitely near points take the following form. Let X be a smooth projective surface defined over k, let x be a closed point on X and X^* the set of infinitely near points of x. For D a curve-germ on X at x and $p \in X^*$, denote by $m_p(D)$ the multiplicity of the strict transform of D at p. If E is another curve on X such that D and E intersect at x and have no common components through x, then the intersection multiplicity of D and E at x is given by

$$(D.E)_x = \sum_{p \in X^*} m_p(D) m_p(E)\, [k(p) : k],$$

where the sum is over the infinitely near points p of x in common to D and E and $k(p)$ is the residue field at the point p. If k is algebraically closed, then each infinitely near point p is rational over k, so the integer $[k(p) : k]$ is equal to 1.

If D is a reduced curve on X with $x \in D$, then the 'double point number' $\delta_x(D)$ is given by

$$\delta_x(D) = \tfrac{1}{2} \textstyle\sum_{p \in X^*} m_p(D)(m_p(D) - 1)[k(p) : k],$$

and we have the proximity relations

$$m_p(D) = \textstyle\sum_{p'} m_{p'}(D)[k(p') : k(p)],$$

where the sum is over all the infinitely near points p' proximate to p.

The real case If B is a single branch admitting an equation $f(x, y) = 0$ with real coefficients, then B admits a real parametrisation. Choose a good parametrisation $\gamma : (\mathbb{C}, 0) \to (B, O)$. Since B is invariant under complex conjugation, this lifts to an anti-holomorphic involution $\sigma : (\mathbb{C}, 0) \to (\mathbb{C}, 0)$. Since σ has 0 as fixed point and its Jacobian matrix at 0 has eigenvalues ± 1, there is a smooth curve of fixed points. If $\phi : (\mathbb{R}, 0) \to (\mathbb{C}, 0)$ parametrises this curve, $\gamma \circ \phi$ gives the required parametrisation.

It follows that the resolution process for B over $\mathbb{R}$ proceeds exactly as in the complex case; each stage of the blowing up produces a copy of the real projective line $P^1(\mathbb{R})$.

For an algebraic curve germ C over $\mathbb{R}$, we can again consider C as a complex curve, take a normalisation $\gamma : \tilde{C} \to C$, and lift complex conjugation to an anti-holomorphic involution σ of the germ $\tilde{C}$. In general, this will interchange some pairs of branches: such branches have no real points other than O. Each branch which is invariant under σ has a real parametrisation as above.

Instead of pro-branches it is more natural over $\mathbb{R}$ to consider semi-branches: a semi-branch is given by a good real parametrisation where the parameter is restricted by $t \geq 0$. An irreducible germ whose equation is defined over $\mathbb{R}$ has 2 real semi-branches. The topology of the set of real points of C is essentially trivial: in a disc neighbourhood of O, the semi-branches are disjoint arcs running from O to the boundary of the neighbourhood – all one can do is count them.

The tangent to a semi-branch has a natural sense, defined by that in which the parameter t increases. The semi-branches belonging to a branch B have tangents in the same sense if and only if the multiplicity $m(B)$ is odd. The exponent of contact has geometrical significance only for two semi-branches with the same oriented tangent. Given two branches, each defining 2 semi-branches, we can ask whether the real curves cross each other at O. A deformation argument shows that this is so if and only if the intersection number of the branches is odd.

4.7 Exercises

Exercise 4.7.1 List all the pro-branches of the curves $y = x^{3/2}$ and $y = x^{3/2} + x^{7/4}$, determine the exponent of contact of each pair of pro-branches, and hence calculate the intersection number of the curves.

Exercise 4.7.2 For each of the following Puiseux characteristics, determine the parameters e_i and $\overline{\beta}_i$, the Herbrand function $H(v)$, and $N(S(B))$:

(a) (6;22,31), (b) (12;15,20), (c) (12;18,32,35) and (d) (72;84,111,160).

Exercise 4.7.3 Classify singularities of multiplicity 3 up to equisingularity by listing all possibilities for Puiseux characteristics and mutual intersection numbers of the branches.

Exercise 4.7.4 Show that the inverse H^{-1} of the Herbrand function of a branch satisfies $H^{-1}(\mathbb{N}) \subseteq \mathbb{N}$.

Exercise 4.7.5 Let B be a branch with good parametrisation $x = t^m$, $y = \psi(t)$. For each $v > 0$ let G_v denote the set of substitutions $t' = \epsilon t$ such that $\epsilon^m = 1$, $\text{ord}\,(\psi(\epsilon t) - \psi(t)) > v$. Show that if $\beta_{i-1} < v < \beta_i$ then $|G_v| = e_{i-1}$, and deduce that the integral (4.14) gives the Herbrand function $H(v)$.

Exercise 4.7.6 Suppose the Newton polygon of f_i consists of a single edge, from $(a_i, 0)$ to $(0, b_i)$, where a_i and b_i are coprime $(i = 1, 2)$; let $b_1 < a_1$ and $a_1 b_2 \neq a_2 b_1$. Determine the intersection number of the curves C_i given by $f_i = 0$ (a) by substituting a parametrisation of C_2 in f_1 and (b) by finding the exponent of contact of the two curves and applying Theorem 4.1.6; check you obtain the same result in all cases.

Exercise 4.7.7 Describe the Eggers trees of each of the following curves:
(a) $(y - x^2)(y^2 - x^7) = 0$;
(b) $(y^3 - x^5)(y^4 - x^7) = 0$;
(c) $y(y^3 - x^5)(y^3 + x^5) = 0$;
(d) $(y^2 - x^3)(y^3 - x^5)(y^3 + x^5) = 0$;
(e) $\{y = x^{3/2} + x^{7/4}\} \cup \{y = x^{3/2} + x^{9/4}\}$;
(f) $\{y = x^{3/2} + x^{9/4}\} \cup \{y = x^{3/2} + x^2 + x^{11/4}\}$.

Exercise 4.7.8 B_1, B_2 are branches at O with respective Puiseux characteristics (6:8,13) and (6;9,13). For each of B_1 and B_2 determine the parameters e_i and $\overline{\beta}_i$, and $N(S)$, and list the positive integers $\leq N(S)$ that do not belong to S.

Exercise 4.7.9 Show that the singular points given by $y^3 + x^6 = 0$ and $y^3 + yx^4 = 0$ are equisingular, but cannot be reduced to each other by a holomorphic change of coordinates. (Hint: expand a general change of coordinates as a power series (you will only need the linear and quadratic terms) substitute in $y^3 + x^6$ and equate successively coefficients of terms of degrees 3, 4 and 5.)

Exercise 4.7.10 Define the Poincaré series of a semigroup $S \subseteq \mathbb{Z}$ to be $G_S(t) := \sum\{t^i \mid i \in S\}$. Show that, under the hypothesis of Lemma 4.3.4, $G_{S_1}(t)$ is equal to $G_S(t)(1 - t^{qb})/(1 - t^b)$. Hence, if S is the semigroup of a branch B, use Lemma 4.3.6 to determine $G_S(t)$ in terms of the $\overline{\beta}_i$ and the e_i.

Exercise 4.7.11 Use (4.6) and the calculation in Theorem 3.5.5 of the Puiseux characteristic of the blown up branch $B^{(1)}$, to calculate the values of $\overline{\beta}_i$ for the blow up of a branch.

Exercise 4.7.12 Use the formula $N(S(B)) = -m + \overline{\beta}_g e_{g-1} - \beta_g$, and the calculation in the preceding Exercise to show that

$$N(S(B)) - N(S(B^{(1)})) = m(m - 1)$$

in all cases.

Exercise 4.7.13 Determine the intersection number of the curves C_1 and C_2 of Exercise 3.8.9.

Exercise 4.7.14 Suppose that a minimal good resolution of a curve C produces infinitely near points and multiplicities $O_0(6)$, $O_1(4)$, $O_2(2)$, $O_2'(2)$, $O_3(1)$, $O_3'(1)$, $O_3''(1)$, $O_4(1)$ where O_i succeeds O_{i-1}, O_2' succeeds O_1 and O_3' and O_3'' both succeed O_2'.

Show that there are two possibilities for the proximity relations between the points, and find in each case the Puiseux characteristics of the three branches and their intersection numbers.

Exercise 4.7.15 Suppose the Newton polygon of f consists of a single edge, from $(a, 0)$ to $(0, b)$, where a and b are coprime. Show that the sequence of multiplicities arising on successive blowing up arises from applying the Euclidean algorithm to a and b; and deduce that the curve is equisingular to $x^a + y^b$.

Exercise 4.7.16 Show that if $O_0, \dots, O_N$ is a sequence of infinitely near points, with each proximate only to its predecessor, there is a smooth curve B passing through them all.

Exercise 4.7.17 Find the multiplicity sequence of the curve $C : y^3 = x^{11}$, and hence determine how many infinitely near points it may have in common with a smooth curve S. For each possibility, determine the intersection number $C.S$ and the exponent of contact $\mathcal{O}(C, S)$. Give examples showing that each case can occur.

Exercise 4.7.18 For curves B and B' with Puiseux characteristics $(4; 10, 13)$ and $(6; 15, 22)$, determine the maximum possible number of infinitely near points they might have in common.

Exercise 4.7.19 Show that for any set of proximity relations between points of the sequence $O_0, \dots, O_N$ satisfying (i) and (ii) of Proposition 3.5.1, there is an irreducible curve C whose infinitely near points satisfy just these conditions. (Hint: use induction on N: let C_1 correspond to the points $O_1, \dots, O_N$: seek C such that blowing up the origin gives C_1.)

Exercise 4.7.20 Let B_1, B_2 and B_3 be the curves given by the respective Puiseux series

$$B_1 : \ y = x^{\frac{3}{2}} + x^{\frac{9}{4}}, \quad B_2 : \ y = x^{\frac{3}{2}} + x^{\frac{11}{4}}, \quad B_3 : \ y = x^{\frac{3}{2}} + x^2 + x^{\frac{11}{4}}.$$

Describe the Eggers trees of $B_1 \cup B_2$ and $B_1 \cup B_3$, giving the 1-chain ν in each case. Hence describe the packets in the decomposition of a transverse polar in each case.

Exercise 4.7.21 Let C be defined by $f(x, y) \cong y^3 + x^8 = 0$. By considering the (transverse) polars given by $\partial f/\partial y + \partial f/\partial x = 0$, $\partial f/\partial y + x\partial f/\partial x = 0$, and $\partial f/\partial y = 0$, show that the numbers of branches of different transverse polars can differ, and that a transverse polar can be non-reduced.

Exercise 4.7.22 Show that any polar curve of the curve C_1 given by $y^3 + x^7 = 0$ has equation of the form $y^2 A(x, y) + x^6 B(x, y) = 0$, and that if the polar is transverse, $A(0, 0) \neq 0$. Deduce that any transverse polar of C_1 is equivalent up to coordinate change with $y^2 = 0$ or $y^2 + x^n = 0$ for some $n \geq 6$.

Show that any transverse polar of the curve C_2 given by $y^3 + x^5 y + x^7 = 0$ has equation of the form $y^2 A(x, y) + x^5 B(x, y) + x^4 y C(x, y) = 0$ with $A(0, 0), B(0, 0) \neq 0$. Deduce that any transverse polar of C_2 is equivalent up to coordinate change with $y^2 + x^5 = 0$.

Show that the curves C_1 and C_2 are equisingular, but *not* equivalent by holomorphic change of coordinates.

Exercise 4.7.23 Show, by considering the leading terms in the parametrisations, that if B, B' are real branches with odd multiplicity, then B and B' cross if and only if $\mathcal{O}(B^+, B'^+)$ has odd numerator (where B, B' have semi-branches $B^\pm$, $B'^\pm$). Show that $\mathcal{O}(B, B') = \mathcal{O}(B^+, B'^+)$, and hence verify that this condition is equivalent to having $B.B'$ odd.

5

Topology of the singularity link

Up to this point we have concentrated on the algebraic side of the description of plane curve singularities. It is even more fascinating to try and visualise them. After a preliminary section on vector fields, in which we recall some standard results from analysis which will give us our main tool for constructing homeomorphisms, we go on to a detailed geometrical description of the local behaviour of a curve at a singular point, which gives in particular a picture of the topology of the link.

We go on to calculate the numerical invariants needed to specify the particular knot or link. Using some basic results about the Alexander polynomial of a knot leads to our main conclusion, that the topology determines the numerical invariants defined earlier.

5.1 Vector fields

In this section we develop our main technique for constructing diffeomorphisms. A first idea is to start with a diffeomorphism somewhere and deform it in a 1-parameter family. We thus define a smooth *isotopy* from X to Y to be a smooth embedding $F : X \times I \to Y \times I$ of the form $F(x,t) = (f_t(x), t)$, so that each f_t is a smooth embedding of X into Y; we also say that the embeddings f_0 and f_1 are isotopic. The fundamental case is when $Y = X$ and we start at the identity map $f_0(x) = x$.

The second idea is to differentiate F with respect to the 'time' variable t. For each $P \in X$, $f_t(P)$ describes a smooth curve in X, which thus has a tangent vector at each point. These tangent vectors form a vector field on X, which is in general time-dependent. Conversely, given a time-dependent vector field, then under fairly mild conditions we can integrate it to give an isotopy. Requirements such as that the isotopy preserve some submanifold, or respect some function on X, can then be translated into equivalent conditions on the vector field.

The third idea is that it is much easier to construct vector fields than maps. In particular, vectors can be added. The situation in practice is that in the neighbourhood of any point we can construct a vector field satisfying the desired conditions, and want to fit these pieces together. This is accomplished using partitions of unity.

A vector in $\mathbb{R}^n$ is specified by its coordinates, say $\xi = (\xi_1, \ldots, \xi_n)$. A *vector field* associates to each point $\mathbf{x}$ of some region $U \subset \mathbb{R}^n$ a vector $\xi(\mathbf{x})$ depending on $\mathbf{x}$; for the vector fields with which we are concerned the dependence will be infinitely differentiable, though it will suffice for the arguments below to have each $\xi_r(\mathbf{x})$ once continuously differentiable. We will call them differentiable, or smooth vector fields. While vectors are conveniently specified by coordinates, one should think of them more geometrically – a vector has a magnitude and a direction – and the coordinates undergo standard transformations under change of coordinates.

A vector field ξ operates on a differentiable function f on U, taking it to $\xi(f) := \sum \xi_r(\mathbf{x})\partial f/\partial x_r$. It is thus convenient to adopt the notation $\xi = \sum \xi_r \partial/\partial x_r$, which has the additional advantage of unifying the transformation formula for change of coordinates with the chain rule for differentiating a function.

We can also think of a vector field as a differential equation. A *solution*, or *integral curve* of ξ is a map $\mathbf{g} : \mathbb{R} \to \mathbb{R}^n$ (not necessarily defined for all values of $t \in \mathbb{R}$) such that, for all t, $\frac{dg_r(t)}{dt} = \xi_r(\mathbf{g}(t))$. Thus for any function f on $\mathbb{R}^n$,

$$\frac{df(g(t))}{dt} = \sum \frac{\partial f}{\partial x_r}(g(t))\frac{dg_r(t)}{dt} = \sum \xi_r \frac{\partial f}{\partial x_r} = \xi(f).$$

The fundamental theorem for (ordinary) differential equations states that equations have unique solutions. The proof only works in a small neighbourhood of the starting point, but does give extra information, as follows.

Theorem 5.1.1 *Let ξ be a smooth vector field defined on a neighbourhood of $O \in \mathbb{R}^n$. Then there is a differentiable map $G : \mathbb{R}^n \times \mathbb{R} \to \mathbb{R}^n$, defined on a neighbourhood $U \times (-\epsilon, \epsilon)$ of $(O, 0)$, such that, for any $\mathbf{x} \in U$, $t \mapsto G(\mathbf{x}, t)$ is the unique solution of the equation for which $g(0) = \mathbf{x}$.*

For a proof see a suitable analysis textbook, e.g. [39] or [52].

For the applications we will make of this Theorem, we will require integral curves which are not defined only for small values of the parameter t. This requires extra conditions as even in the case $n = 1$ there are examples such as $(1 + x^2)\partial/\partial x$ with integral $x = \tan(t)$ which goes to infinity as $t \to \frac{\pi}{2}$. However, once we have started to integrate, we can continue the integral curves using the identity $G(\mathbf{x}, s+t) = G(G(\mathbf{x}, s), t)$, which results from the uniqueness of the integral curve, as long as we stay inside U.

Since the theorem refers only to what happens in a neighbourhood of a point, we can transpose it to a statement about a vector field on any manifold M. It follows at once that if ξ is a tangent vector field on M, there is a map G defined on a neighbourhood V of $M \times 0$ in $M \times \mathbb{R}$ and with values in M and giving integral curves of ξ. The easiest case is when M is compact.

Corollary 5.1.2 *Let ξ be a smooth vector field on a compact manifold M. Then there is a map $G : M \times \mathbb{R} \to M$ which is the unique solution of $\frac{\partial}{\partial t} f(G(x,t)) = \xi(f)$ with $G(x, 0) = \mathbf{x}$.*

For each $s \in \mathbb{R}$, the map $g_s : M \to M$ defined by $G(x, s) = g_s(x)$ is a diffeomorphism.

Proof The theorem shows that each $(x, 0)$ has a neighbourhood $U_x \times (-\epsilon_x, \epsilon_x)$ on which G is defined. Since M is compact, we can choose a finite collection of the neighbourhoods U_x which cover the whole of M, and then take ϵ as the smallest of the corresponding ϵ_x. This gives a map G defined on $M \times (-\epsilon, \epsilon)$.

But now we can extend the integral curves as noted above, since we cannot leave M and can always extend the parameter value by a fixed amount: thus we reach $t = K$ after at most $\lceil \frac{K}{\epsilon} \rceil$ steps.

We have constructed a differentiable map g_s; a differentiable inverse is provided by g_{-s} in view of the uniqueness of solution curves. □

We will require corresponding statements when the manifold M has a boundary. First consider the local situation at $O \in \mathbb{R}^n$, and the manifold $\mathbb{R}^n_+$ is defined by $x_n \geq 0$. A vector field defined near O on $\mathbb{R}^n_+$ is said to be smooth if it extends to a smooth vector field defined on a neighbourhood of O in $\mathbb{R}^n$. Thus we can use Theorem 5.1.1 to obtain an integral. However, the integral curves may cross in and out of the boundary.

We thus restrict to one of two cases. We say that the vector field $\xi(x) = \sum_i a_i(x)\partial/\partial x_i$ is *tangent to the boundary* if $a_n(x) = 0$ whenever $x_n = 0$. We say that it *points outwards* if $a_n(x) < 0$ for $x_n = 0$. In general, the vector field ξ on M is tangent to the boundary (or points outwards, respectively) if this condition holds in local coordinates at each point of the boundary. We leave it to the reader to verify (see Exercise 5.7.1) that this condition is independent of the choice of local coordinates.

In the former case, Corollary 5.1.2 extends as follows.

Corollary 5.1.3 *Let ξ be a smooth vector field on a compact manifold M with boundary, which is tangent to the boundary. Then there is a map $G : (M, \partial M) \times \mathbb{R} \to (M, \partial M)$ which is the unique solution of $\frac{\partial}{\partial t} f(G(x,t)) = \xi(f)$ with $G(x,0) = \mathbf{x}$.*

For each $s \in \mathbb{R}$, the map g_s defined by $G(x,s) = g_s(x)$ is a diffeomorphism of M keeping the boundary ∂M invariant.

Proof As ξ is tangent to the boundary, in local coordinates its restriction to $\mathbb{R}^{n-1}$ is a tangent vector field to $\mathbb{R}^{n-1}$, so any integral curve through a point of $\mathbb{R}^{n-1}$ remains in $\mathbb{R}^{n-1}$ (so long as we stay in a suitable neighbourhood of O). Hence no integral curve starting in the half-space $x_n > 0$ can reach the boundary. Thus Theorem 5.1.1 gives us a smooth map $G : \mathbb{R}^n_+ \times \mathbb{R} \to \mathbb{R}^n_+$ defined on a neighbourhood $U \times (-\epsilon, \epsilon)$ of $(O, 0)$.

The result now follows by the same argument as for Corollary 5.1.2. □

If ξ is outward pointing, then in the local situation we can find K with $a_n \leq K$ on some neighbourhood of O. Thus x_n decreases along each integral curve with $\frac{\partial}{\partial t} x_n \leq K$. Thus an integral curve in $\mathbb{R}^n_+$ which goes out to the boundary $\mathbb{R}^{n-1}$ must then leave the manifold. Globally, we infer

Corollary 5.1.4 *Let ξ be a smooth vector field on a compact manifold M with boundary, which is outward pointing on the boundary. Then there is a map $G : N \to M$, defined on a submanifold N (with boundary) of $M \times \mathbb{R}$, which is the unique solution of $\frac{\partial}{\partial t} f(G(x,t)) = \xi(f)$ with $G(x,0) = \mathbf{x}$. For each $x \in M$, either $G(x,t)$ is defined for all $x \in \mathbb{R}$ or, for some κ_x, it is defined for $-\infty < t \leq \kappa_x$, and $G(x, \kappa_x) \in \partial M$.*

The pattern of our application of these results will be to construct a vector field having certain properties, and infer properties of the map G defining the integral.

Given a topological space X, and a cover $\{U_\alpha\}$ of X by open sets, a *partition of unity* subordinate to the given cover is a set of continuous functions $\phi_\alpha : X \to \mathbb{R}$ such that ϕ_α is non-negative, and equal to 0 except in U_α, and for each $x \in X$, $\sum_\alpha \phi_\alpha(x) = 1$. We may also require that near each point $\mathbf{x}$ of X only finitely many of the ϕ_k are non-zero, and those ones add up to 1. In the case when X is compact, one can always construct a partition of unity, which consists of a finite set of functions. If X is a smooth manifold, one can (see e.g. [89]) construct a partition of unity with each ϕ_α differentiable (of class C^∞), and equal to 0 near the frontier of U_α.

Theorem 5.1.5 *Let X be a smooth manifold; suppose given at each point $x \in X$ a convex subset A_x of the space T_xX of tangent vectors to X at x. Suppose also that there exist an open cover $\{U_k\}$ of X, and smooth tangent vector fields ξ_k defined on U_k, such that for each $x \in U_k$, $\xi_k(x) \in A_x$. Then there exists a vector field ξ on X such that, for all $x \in X$, $\xi(x) \in A_x$.*

Proof Choose a differentiable partition of unity $\{\phi_k\}$ such that ϕ_k is zero outside, and in a neighbourhood of the boundary of, U_k. Then the product $\phi_k\xi_k$ is a differentiable vector field on U_k, vanishing near the frontier, and so can be extended to a differentiable vector field on X by taking it to be 0 outside U_k. The sum $\sum \phi_k\xi_k$ is defined and differentiable, since near any point $\mathbf{x}$ the sum is finite, so there are no convergence problems. And since each $\xi_k(\mathbf{x})$ with $\phi_k(\mathbf{x}) \neq 0$ lies in A_x, and we have added non-negative multiples with sum 1, the result also lies in A_x. □

A typical example of the type of convex sets that will occur, and how to apply the theorem, is the following. Suppose we are given a finite list of functions $f_1, f_2, \ldots, f_r$ on X. Then define $\xi \in A_x$ if, for each i, we have $\xi(f_i) > 0$. Suppose that $A_x \neq \emptyset$ for each $x \in X$. Choose $\xi_x \in A_x$, and extend ξ_x to a vector field defined in some neighbourhood of x. Each condition $\xi_x(f_i) > 0$ holds on an open set, so all will hold in some smaller neighbourhood U_x of x. We can now apply the theorem to construct a vector field ξ on X such that all the conditions hold on all of X. Thus along any integral curve of ξ all the functions f_i are strictly increasing.

The method is illustrated by the following useful result.

Lemma 5.1.6 *(Isotopy Extension Theorem) Suppose X and Y compact smooth manifolds, and $F : X \times I \to Y \times I$ a smooth isotopy: write $F(x,t) = (f_t(x), t)$. Then there is a diffeomorphism H of $Y \times I$ of the form $H(y,t) = (h_t(y), t)$ such that, for each $(x,t) \in X \times I$, $f_t(x) = h_t(f_0(x))$.*

Thus the homeomorphism h_1 takes the embedding f_0 to f_1.

Proof We argue by constructing a vector field. We want the paths $\{f_t(x)\}$ to be among the integral curves. Thus we require a vector field ξ on $Y \times I$ such that

(i) the second component of ξ is $\partial/\partial t$ and
(ii) for all $\mathbf{x} \in X$ and $0 \leq t \leq 1$ we have $\xi(f_t(\mathbf{x})) = DF(\partial/\partial t)$.

Then by (i) the integral H of ξ is of the form $H(\mathbf{y}, t) = (h_t(\mathbf{y}), t)$, and by (ii) h_s has the desired property.

Thus ξ is assigned along the image of $X \times I$ and is unrestricted elsewhere. By Theorem 5.1.5 it suffices to construct a suitable vector field in the neighbourhood of each point. For a point not in $F(X \times I)$ we can take $\xi = \partial/\partial t$.

At a point in the image, we take local coordinates $(y_1, \ldots, y_{p+q})$ on $Y \times I$ in which $F(X \times I)$ is given by $y_{p+1} = \ldots = y_{p+q} = 0$. The given vector field along $F(X \times I)$ thus has the form $\sum_{i=1}^{p} a_i(y_1, \ldots, y_p)\partial/\partial y_i$. We use the same formula to define a vector field ξ' near the point in question. This satisfies condition (ii), but not yet (i). However, its second component $b(y)\partial/\partial t$ with respect to the product decomposition $Y \times I$ is such that $b(y)$ is equal to 1 at the point in question, and so is non-zero in some neighbourhood. We may thus take $\xi := b(y)^{-1}\xi'$. □

5.2 Knots and links

For each curve germ C defined at $O \in \mathbb{C}^2$ we can define a link in the 3-sphere S^3. This consists of a disjoint union of embedded copies of the circle S^1, one for each branch of C at O. We show that this link is defined uniquely up to isotopy.

Consider the germ at $O \in \mathbb{C}^2$ of a curve C. Take coordinates (x, y) in $\mathbb{C}^2$ and write D_ϵ for the disc with centre O and radius ϵ, defined by $|x|^2 + |y|^2 \leq \epsilon^2$, and S_ϵ for its boundary sphere $|x|^2 + |y|^2 = \epsilon^2$. We may suppose ϵ small enough so that C is defined in the neighbourhood D_ϵ of O. We will describe the intersection $C \cap D_\epsilon$: the first point is that this is essentially independent of ϵ provided that ϵ is small enough.

Lemma 5.2.1 *For ϵ sufficiently small, $K = C \cap S_\epsilon$ is a 1-manifold smoothly embedded in S_ϵ, and there is a homeomorphism of the pair $(D_\epsilon, C \cap D_\epsilon)$ to the cone on $(S_\epsilon, C \cap S_\epsilon)$, which may be chosen compatible with the natural projections on $[0, \epsilon]$.*

Proof We use the technique of Section 5.1. For suitable ϵ we will construct a vector field ξ on $D_\epsilon - \{O\}$ such that (i) at all points, ξ has inner product with the radius vector equal to 1, and (ii) at points of C, ξ is tangent to C.

After constructing ξ we integrate the vector field $-\xi$. Since the inner product of ξ with the radius vector is 1, if $r = \sqrt{(\| x^2 \| + \| y^2 \|)}$, then $\xi(r) = 1$ so that if we integrate on the compact set $\epsilon_1 \leq r \leq \epsilon$, the integral $G : (S_\epsilon \times \mathbb{R}) \to D_\epsilon$ will be defined for all $t \leq \epsilon - \epsilon_1$ and at $t = \epsilon - \epsilon_1$ it will take values on the sphere S_{ϵ_1}. As $t \to \epsilon$ these converge uniformly to the origin, so G gives a continuous map $S_\epsilon \times [0, \epsilon] \to D_\epsilon$. Each integral curve meets each concentric sphere S_{ϵ_1} in just one point, and there is just one integral curve through each point on each such sphere, so our map is bijective except that $S_\epsilon \times \epsilon$ is mapped to the origin. Hence it induces a homeomorphism of the cone on S_ϵ onto D_ϵ. Finally, since ξ is tangent to C, each integral curve that meets C stays within C so that our homeomorphism does indeed take the cone on $S_\epsilon \cap C$ to $D_\epsilon \cap C$.

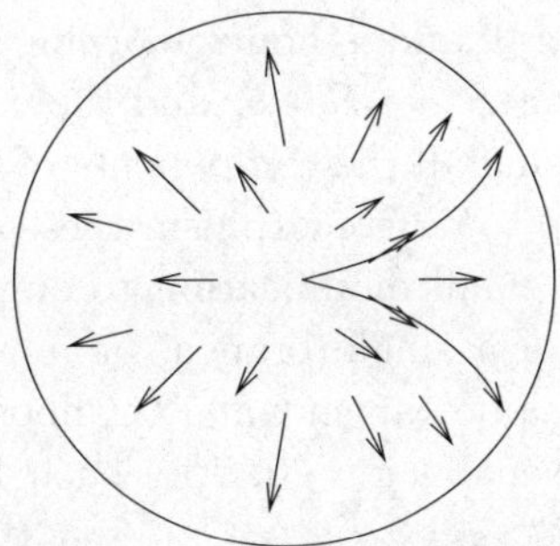

Fig. 5.1. Vector field on a cone

It remains to construct the vector field. Since conditions (i) and (ii) on ξ define a convex subset of the possible vectors at each point, it follows from Theorem 5.1.5 that it is enough to construct at each point a suitable vector field in the neighbourhood of that point. For points $\mathbf{x}$ not on C, we can simply choose the radius vector. For a point $\mathbf{x}$ on C where the radius vector is not perpendicular to all (real) tangent vectors of C we proceed as follows. For $\mathbf{y}$ in a neighbourhood of $\mathbf{x}$, take the orthogonal projection of the radius vector $\mathbf{y}$ on the tangent space at $\mathbf{y}$, and replace by a suitable multiple to normalise the inner product to 1.

The crucial geometrical condition is thus that the radius vector is not perpendicular to all tangent vectors of C at $\mathbf{x}$. We have seen in

Section 2.3 that each branch of C has a well defined complex tangent and that its tangent spaces converge to this as the points tend to O. If this is, say, $y = 0$ then as points on the branch converge to O the tangent plane is close to this direction, and so is the (radius vector of) the point. Thus in a neighbourhood of O, the two are not orthogonal. We may take any such neighbourhood as D_ϵ and then the above argument applies. □

The final clause of the lemma means that the points with $|x|^2 + |y|^2 = \delta^2$ are mapped to the points on the cone at height δ, so that the pairs $(S_\epsilon, C \cap S_\epsilon)$, $(S_\delta, C \cap S_\delta)$ are homeomorphic: this is what we mean by 'essentially independent of ϵ'.

To analyse $C \cap S_\epsilon$, consider a branch B of C. We may suppose B tangent to $y = 0$ and take a parametrisation $x = t^m$, $y = \sum a_r t^r$ with $a_r = 0$ if $r \leq m$. The vector field, restricted to B, defines a vector field in the t plane. Since for t small the radius vector is nearly in the plane $y = 0$, we see that (in a small neighbourhood of O) $|t|$ increases along integral curves of the vector field. Thus $B \cap S_\epsilon$ corresponds to a curve in the t plane encircling 0, and is homeomorphic to a circle.

Using diffeomorphisms $S^1 \to B \cap S_\epsilon$ and the central projection $S_\epsilon \to S^3$ we obtain a knot; using all the components of $C \cap S_\epsilon$, we have a link. We have shown that up to diffeomorphism these depend only on C.

We now introduce a small modification to our picture. Consider the case when C consists of a single branch, or more generally of several branches all with the same tangent at O. Choose coordinates so that this tangent line is given by $y = 0$. Any such branch has a Puiseux parametrisation $x = t^m, y = a(t) = \sum_{r=n}^{\infty} a_r t^r$, with $n > m$. Since $t^{-m}y$ tends to 0 with t, we can choose $\epsilon > 0$ such that $|x| < \epsilon$ implies $|y| < \frac{\epsilon}{100}$, say. We may replace the sphere S_ϵ by a nearby manifold S'_ϵ which coincides with $|x| = \epsilon$ in the region where $|y| < \frac{|x|}{100}$. This has the advantage that taking $|t| = \epsilon^{1/m}$ or setting $t = \epsilon^{1/m} e^{2\pi i\theta}$ gives an explicit parametrisation for the knot $C \cap S'_\epsilon$. Moreover the change is inessential.

Lemma 5.2.2 *For ϵ small enough, the knots $(S_\epsilon, C \cap S_\epsilon)$ and $(S'_\epsilon, C \cap S'_\epsilon)$ are homeomorphic.*

We can use the method of proof of Lemma 5.2.1: the only difference is that we replace the radius vector (x, y) by the vector field $(x, 0)$.

For any two embeddings (or even maps) of S^1 in S^3 with disjoint images K_1, K_2, the *linking number* is defined as follows. Span K_2 by an oriented surface X (there is no need for X either to be the image of the disc D^2 or to be embedded, but we do need it to meet K_1 transversely) and count the number of intersection points of K_1 with X, with appropriate signs. The result is an integer, which we denote $Lk(K_1, K_2)$. This does not depend on the choice of X, since if X' is another such surface, the union of X and X' (with orientation changed) is a closed surface in S^3, which necessarily has zero intersection number with the closed curve K_1.

We may also span both K_1 and K_2 by surfaces in the disc D^4: these also may be supposed transverse, and we can again count their (signed) intersections: as above, we see that the resulting intersection number does not depend on the choice of the surfaces. If the above surface X is deformed slightly into the interior of D^4, we see that the intersections in S^3 with K_1 move to intersections in D^4 with the surface spanning K_1. Hence this intersection number is equal to the linking number $Lk(K_1, K_2)$. We thus have

Lemma 5.2.3

(i) *Given two branches B, B' at O, their intersection number at O is equal to the linking number of the corresponding knots K and K'.*

(ii) $Lk(K, K') = Lk(K', K)$.

(ii) follows since the intersection number is symmetric.

5.3 Description of the geometry of the link

In this section, we give an explicit geometrical model for the links we have just defined, and show that the isotopy class of the link depends only on the equisingularity class of C. The model resembles Ptolemy's description of the solar system in that it consists of circles with their centres on other circles, all revolving together.

We begin by considering the example $y = x^{3/2} + x^{7/4}$: however, first look at the simpler curve $y = x^{3/2}$. For the corresponding knot in S'_ϵ we have a parametrisation $(x, y) = (\epsilon e^{2i\theta}, \epsilon^{3/2} e^{3i\theta})$. Each value of x determines 2 values of y, which lie on the circle with centre 0 and (small) radius $\epsilon^{3/2}$; as x moves round the circle $|x| = \epsilon$ once, the values of y move round this circle $\frac{3}{2}$ times, getting interchanged in the process.

For the given curve, the obvious parametrisation leads to the knot $(x, y) = (\epsilon e^{4i\theta}, \epsilon^{3/2} e^{6i\theta} + \epsilon^{7/4} e^{7i\theta})$. For each value of x we have 4 values of y. Since $\epsilon^{7/4}$ is small compared to $\epsilon^{3/2}$, it is natural to think of these as being near the previous points: indeed, we may draw circles of radius

$\epsilon^{7/4}$ centred at the points corresponding to the other curve: then we have two points on each of these circles, and as x moves round the circle once, as well as our 'first order points' moving round the circle $|y| = \epsilon^{3/2}$, these points move round these auxiliary circles $\frac{7}{4}$ times.

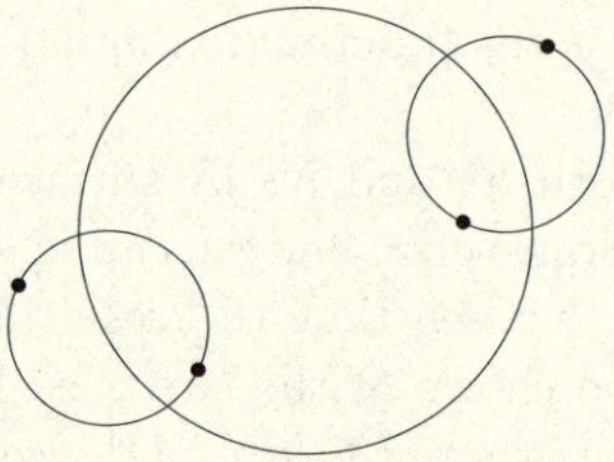

Fig. 5.2. The carousel

The picture of a hierarchy of circles, each with centre on a circle at the preceding level, and with radii each small compared to earlier ones, and all rotating at appropriate rates: has been appropriately dubbed a *carousel.* It is qualitatively correct for any branch B, as we show next.

Take a Puiseux parametrisation $x = t^m$, $y = a(t) = \sum_{r=n}^{\infty} a_r t^r$ of B, and use the usual notation for the Puiseux characteristic. We define a sequence of branches B_k $(k \geq n)$ by the parametrisations (not necessarily good) $x = t^m$, $y = \sum_{r=n}^{k} a_r t^r$ and consider what happens to B_k or rather, to the corresponding knot $K_k := B_k \cap S'_\epsilon$, as we increase k.

If $k < \beta_1$, y_k is a polynomial in x, so we have a unique value of y_k for each value of x. Thus K_k is unknotted; there is no merit in decomposing this as a sum of points on different circles.

If $k = \beta_1$, we describe a circle of radius $|a_k|\epsilon^{\frac{\beta_1}{m}}$ with centre y_{k-1}. The values of y_k give m/e_1 points equally spaced round this circle, and as x moves round the circle $|x| = \epsilon$, they rotate uniformly: this rotation is superimposed on the motion of the centre y_{k-1} of the circle. Each of these points corresponds to e_1 values of x.

For $\beta_1 < k < \beta_2$ we have a similar situation to the case $k < \beta_1$: the further terms form a polynomial in $t^{e_1} = x^{\frac{e_1}{m}}$, so we still have m/e_1 points and the effect, since the later terms are much smaller than the radius of the above circle (for ϵ small enough) is to move these points by distances which are small compared to their distances apart.

More precisely, these knots K_k are all isotopic to the knot K_{β_1}. Indeed, multiplying the other coefficients by s provides an isotopy. Explicitly, an isotopy is given, with s as the isotopy parameter, by

$$x = \epsilon e^{ie_1\theta}, \; y = \sum_{r=n}^{\beta_1} a_r e^{ire_1\theta/m} \epsilon^{r/m} + \sum_{\beta_1+1}^{k} s a_r e^{ire_1\theta/m} \epsilon^{r/m}.$$

The general pattern is now clear: as k increases by 1, in general K_k is changed by an isotopy, but when k attains a value β_q each point splits into e_{q-1}/e_q points lying on a small circle surrounding it. We proceed to a more formal treatment.

Proposition 5.3.1 *Two branches with the same Puiseux characteristic determine isotopic knots in S'_ϵ for ϵ small enough.*

Proof Consider a branch B; introduce notation as above. The knot K is given by the parametrisation $x = \epsilon e^{im\theta}$, $y = \sum_{r=n}^{\infty} a_r \epsilon^{r/m} e^{ir\theta}$.

We claim that the following deformation (for $(0 \le s \le 1)$) defines an isotopy of K, provided ϵ is small enough:

$$x = \epsilon e^{im\theta}, \quad y = \sum_{r=n}^{\infty} s_r a_r \epsilon^{r/m} e^{ir\theta},$$

where $s_r = 1$ if $r = \beta_q$ for some q and $s_r = s$ otherwise.

This formula certainly defines a smooth 1-parameter family of maps from S^1 to S^3. That it is an isotopy means that each map in the family is a smooth imbedding - i.e. is injective and has nowhere zero derivative. The latter condition is clear since already $dx/d\theta \neq 0$.

It remains to show that θ and $\theta + \frac{2\pi k}{m}$ lead to different values of y unless $m|k$. It follows from the geometrical description above that it is enough to show that for each q, the sum of the contributions of terms between β_q and β_{q+1} is small compared to the distance apart of two points on the qth small circle. Let us consider the final case $q = g$ (the rest are similar but easier). We may choose R so that $|a_r|R^r$ is uniformly bounded, by M, say. Then the relevant sum is majorised by

$$\sum_{r=\beta_g+1}^{\infty} |a_r|\epsilon^{r/m} \le \sum_{r=\beta_g+1}^{\infty} MR^{-r}\epsilon^{r/m} = \frac{M\epsilon^{(\beta_g+1)/m}}{R^{\beta_g}(R-\epsilon^{1/m})},$$

and as $\epsilon \to 0$, this is of higher order than $\epsilon^{\beta_g/m}$, as required.

We have shown that the knot corresponding to B is isotopic to that corresponding to the branch B' with parametrisation $x = t^m$, $y = \sum_{q=1}^{g} c_q t^{\beta_q}$, where $c_q = a_{\beta_q}$. We will reduce the c_q to 1 by a further isotopy; this will complete the proof. Write $c_q = e^{l_q}$, and consider the deformation

$$x = t^m, \quad y = \sum_{q=1}^{g} e^{sl_q} t^{\beta_q}, \text{ for } 0 \le s \le 1.$$

It again follows, since the later circles are uniformly small compared to earlier ones, that for small enough ϵ, the corresponding deformation of knots is an isotopy. □

Using the same methods we can obtain an important extension of this result.

Proposition 5.3.2 *Let C and C' be two curve-germs; suppose C and C' are equisingular. Then C and C' determine isotopic links in S^3.*

Proof By the definition of equisingularity, we have a bijection between the components B_i of C and the components B'_i of C' (where $1 \le i \le k$, say) such that

(a) for each i, B_i and B'_i have the same Puiseux characteristic, and
(b) for each pair (i, j), the exponent of contact $c_{i,j}$ of B_i and B_j equals the exponent of contact of B'_i and B'_j.

Choose coordinates with the line $x = 0$ not tangent to C (or to C'), so that each B_i admits a Puiseux series $y = \sum a_{r,i} x^r$. Consider a sector $|\arg x| < \frac{1}{2}\pi$, and choose the fractional powers x^r to be positive when x is, so the choices of Puiseux series correspond to choices of pro-branches γ_i of B_i.

We claim that we can choose all the γ_i simultaneously so that, for $i \ne j$, $\mathcal{O}(\gamma_i, \gamma_j) = \mathcal{O}(B_i, B_j)$. We verify this by induction on the number k of components. Thus we may suppose γ_i already chosen for $1 \le i \le k - 1$. Let j be such that $\mathcal{O}(B_j, B_k) = \max\{\mathcal{O}(B_i, B_k) \,|\, 1 \le i < k\}$. By Lemma 4.1.2 we can choose γ_k so that $\mathcal{O}(\gamma_k, \gamma_j) = \mathcal{O}(B_k, B_j)$. For any $i \ne j, k$ we have $\mathcal{O}(\gamma_i, \gamma_k) = \mathcal{O}(\gamma_i, \gamma_j)$ by (4.2), since $\mathcal{O}(\gamma_j, \gamma_k) = \mathcal{O}(B_j, B_k) \ge \mathcal{O}(B_i, B_k) \ge \mathcal{O}(\gamma_i, \gamma_k)$; now $\mathcal{O}(\gamma_i, \gamma_j) = \mathcal{O}(B_i, B_j)$ by our inductive hypothesis, and $\mathcal{O}(B_i, B_j) = \mathcal{O}(B_i, B_k)$ by (4.3). Our claim is proved.

The same considerations apply to the curve C'.

We next seek deformations $a_{r,i}(s)$ $(0 \le s \le 1)$ of the coefficients $a_{r,i}(0) = a_{r,i}$ for the curve C to those $a_{r,i}(1) = a'_{r,i}$ for C'. These must satisfy the conditions

(i) the Puiseux characteristic of the curve $B_i(s)$ given by $y = \sum_r a_{r,i}(s)x^r$ is independent of s, for each i;

(ii) the exponent of contact of $B_i(s)$ and $B_j(s)$ is independent of s.

We have seen in the proof of Proposition 5.3.1 how to satisfy (i) for a single branch. Here it is not convenient to write explicit formulae. Instead, we again argue by induction on k. Thus suppose deformations $a_{r,i}(s)$ already chosen for $1 \leq i < k$ satisfying (i) and (ii). Write κ_k for the maximum $\mathcal{O}(B_j, B_k)$ as above. Note that there may be several branches B_j realising the maximum. Then we must take the coefficients $a_{r,k}(s) = a_{r,j}(s)$ for $r < \kappa_k$ and j realising the maximum.

It will be convenient for the next argument temporarily to introduce the following terminology. Let B be a branch with Puiseux characteristic $(m, \beta_1, \ldots, \beta_g)$. We say that r is a characteristic exponent for B if, for some q with $1 \leq q \leq g$, $r = \alpha_q$. Say r is a free exponent if, for some q with $1 \leq q \leq g+1$, $\alpha_{q-1} < r < \alpha_q$ and $mr/e_{q-1} \in \mathbb{Z}$ (here $e_0 = m$ and $\beta_{g+1} = \infty$). Say r is a forbidden exponent otherwise. A Puiseux series has the indicated characteristic if and only if coefficients corresponding to forbidden exponents are zero and those corresponding to characteristic exponents are non-zero. This is essentially the same as the definition of the ring $\mathcal{O}_S$ in the proof of Lemma 3.5.4.

If $r = \kappa_k$ is a forbidden exponent for B_k, it must be characteristic for each B_j. Thus $a_{j,k}(s) \neq 0$ for each j and s and we can safely take $a_{r,k}(s) = 0$ for all s. Otherwise we may argue as follows. Choose ϵ small enough so that none of the coefficients $a_{r,j}(s)$ moves more than a small distance for $s \in [0, \epsilon]$ and $s \in [1-\epsilon, 1]$. Choose a path $a_{r,k}(s)$ for $0 \leq s \leq \epsilon$ which goes from the given value of $a_{r,k}$ to some very large value, avoiding the $a_{r,j}$ (which haven't moved very far). Then keep $a_{r,k}(s)$ constant for $\epsilon \leq s \leq 1-\epsilon$: it will be out of the way of the paths $a_{r,j}(s)$. Reverse the argument at the end to come back to $a'_{r,k}$. If moreover r is characteristic for B_k the paths must also avoid 0.

Once we have passed $r = \kappa$, the exponents of contact with all the other branches are already fixed, and all we need do is join $a_{r,k}$ and $a'_{r,k}$ by a path – which could be a straight line – or, if the exponent r is characteristic, by a path avoiding 0.

Intersecting these deformations with a small enough sphere provides the desired isotopy. This follows from essentially the same argument as before: in each 'slice' where x is constant we have finitely many values of y for each branch, and we have to ensure that no two of these coincide during the deformation. But the distance between any two of them is measured by a convergent power series, and the above conditions determine the orders of all these series. The result thus follows. □

Corollary 5.3.3 *Any curve C is equisingular to a curve C' whose branches have parametrisations with real coefficients; thus the singularity links of C and C' are isotopic.*

Proof We can construct C' by writing down parametrisations with the same characteristic exponents and orders of contact with all the coefficients having real, or even integer values (indeed, those not involved in these characteristic exponents and orders can be taken to be zero). The proposition then provides an isotopy. □

One of the major aims of this chapter is to prove the converse of Proposition 5.3.2, but to achieve this we need new tools.

5.4 Cable knots

The carousel pictured in the preceding section gave a picture of the possible values of y for a given x: we must integrate this to give an effective description of the knot. If C is a single branch, the link is a knot, and the description in the preceding section can be re-stated in terms of the knot construction known as cabling. In order to apply properties of this construction we need to evaluate the relevant parameters in terms of the Puiseux characteristic of the branch.

Before starting we observe, however, that the carousel description is already complete and fits into one of the standard approaches to knot theory: the theory of braids. Indeed, the definition of a braid is that we have a set of m points in a disc D – which may be taken as $\{y \in \mathbb{C} \mid |y| < \delta\}$ for any fixed δ – varying continuously as functions of θ with $0 \leq \theta \leq 2\pi$, and having the same points at both ends $\theta = 0$ and $\theta = 2\pi$ (though perhaps in a different order). This is converted to a knot by identifying the two ends $\theta = 0$ and $\theta = 2\pi$, thereby giving one or more closed curves in $D \times S^1$, and embedding this in turn in S^3 in a standard way. But this is precisely the procedure above.

For our present purpose, however, we use a different approach. Since by Proposition 5.3.1 we may choose any branch with the given Puiseux characteristic to get an equivalent knot, we may take our knot as parametrised by $x = e^{im\theta}$, $y = \sum_{q=1}^{g} \epsilon_q e^{i\beta_q\theta}$, where $\epsilon_0 = 1$ and each ϵ_q is very small compared to its predecessor – e.g it will suffice to take

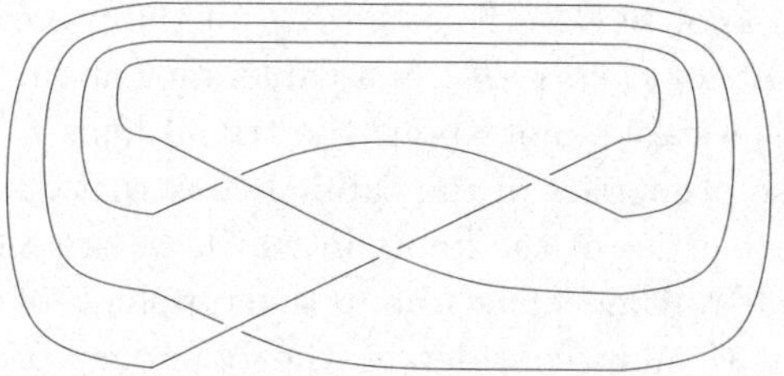

Fig. 5.3. A braid

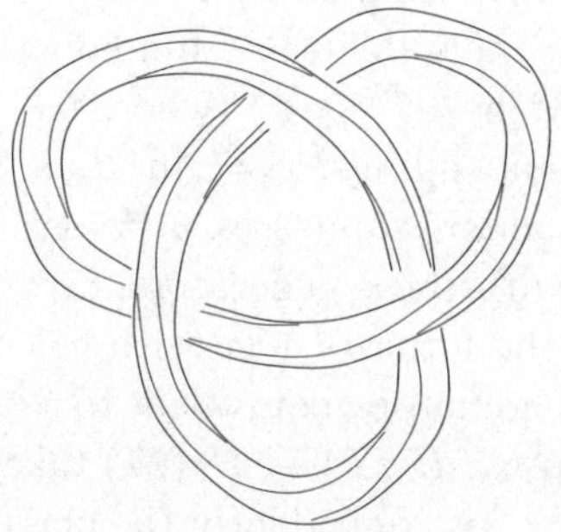

Fig. 5.4. A cable knot

$\epsilon_q = 10^{-\beta_q}$. With this choice the unit circle $|x| = 1$ is sufficiently small in the sense of the proposition.

Recall that the first term leads to a small circle $|y| = \epsilon_1$. In the 3-sphere S'_1, this sweeps out a torus $|x| = 1,\ |y| = \epsilon_1$, which is the boundary of the solid torus $|x| = 1,\ |y| \le \epsilon_1$ which is a neighbourhood of the circle $y = 0$. Similarly the next term leads us to consider the solid torus which is the union of the discs $x = e^{im\theta},\ |y - \epsilon_1 e^{i\beta_1\theta}| \le \epsilon_2$.

The general construction is the following. For any knot K given by an embedding $f : S^1 \to S^3$, we can find neighbourhoods given by embeddings $F : S^1 \times D^2 \to S^3$ such that $F(x, 0) = f(x)$ for each $x \in S^1$. Such an embedding, which is called a *tubular neighbourhood* of the knot K, is almost unique up to isotopy: we will have to consider the extent of uniqueness below. Any simple closed curve traced on $S^1 \times S^1$ is mapped by F to the boundary of the neighbourhood, and so defines by composition a further knot. Such a knot is said to be a *cable knot* about K: see Figure 5.4. The branch B^-_{q+1} was defined in the proof of (ii) of Lemma 4.3.1 by the parametrisation $x = t^m,\ y = \sum_{1 \le r < \beta_{q+1}} a_r t^r$, so meets S'_1 in the knot parametrised by $x = e^{im\theta},\ y = \sum_{s=1}^{q} \epsilon_s e^{i\beta_s\theta}$. We now denote this knot by K^q (it is the K_{β_q} of the preceding section). It is immediate that for each q, K^q is a cable knot about K^{q-1} – thus in particular, K^1 is a cable knot about the trivial knot K^0 ($y = 0$). Thus if we can analyse properties of the cabling construction, we will be able to obtain the properties of the knots in which we are interested.

We need some routine facts about the topology of the torus $T = S^1 \times S^1$, which we may consider as the boundary of the solid torus $M = S^1 \times D^2$; we write $i : T \to M$ for the inclusion. The circle S^1 carries a standard orientation, which we may consider as being defined by the positive sense of rotation or equivalently by the parametrisation $z = e^{i\theta}$, $0 \leq \theta \leq 2\pi$: this defines a fundamental homology class in $H_1(S^1)$. Parametrise M by $(e^{i\theta}, \rho e^{i\phi})$ (where $0 \leq \rho \leq 1$)

Under the natural embeddings of S^1 in T as $S^1 \times 1$ and as $1 \times S^1$, the fundamental class maps to classes u, v say in $H_1(T)$. The group $H_1(T)$ is free abelian with these generators: every element of it may be uniquely expressed in the form $au + bv$ with $a, b \in \mathbb{Z}$.

The solid torus M is homotopy equivalent to S^1, and $H_1(M)$ is infinite cyclic; the natural map $i_* : H_1(T) \to H_1(M)$ takes u to a generator and v to 0. The curve $1 \times S^1$, or equivalently the class v which it represents, is called a *meridian* of the torus T: it is characterised up to sign as generating the kernel of i_*.

The choice of the class u is more arbitrary, and indeed we have self-homeomorphisms of M given by $h_r(w, z) = (w, w^r z)$ for $r \in \mathbb{Z}$: we may say that h_r twists M r times. We have $h_{r*}(u) = u + rv$: any of the classes $u + rv$ may be called a latitude; choosing a particular one among them will be of concern to us below.

Intersection numbers on the manifold T (with the product orientation) are easily determined: we have

$$u.u = 0, \quad u.v = 1, \quad v.u = -1, \quad v.v = 0;$$

since intersections are bilinear, $(au + bv).(a'u + b'v) = ab' - ba'$. If a and b are coprime, $(e^{ia\psi}, e^{ib\psi})$ $(0 \leq \psi \leq 2\pi)$ parametrises a simple closed curve on T with homology class $au + bv$.

Lemma 5.4.1

(i) *If j is an embedding of the torus T in S^3 with image disjoint from the knot K, taking linking numbers with K defines a homomorphism $\phi : H_1(T) \to \mathbb{Z}$.*

(ii) *If $j : M \to S^3$ is an embedding of the solid torus, K is the knot $j(S^1 \times 0)$, and v is a meridian, then $\phi(v) = 1$.*

Proof For each curve x lying on the surface $j(T)$ we may count its linking number with K: recall that we choose a spanning surface X, so that $\partial X = x$, and count the intersections $X \cap K$. If two curves x and y are homologous on the torus, their difference bounds a surface Y, say, lying on $j(T)$: $\partial Y = y - x$. Thus if X is a surface which spans x, then $\partial(X + Y) = y$, so $X + Y$ spans y. Since Y is disjoint from K, $(X \cup Y) \cap K = X \cap K$, and the signs are the same, so $Lk(x, K) = Lk(y, K)$.

Thus the construction defines a map $\phi : H_1(T) \to \mathbb{Z}$. This is a homomorphism, since if the curves x, x' represent two classes in $H_1(T)$, their union represents the sum. If $\partial X = x$ and $\partial X' = x'$ we have $\partial(X + X') = x + x'$, and the intersections $(X \cup X') \cap K = (X \cap K) \cup (X' \cap K)$. Moreover the sign attached to each point is the same in both cases.
The knot K is disjoint from $j(T)$, so (i) defines a homomorphism $\phi : H_1(T) \to \mathbb{Z}$. If v is a meridian, the surface X can be taken to be the disc $j(1 \times D^2)$, and this meets K just once, transversely, in the point $j(1, 0)$. □

In the situation of (ii), if $\phi(u) = s$ there is a unique value of r, namely $r = -s$, such that $\phi(u + rv) = 0$. The corresponding class $u + rv$, or a representative curve on $j(T)$, is called a *longitude* of the torus in S^3. This is characterised by having zero linking number with the knot K. Indeed it follows that, if u is a longitude, then for any curve C on $j(T)$ having homology class $[C] = au + bv$, $Lk(C, K) = \phi([C]) = b$.

Fix u to be a longitude. A knot K' which is cabled about K is the image by j of a simple closed curve on T. If this curve has homology class $mu + pv$, then m is called the *winding number* of K' about K: the pair (m, p) we will call the *cabling invariant.* Another well known property of the torus states that the class $mu+pv$ determines the curve up to isotopy on T. Since the knots in which we are interested are formed by iterating this cabling construction, they are determined by the corresponding cabling invariants.

Theorem 5.4.2 *The linking number* $Lk(K^q, K^{q-1}) = \overline{\beta}_q/e_q$.

Proof The knot K^q is given by $x = e^{im\theta}$, $y = \sum_{r=1}^{q} \epsilon_r e^{i\beta_r\theta}$ (here a good parameter is $e^{ie_q\theta}$). This lies on the torus T_{q-1} (a tubular neighbourhood of K^{q-1}) given by

$$x = e^{im\theta}, \quad \left| y - \sum_{r=1}^{q-1} \epsilon_r e^{i\beta_r\theta} \right| = \epsilon_q.$$

We parametrise T_{q-1} by setting $x = e^{im\theta/e_{q-1}}$, $y = \sum_{r=1}^{q-1} \epsilon_r e^{i\beta_r\theta/e_{q-1}} + \epsilon_q e^{i\phi}$. This gives consistent orientations: for ϕ fixed, increasing θ gives the positive sense of rotation in the x plane, and for x fixed, increasing ϕ gives the positive sense of rotation in the y plane.

Taking $\theta = 0$ gives the circle $x = 1$, $\left| y - \sum_{r=1}^{q-1} \epsilon_r \right| = \epsilon_q$. This is a meridian of T_{q-1}, with class v_{q-1} say, since it spans the disc inside this torus given by $x = 1$, $\left| y - \sum_{r=1}^{q-1} \epsilon_r \right| \le \epsilon_q$. The curve K^q itself has a good parametrisation obtained by setting $(\theta, \phi) = (\frac{e_{q-1}}{e_q}\psi, \frac{\beta_q}{e_q}\psi)$ (with $0 \le \psi \le 2\pi$). For a curve to represent a class u_{q-1}, the natural choice is given by setting $\phi = 0$. We also define w_{q-1} as the curve

$$x = e^{im\theta}, \; y = \sum_{r=1}^{q-2} \epsilon_r e^{i\beta_r\theta} + (\epsilon_{q-1} - \epsilon_q)e^{i\beta_{q-1}\theta},$$

which has the good parametrisation $\phi = \pi + \frac{\beta_{q-1}}{e_{q-1}}\theta$. See Figure 5.5; here the torus T_{q-1} meets the plane of the diagram transversely and w_{q-1} is the indicated curve of intersection.

The homology classes of these curves on T_{q-1} are thus related by

$$[K^q] = \tfrac{e_{q-1}}{e_q}[u_{q-1}] + \tfrac{\beta_q}{e_q}[v_{q-1}], \qquad [w_{q-1}] = [u_{q-1}] + \tfrac{\beta_{q-1}}{e_{q-1}}[v_{q-1}].$$

We prove the theorem by induction on q. To start the induction, take K^0, which is the knot $y = 0$, and K^1, which lies on T_0. Since the linking number of u_0 and K^0 is zero (the two curves are parallel circles), the above calculation of the homology class determines the cabling numbers $(\frac{m}{e_1}, \frac{\beta_1}{e_1})$ and in particular the linking number $Lk(K^1, K^0) = \beta_1/e_1 = \overline{\beta}_1/e_1$.

For the induction step, we first observe that the linking number of v_{q-1} with K^{q-1} is 1, since v_{q-1} is a meridian. We determine the linking number of w_{q-1} with K^{q-1}. Observe that the union of the curves

$$x = e^{im\theta}, \; y = \sum_{r=1}^{q-2} \epsilon_r e^{i\beta_r\theta} + s(\epsilon_{q-1} - \epsilon_q)e^{i\beta_{q-1}\theta}$$

as s runs from 1 to 0 gives a surface X_q with one boundary on w_{q-1}

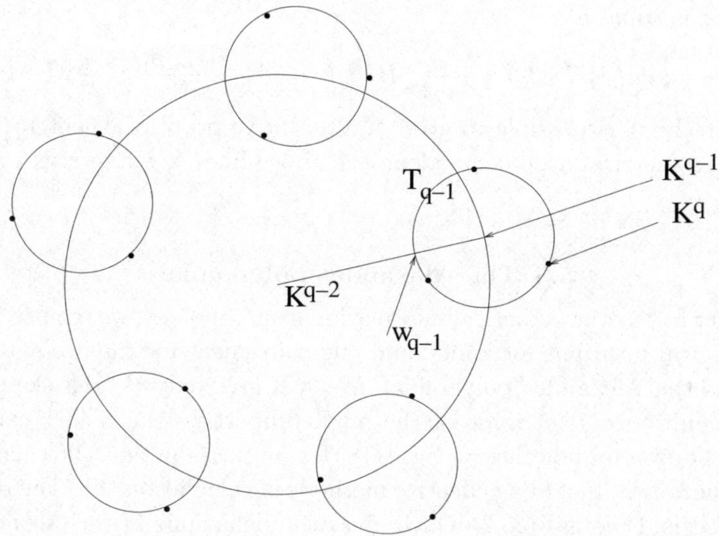

Fig. 5.5. Construction of curve to represent w_{q-1}

and the other on K^{q-2}, taken e_{q-2}/e_{q-1} times. As this surface does not intersect K^{q-1}, we have

$$Lk(w_{q-1}, K^{q-1}) = \tfrac{e_{q-2}}{e_{q-1}} Lk(K^{q-2}, K^{q-1})$$

which, by inductive hypothesis, equals $\frac{e_{q-2}}{e_{q-1}} . \frac{\overline{\beta}_{q-1}}{e_{q-1}}$.

Now $[K^q] = \frac{e_{q-1}}{e_q}[w_{q-1}] + \frac{\beta_q - \beta_{q-1}}{e_q}[v_{q-1}]$, so

$$\begin{aligned} Lk(K^q, K^{q-1}) &= \tfrac{e_{q-1}}{e_q} Lk(w_{q-1}, K^{q-1}) + \tfrac{\beta_q - \beta_{q-1}}{e_q} Lk(v_{q-1}, K^{q-1}) \\ &= \tfrac{e_{q-1}}{e_q} . \tfrac{e_{q-2}}{e_{q-1}} . \tfrac{\overline{\beta}_{q-1}}{e_{q-1}} + \tfrac{\beta_q - \beta_{q-1}}{e_q}, \end{aligned}$$

and by (4.6), this is equal to $\frac{\overline{\beta}_q}{e_q}$. □

Remark An alternative proof – which thus, in particular, checks the signs in the above – can be obtained from Theorem 4.1.6. For (if we suppose the only exponents with non-zero coefficients in the Puiseux series are the α_q) K^q corresponds to the branch B^-_{q+1} of Section 4.3, and thus the linking number $Lk(K^q, K^{q-1})$ equals the intersection number of B^-_{q+1} and B^-_q. We have $m(B^-_q) = m/e_{q-1}$, $m(B^-_{q+1}) = m/e_q$; the exponent of contact is β_q/e_{q-1} and up to this point the Herbrand function H' for B_q equals that for B divided by e_q. Hence the intersection number is equal to

$$m(B_q^-)H'(\beta_q) = \tfrac{m}{e_{q-1}e_q}H(\beta_q) = \tfrac{m}{e_{q-1}e_q}.\tfrac{e_{q-1}\overline{\beta}_q}{m} = \tfrac{\overline{\beta}_q}{e_q}.$$

Conversely, it is possible to give an alternative proof of Theorem 4.1.6 using geometrical arguments along the above lines: see Exercise 5.7.3.

5.5 The Alexander polynomial

In order to recover numerical information from topology, we require some topological invariant for knots, and one convenient for the present purpose is the Alexander polynomial. We will give only a brief sketch of the definition and of some of the main properties: the reader can refer to books on knot theory, e.g. [118] for an introduction, [96], [26] for fuller accounts, including effective methods of calculation. The key result we need is Theorem 5.5.2, which gives the behaviour of the Alexander polynomial under the cabling construction. This leads easily to the calculation of the polynomial for our knots, and the conclusion that the equisingularity class of a branch is determined by this polynomial, and hence by the isotopy class of the knot. This rounds off the whole discussion of equisingularity.

For any knot K in S^3, all invariants start from the fundamental group G of its complement: $G = \pi_1(S^3 - K)$. If G is made into an abelian group G^{ab} by factoring out its commutator subgroup G' we obtain the homology group

$$G^{ab} = G/G' \cong H_1(S^3 - K) \cong \mathbb{Z}.$$

Since this holds for any knot, it tells us nothing about K, so next we consider G' made abelian: $G^{'ab} = G'/G''$. Choose an element $\tau \in G$ which projects to $1 \in G/G' \equiv \mathbb{Z}$. Then conjugation by τ gives an inner automorphism $\lambda \mapsto \tau^{-1}\lambda\tau$ of G which induces an automorphism of G' and hence one of $G^{'ab}$.

Write V for $G^{'ab}$ made into a rational vector space by introducing denominators: $V = G^{'ab} \otimes \mathbb{Q}$: then we also have an induced automorphism A_t of V, i.e. a linear map of V into itself. A basic result (see e.g. [26]) states that V is always finite dimensional. We define the *Alexander polynomial* $\Delta^K(t)$ to be the characteristic polynomial of A_t. Other, more computational approaches to the definition may also be given; we will discuss Seifert surfaces in Chapter 10.

For the case of links, one may proceed as follows. Write G for the fundamental group of the complement of the link. Taking linking numbers with the several components of the link defines homomorphisms $G \to \mathbb{Z}$, which we assemble to give $\pi : G \to \mathbb{Z}^r$. One can show that $G^{ab} = \mathbb{Z}^r$. We can thus identify $\operatorname{Ker}\pi$ with the fundamental group G' of the universal abelian cover $\tilde{\Sigma}$ of the link complement $S^3 \setminus L$. Thus $\mathbb{Z}^r$ acts on the cover, and hence on its fundamental group G' (by outer automorphisms, as above) and on its homology group $H_1(\tilde{\Sigma}) = G'^{ab}$. We may thus regard the latter as a module over the group $\mathbb{Z}^r$, and over its (integer) group ring $R := \mathbb{Z}[t_1, t_1^{-1}, \ldots, t_r, t_r^{-1}]$.

Take a presentation of this module by generators g_i $(1 \le i \le g)$ and relations $r_j = \sum c_{i,j} g_i$. The Fitting ideal $F \lhd R$ is defined to be the ideal in R generated by all $(g-1) \times (g-1)$ minors of the matrix $(c_{i,j})$. It is shown in [65] that if I denotes the augmentation ideal of R (the kernel of the map $R \to BZ$ in which $T_i \mapsto 1$) then for some $\Delta_*^L \in R$, $F = \Delta_*^L.I$. The greatest common divisor of elements of F is thus Δ_*^L if $r > 1$ and $\Delta^L(t_1 - 1) = \Delta_*^L(t_1 - 1)$ if $r = 1$. The element Δ_*^L is unique up to multiplication by a unit of R, i.e. an element of the form $\pm t_1^{a_1} \cdots t_r^{a_r}$. (This ambiguity does not occur in the definition of $\Delta^K(t)$ which is a monic polynomial with non-zero constant term.) The following properties are known.

Lemma 5.5.1 *If the link L has components K_i $(1 \le i \le n)$ and the linking number of K_i and K_j is $\ell_{i,j}$, then if L' is obtained by omitting K_n from K we have*

$$\Delta_*^L(t_1, \ldots, t_{n-1}, 1) = (t_1^{\ell_{1,1}} t_2^{\ell_{1,2}} \cdots t_{n-1}^{\ell_{1,n-1}} - 1)\Delta_*^{L'}(t_1, \ldots, t_{n-1}).$$

If also $n = 2$, then $\Delta_*^L(1,1) = |\ell_{1,2}|$.

An example which is crucial for us is the case of torus knots. Consider the standard torus $T = \{(z,w) \in \mathbb{C}^2 | \, |z| = |w| = \frac{1}{\sqrt{2}}\}$: then for any mutually coprime integers m, p we have a knot parametrised by $z = \frac{1}{\sqrt{2}} e^{im\theta}$, $w = \frac{1}{\sqrt{2}} e^{ip\theta}$. In this case, it is known that the fundamental group may be presented as $\langle x, y \mid x^m = y^p \rangle$. The projection to $\mathbb{Z}$ takes x to p and y to m. Its kernel G' is generated by the elements $\sigma_{a,b} = x^a y^b x^{-a} y^{-b}$. The element $\sigma_{a,b}$ depends only on $a \pmod m$ and on $b \pmod p$, and is trivial if either is zero. It can be shown that the elements $\{\sigma_{a,b} \mid 1 \le a < m, 1 \le b < p\}$ map to a base of V, and then the inner automorphism by τ can be calculated. However, the usual method of calculation is the free differential calculus: this example is treated by this method in [96]p 266. The end result is the polynomial

$$\Delta_{(m,p)}(t) = \frac{(t^{mp} - 1)(t - 1)}{(t^m - 1)(t^p - 1)}.$$

This is a key ingredient in the following.

Theorem 5.5.2 *Suppose the knot K' is a cable knot about K with cabling invariant (m,p). Then*

$$\Delta(K',t) = \Delta(K,t^m)\Delta_{(m,p)}(t).$$

For proofs see [26], or Theorem 9.7.2, which is copied from [65].

We return to our singularity link K. We recall that we have a sequence K^0 (unknotted), $K^1, \dots, K^g = K$, and for each q, K^q cables around K^{q-1} with cabling invariant $(\frac{e_{q-1}}{e_q}, \frac{\overline{\beta}_q}{e_q})$, the latter parameter having been given by Theorem 5.4.2. We may thus use the above theorem to calculate $\Delta(K)$ inductively.

To express the result in a convenient form we introduce some notation. Write $P_z(n)$ for the factor $(t^n - 1)$, and regard the symbols $P_z(n)$ as independent generators of an additive group, where addition corresponds to multiplying the polynomials. We first observe that there are no non-trivial relations between these polynomials, so that the symbol, when it exists, is uniquely determined.

Lemma 5.5.3 *If the relation $\prod_1^N (t^n-1)^{a_n} = \prod_1^N (t^n-1)^{b_n}$ holds identically in t, then $a_n = b_n$ for all n.*

Proof We proceed by induction on N: the result is vacuous for $N=0$. Cancelling a power of $t^N - 1$ and interchanging left and right hand sides, if necessary, we may suppose that $a_N \geq b_N = 0$. Substitute $t = e^{2\pi i/N}$: the left hand side becomes zero, so the right must also vanish. Hence $a_N = 0$. The conclusion follows by the inductive hypothesis. □

Proposition 5.5.4 *The symbol of $\Delta(K^q,t)$ is*

$$\sum_{r=1}^{q} P_z\left(\frac{e_{r-1}\overline{\beta}_r}{e_r e_q}\right) - \sum_{r=1}^{q} P_z\left(\frac{\overline{\beta}_r}{e_q}\right) - P_z\left(\frac{m}{e_q}\right) + P_z(1).$$

Proof For $q=0$, K^0 is unknotted; the group G' is trivial, and the Alexander polynomial is equal to 1; since we have to interpret e_0 as m, the formula in the statement reduces to $-P_z(\frac{m}{e_0}) + P_z(1) = -P_z(1) + P_z(1) = 0$, confirming its correctness.

Inductively assume the formula true as stated, and consider K^{q+1}: to pass from $\Delta(K^q,t)$ to $\Delta(K^{q+1},t)$ we have first to substitute $t^{e_q/e_{q+1}}$ for t, which amounts to multiplying the n in each symbol $P_z(n)$ by e_q/e_{q+1}, and secondly to multiply the polynomial by the appropriate term $\Delta_{(m,p)}$, which involves adding $P_z(mp) - P_z(m) - P_z(p) + P_z(1)$ to the symbol. We thus obtain

$$\sum_{r=1}^{q} P_z\left(\tfrac{e_{r-1}\overline{\beta}_r}{e_r e_{q+1}}\right) - \sum_{r=1}^{q} P_z\left(\tfrac{\overline{\beta}_r}{e_{q+1}}\right) - P_z\left(\tfrac{m}{e_{q+1}}\right) + P_z\left(\tfrac{e_q}{e_{q+1}}\right)$$

$$+ P_z\left(\tfrac{e_q}{e_{q+1}}.\tfrac{\overline{\beta}_{q+1}}{e_{q+1}}\right) - P_z\left(\tfrac{e_q}{e_{q+1}}\right) - P_z\left(\tfrac{\overline{\beta}_{q+1}}{e_{q+1}}\right) + P_z\left(1\right),$$

and here one pair of terms cancels and two more terms can absorbed into the summations, leading to the desired formula for $\Delta(K^{q+1}, t)$. The result thus follows by induction. □

Corollary 5.5.5 $\deg \Delta^{K^q}(t) = (e_{q-1}\overline{\beta}_q - \beta_q - m + e_q)/e_q$.

Proof Since $t^n - 1$ has degree n we deduce that

$$e_q \deg\ \Delta(K^q, t) = \sum_{r=1}^{q} \frac{e_{r-1}\overline{\beta}_r}{e_r} - \sum_{r=1}^{q} \overline{\beta}_r - m + e_q$$

$$= e_{q-1}\overline{\beta}_q/e_q + \sum_{r=1}^{q-1}\left(\frac{e_{r-1}\overline{\beta}_r}{e_r} - \overline{\beta}_{r+1}\right) - \overline{\beta}_1 - m + e_q$$

$$= e_{q-1}\overline{\beta}_q + \sum_{r=1}^{q-1}(\beta_r - \beta_{r+1}) - \beta_1 - m + e_q = e_{q-1}\overline{\beta}_q - \beta_q - m + e_q.$$

□

In particular, since $e_g = 1$ we have

Corollary 5.5.6 *The symbol of* $\Delta^K(t)$ *is*

$$\sum_{q=1}^{g} P_z\left(\tfrac{e_{q-1}\overline{\beta}_q}{e_q}\right) - \sum_{q=1}^{g} P_z\left(\overline{\beta}_q\right) - P_z\left(m\right) + P_z\left(1\right),$$

and its degree is $e_{g-1}\overline{\beta}_g - \beta_g - m + 1 = N(S(B)) + 1$.

We are now ready to apply these calculations to obtain our main conclusions.

Theorem 5.5.7 *The Alexander polynomial* $\Delta^K(t)$ *and hence the fundamental group* $G = \pi_1(S^3 - K)$ *determines the Puiseux characteristic of the branch.*

Proof We first observe that in the expression for $\Delta^K(t)$ in Corollary 5.5.6, there can be no further cancellation. For if, for example, $\frac{e_{q-1}\overline{\beta}_q}{e_q} = \overline{\beta}_r$, it would follow that $\overline{\beta}_r$ was a divisor of $\overline{\beta}_q$, contradicting (ii) of Theorem 4.3.5. Thus the terms with negative coefficients are just the $\{\overline{\beta}_q \,|\, 0 \leq q \leq g\}$. The result thus follows by Proposition 4.3.8. □

Although our expression for $\Delta^K(t)$ involves denominators, since it is a polynomial, these must cancel out. We briefly discuss how to do this explicitly.

We recall that a complex number t is said to be a primitive d^{th} root of unity if d is the least positive integer such that $t^d = 1$. If Φ_d is the polynomial with these numbers as roots, then since $t^n = 1$ if and only if the least d such that $t^d = 1$ is a divisor of n, we have $t^n - 1 = \prod_{d|n} \Phi_d(t)$, where the product is extended over all divisors d of n (including 1 and n itself). If these factorisations are inserted in the expression for $\Delta^K(t)$ then indeed the denominator will cancel.

Example 5.5.1 Consider Example 2.1.1 from the present viewpoint. We have $\beta = (4; 6, 7)$, $e = (4; 2, 1)$, $\overline{\beta} = (4; 6, 13)$; the semigroup is

$$S(B) = \{0, 4, 6, 8, 10, 12, 13, 14, 16, \dots\},$$

$$P_z(26) + P_z(12) - P_z(13) - P_z(6) - P_z(4) + P_z(1)$$

and degree 16. We write it as a product of cyclotomic polynomials $\Phi_k(t)$; after cancellation, we obtain $\Delta^K(t) = \Phi_{26}(t)\Phi_{12}(t) = (t^{12} - t^{11} + t^{10} - t^9 + t^8 - t^7 + t^6 - t^5 + t^4 - t^3 + t^2 - t + 1)(t^4 - t^2 + 1)$.

Theorem 5.5.7 completes the circle of equivalences between the different notions of equisingularity, and thus rounds off the first part of this book. First we consider the case of a single branch.

Theorem 5.5.8 *The following conditions on a pair of branches B, B' are equivalent:*

(i) *B and B' are equisingular;*
(ii) *The corresponding knots K and K' have the same Alexander polynomial;*
(iii) *The fundamental groups $\pi_1(S^3 \backslash K)$ and $\pi_1(S^3 \backslash K')$ are isomorphic;*
(iv) *The knots K and K' are isotopic;*
(v) *The pairs $(D^4_\epsilon, D^4_\epsilon \cap B)$, $(D^4_\epsilon, D^4_\epsilon \cap B')$ are topologically equivalent for small enough ϵ.*

Proof The implications (v) $\Rightarrow$ (iv) $\Rightarrow$ (iii) $\Rightarrow$ (ii) are immediate. By Proposition 5.3.1, (i) implies (iv), and by Theorem 5.5.7, (ii) implies (i). Finally, (iv) implies, by Lemma 5.1.6, that the pairs $(S^3_\epsilon, S^3_\epsilon \cap B)$ $(S^3_\epsilon, S^3_\epsilon \cap B')$ are homeomorphic for small ϵ, and by Lemma 5.2.1 this in turn implies (v). □

To state the result for curves with several branches, define the *peripheral classes* of a link $L \subset S^3$ to be the collection of conjugacy classes in $\pi_1(S^3 \setminus L)$ consisting of the classes of meridians in tori bounding regular neighbourhoods of the several components of L.

Theorem 5.5.9 *Suppose C and C' are plane curve singularities defining links L, L'. Then the following are equivalent:*

(i) *C and C' are equisingular.*
(ii) *There is an isomorphism $\pi_1(S^3 \setminus L) \cong \pi_1(S^3 \setminus L')$ taking the peripheral classes of L to those of L'.*
(iii) *There is an isotopy of L on L'.*
(iv) *The pairs $(D^4_\epsilon, D^4_\epsilon \cap C)$, $(D^4_\epsilon, D^4_\epsilon \cap C')$ are topologically equivalent for small enough ϵ.*

Proof The implications (iv) $\Rightarrow$ (iii) $\Rightarrow$ (ii) are immediate. By Proposition 5.3.2, (i) implies (iii). As above, (iii) implies, by Lemma 5.1.6, that the pairs $(S^3_\epsilon, S^3_\epsilon \cap C)$ $(S^3_\epsilon, S^3_\epsilon \cap C')$ are homeomorphic for small ϵ, and by Lemma 5.2.1 this in turn implies (iv).

Now suppose (ii) holds. Write B_j for the components of C, K_j for the corresponding components of L, and label the components of C', L' as B'_j, K'_j using the bijection given by the hypothesis of (ii). Since $\pi_1(S^3 \setminus K_i)$ is obtained from $\pi_1(S^3 \setminus L)$ by killing the classes of small loops round the remaining components K_r, it is isomorphic to $\pi_1(S^3 \setminus K'_i)$; hence by Theorem 5.5.8, B_j and B'_j are equisingular. Now since $\pi_1(S^3 \setminus (K_i \cup K_j))$ is obtained from $\pi_1(S^3 \setminus L)$ by killing the classes of small loops round the remaining components K_r, it is isomorphic to $\pi_1(S^3 \setminus (K'_i \cup K'_j))$; hence their Alexander polynomials are equal. It now follows from Lemma 5.5.1 that $B_i.B_j = B'_i.B'_j$. As this holds for all i and j, C and C' are equisingular. □

We will discuss Alexander polynomials of curves with several branches (particularly the one-variable polynomial) further in Chapter 10, especially in Section 10.3.

5.6 Notes

Section 5.1 The results of this section are routine in differential topology. One text on this area is [89], but this does not develop vector fields sufficiently far for our needs.

Section 5.2 The arguments in this section apply to the neighbourhood of a point in an algebraic variety of any dimension: the cone property and the identification of intersection and linking numbers hold very generally.

Section 5.3 The origin of the ideas in this chapter was the study of the fundamental groups of the knots and links arising from singularities: according to [65], the problem was given to Brauner by his thesis adviser Wirtinger, who had spoken of it as early as 1905. Brauner's paper [21] described the links geometrically using stereographic projection and gave a presentation for the fundamental group. The simplified description in Lemma 5.2.2 is due to Kähler [95].

Section 5.4 Much of this section is a presentation of tools familiar to knot theorists (see e.g. Schubert [160], Burde and Zieschang [26]). For the approach to Theorem 5.4.2, I am indebted to my colleague Hugh Morton.

Section 5.5 The Alexander polynomial is due to Alexander [8]. Its application to distinguish topologically between knots arising from singularities with distinct Puiseux characteristics (thus proving Theorem 5.5.7) was effected by Burau [24], and independently by Zariski [203], in 1932. Much of the development in this chapter – the description by the carousel, the Alexander polynomial for torus knots, and the formula Theorem 5.5.2 for cable knots – goes back to these references. It was shown by Rapaport [154] and also by Crowell [41] that for *any* knot group G, the group G'^{ab} is torsion-free of finite rank, equal to the degree of the Alexander polynomial.

The free differential calculus was invented by R. H. Fox: for the original references see [72]; a brief summary is given in [96] Chapter 9. Another standard method of calculation is to use a Seifert matrix: we will mention these in Chapter 10.

The situation for singularities with several branches was somewhat less clear: the case with 2 branches was again decided by Burau [25] using the Alexander polynomials for the two branches and their intersection number. The old results were reviewed by Reeve [155], but it seems that the first complete proofs of Theorem 5.5.9 were given by Lejeune [114] and Zariski [208].

The multi-variable Alexander polynomial is due to Fox. It is found in few references, e.g. Torres [178], Fox [72], Milnor [132] and Eisenbud

& Neumann [65], the latter of which is our main source for its properties. The fact that the multi-variable Alexander polynomial suffices in all cases to distinguish equisingularity classes of singularities was proved much later: see Evers [69] and Yamamoto [200]. The multi-variable Alexander polynomial is trivial whenever there is an embedded sphere in S^3 with some components of the link on each side. It follows that there is no universal procedure for obtaining linking numbers from the polynomial. The situation when the linking numbers are all non-zero is unclear: see Exercise 5.7.9 for the case with 3 components.

The original paper of Alexander defined a series of 'Alexander invariants', which can be calculated as Fitting ideals of a Seifert matrix. Another generalisation of the original concept is to Alexander invariants of a complete curve in the projective plane.

It may be noted that the Jones polynomial is much less convenient for our purpose: indeed, although there are several results on calculation of polynomials for cabled knots, there is no simple formula for the Jones (or any other new) knot polynomial for the class of knots considered here.

There are various notations for the cyclotomic polynomials. Some authors write $((n))$ to denote the factor $(t^n - 1)$, and regard the symbols $((n))$ as independent generators of an additive group. Also often-used is the ATLAS notation 'Frame shapes', where $\prod_i m_i^{r_i}$ denotes $\prod_i (1-t^{m_i})^{r_i}$. Our notation $P_z(n)$ was chosen to avoid confusion.

The result of Exercise 5.7.7 is due to Campillo et. al. [28], who have also obtained a version for curves with several branches using the extended semigroup of [29].

5.7 Exercises

Exercise 5.7.1 Verify that the conditions on a vector field to be tangent to the boundary, or to point outwards, are independent of the choice of local coordinates.

Exercise 5.7.2 Draw the intersections of a sphere of small radius with the curve $x = t^6$, $y = t^9 + t^{11}$ with the subspaces (a) x real and positive, (b) $\arg x = \pi/6$.

Exercise 5.7.3 Use Theorem 5.4.2 to give an alternative proof of Theorem 4.1.6.

Exercise 5.7.4 Find the Alexander polynomial for a curve with an E_8 singularity.

Exercise 5.7.5 Calculate the symbol of Δ, and the factorisation of the Alexander polynomial, for curves with Puiseux characteristic (i) $(8;11)$, (ii) $(6;15,20)$, (iii) $(4;10,13)$, (iv) $(12;18,22,25)$.

Exercise 5.7.6 Verify that for any branch B, $\Delta^B(1) = 1$.

Exercise 5.7.7 Verify, using Exercise 4.7.10, that for a branch B with knot K and semigroup S we have $\Delta^K(t) = (1-t)G_S(t)$.

Exercise 5.7.8 Show that the non-zero coefficients in the expansion of the Alexander polynomial $\Delta_K(t)$ in ascending powers of t are equal to ± 1 and alternate in sign. (Hint: use the preceding exercise.)

Verify this conclusion for the polynomials calculated in Exercises 5.7.4 and 5.7.5 (iii).

Exercise 5.7.9 Let L be a link with 3 components. Show that the expansion of Δ^L about the point $(1,1,1)$ begins $\ell_{1,2}\ell_{1,3}(t_1-1) + \ell_{2,3}\ell_{2,1}(t_2-1) + \ell_{3,1}\ell_{3,2}(t_3-1)$, and hence that Δ_L determines the mutual linking numbers of the branches.

6

The Milnor fibration

We begin this chapter by explaining what a fibration is, and giving a method of establishing that certain maps are fibrations. Then we show that two maps defined explicitly in terms of an isolated curve singularity give equivalent fibrations: each of these is termed 'the Milnor fibration'. All the more delicate topology of C is encoded in the Milnor fibration, and studying its geometry gives a very close insight into the topology and geometry associated to the singularity. In this chapter, we give some elementary properties, leading to various calculations of the Betti numbers of the fibre. A detailed study will be made in Chapter 10.

6.1 Fibrations

A fibration is a sort of twisted product. More precisely, a map $\pi : E \to B$ is (the projection of) a *fibration* with *fibre* F if each point $b \in B$ has a neighbourhood U such that there is a homeomorphism ϕ of $\pi^{-1}(U)$ onto $F \times U$ whose second component is the restriction of π. Thus to construct the homeomorphism one needs only the first component, a map onto F, which may well be defined via a map onto $F_b = \pi^{-1}(b)$.

Properties of fibrations are derived in full in textbooks of algebraic topology, e.g. [169], and we content ourselves here with citing those we shall require.

If U is any contractible subset of B, then a homeomorphism ϕ as above may be constructed over U. In particular, if $B = S^1$, we may take U to be either the upper or the lower semicircle: U^+, U^- say. We thus have homeomorphisms

$$\phi^+ : \pi^{-1}(U^+) \to F \times U^+, \quad \phi^- : \pi^{-1}(U^-) \to F \times U^-.$$

The sets U^+, U^- intersect in their common end-points ± 1. Each of $\phi^\pm$ gives a homeomorphism $h^\pm$ of $\pi^{-1}(-1)$ onto F. If we replace ϕ^- by $((h^+ \circ (h^-)^{-1}) \times 1) \circ \phi^-$, the two homeomorphisms of $\pi^{-1}(-1)$ onto F agree. It is not in general possible also to arrange the two homeomorphisms of $\pi^{-1}(1)$ onto F to agree. The final picture is thus as follows. We have the product of F by an interval $[0, 2\pi]$, and will identify, for each $x \in F$, the point $(x, 2\pi)$ with $(h(x), 0)$ for a suitable homeomorphism h. The resulting space X_h is called the *mapping torus* of h; we map it to S^1 by taking (x, θ) to $e^{i\theta}$. The map h is called the *monodromy* of the fibration.

Lemma 6.1.1 *The monodromy of a fibration over S^1 is determined uniquely up to isotopy.*

Proof If h and h' are both possible choices for the monodromy, there is a homeomorphism of X_h on $X_{h'}$ which is the identity on F and respects the projection on S^1. This lifts to a homeomorphism K of $F \times I$ on itself of the form $K(x, \theta) = (k_\theta(x), \theta)$ with k_0 the identity. To obtain X_h we identify $(x, 2\pi)$ with $(h(x), 0)$. In the target, $(k_{2\pi}(x), 2\pi)$ must be identified with $(k_0(h(x)), 0) = (h(x), 0)$ but also, to obtain $X_{h'}$, with $(h'(k_{2\pi}(x)), 0)$. Hence $h \cong h' \circ k_{2\pi}$, and $h_\theta = h' \circ k_\theta$ gives the required isotopy. □

To establish that a differentiable map $\pi : E \to S^1$ is a fibration, the following method is available. Suppose first that E is a compact manifold without boundary. Then we seek a vector field ξ on E which is mapped by π to the vector field $\frac{\partial}{\partial t} = i\theta \frac{\partial}{\partial \theta}$ on S^1. Since E is compact, we can then integrate ξ, giving a map $\Phi : E \times \mathbb{R} \to E$. The restriction of Φ to $F \times [0, 2\pi]$, where $F_i = \pi^{-1}(0)$, is surjective, and the monodromy is given by $h(x) = \Phi(x, 2\pi)$.

If E has a boundary ∂E, we also require the restriction of π to ∂E to be a submersion. We then choose a vector field on ∂E mapped by π to $\frac{\partial}{\partial t}$, and extend to a vector field on E (which is thus tangent to the boundary) with the same property. We can then apply Corollary 5.1.3,

In general we have

Theorem 6.1.2 *(Ehresmann fibration theorem) If E is a manifold, and $\pi : E \to B$ a map such that E is compact, or more generally π is proper, and both π and (if ∂E is non-empty) $\pi \,|\, \partial E$ are submersions, then π is a fibration.*

Proof Since the definition of fibration refers to the preimage of a neighbourhood in B, it is enough to restrict to such a neighbourhood. Take coordinates to express this as a subset – say $[-\epsilon, \epsilon]^n$ – of $\mathbb{R}^n$, and proceed by induction on n.

The crucial case is $n = 1$. To construct a vector field with the desired property near a point not on the boundary, take any vector ξ at that point with $\pi_*\xi \neq 0$, and any vector field extending ξ. For a point on the boundary, first proceed as above for the restriction to the boundary; then extend to a neighbourhood in E. Then the projection is non-zero provided we restrict to some neighbourhood of the point, and dividing by the length of the projection we obtain a vector field near the point with the desired property. We may use a partition of unity, as described in Section 5.1, to piece together these vector fields defined on small patches to obtain a vector field defined everywhere and with the desired properties. Integrating this gives the desired homeomorphism.

Suppose the result proved for $n - 1$. The projection $\pi_1 : \mathbb{R}^n \to \mathbb{R}$ on the last coordinate is a submersion, hence so is $\pi_1 \circ \pi$. We restrict to the above neighbourhood. Applying the result for $n = 1$ to this case, we obtain a homeomorphism from $\pi^{-1}[-\epsilon, \epsilon]^n$ to $\pi^{-1}([-\epsilon, \epsilon]^{n-1} \times 0) \times [-\epsilon, \epsilon]$ which is compatible with projection on the final coordinate. The induction hypothesis provides a homeomorphism from $\pi^{-1}([-\epsilon, \epsilon]^{n-1} \times 0)$ to $\pi^{-1}(O) \times [-\epsilon, \epsilon]^{n-1}$ which is compatible with projection on the second factor. Combining these, the result follows. □

6.2 The Milnor fibration

For each curve singularity there are two fibrations, equivalent to each other, describing how the defining function f behaves in a neighbourhood. Here we construct the fibrations and prove their equivalence: the basic technique is the use of vector fields.

Consider the germ at O of an equation $f(x, y) = 0$ such that f has no repeated factor. Then O is an isolated point of the intersection $\{(x, y) \mid \partial f/\partial x = \partial f/\partial y = 0\}$. For if this intersection contained a curve D, then since both partial derivatives of f vanish along D, f would take a constant value along D, and hence be zero. But then any branch of D would correspond to a repeated factor of f.

Consider the discs $B_\epsilon := \{(x, y) \mid |x|^2 + |y|^2 \leq \epsilon^2\}$ in $\mathbb{C}^2$, with boundary sphere S_ϵ and $D_\eta := \{z \mid |z| \leq \eta\}$ in the complex plane, with boundary S_η. Write D_η^* for the 'punctured disc' obtained by removing the point

0 from D_η. Following Milnor [132] we will define two closely related fibrations.

Theorem 6.2.1 *If ϵ is small enough, we can find η_0 such that for $\eta < \eta_0$, the map*

$$f_1 : B_\epsilon \cap f^{-1}(D^*_\eta) \to D^*_\eta,$$

defined by restriction of f, is the projection of a smooth fibration.

Proof Since, by the above remark, df does not vanish on B_ϵ (except at O), the map is a submersion. We next show that the restriction of f to $S_\epsilon \cap f^{-1}(D^*_\eta)$ is also a submersion. Since the fibres are closed in $B_\epsilon \cap f^{-1}D_\eta$, and so are compact, the conclusion will follow from the fibration theorem 6.1.2.

At a point of $S_\epsilon \cap f^{-1}(0)$, the restriction of f to S_ϵ will fail to be a submersion if and only if $f^{-1}(0)$ is tangent to S_ϵ, i.e. is perpendicular to the radius vector. We showed in the proof of Lemma 5.2.1 that, for ϵ small enough, this does not occur. It follows by openness of the submersion condition that the restriction is also a submersion on some neighbourhood of $S_\epsilon \cap f^{-1}(0)$ in S_ϵ. But if η_0 is small enough, $S_\epsilon \cap f^{-1}(D^*_{\eta_0})$ is contained in this neighbourhood. The result thus follows. □

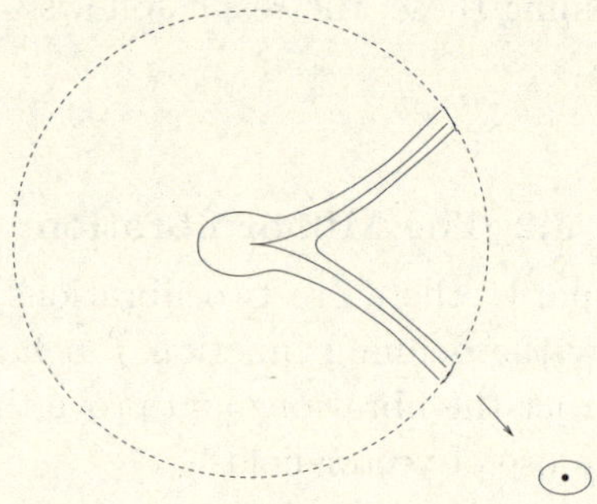

Fig. 6.1. The first Milnor fibration

Write $K := f^{-1}(0) \cap S_\epsilon$, with tubular neighbourhood

$$N(K) := f^{-1}(D_\eta) \cap S_\epsilon,$$

which has boundary $\partial N(K) = f^{-1}(S_\eta) \cap S_\epsilon$. This boundary divides the sphere S_ϵ into two parts: one is $N(K)$; let us denote the closure of the other by W. This is called the *closed complement* of the link. Our use here of the symbol K does not exclude the case that K is a link with several components.

Theorem 6.2.2 *The map f_2 defined by $f/|f|$ from W to S^1 is also a fibration, equivalent to the restriction f_1 of f to $B_\epsilon \cap f^{-1}(S_\eta)$.*

Proof The proof will again be effected by constructing, and then integrating vector fields, but the arguments here are more delicate.

For the first assertion of the theorem we need to lift the vector field on S^1 to one tangent to W. i.e. to S^3. Thus we require a vector field ξ which at each point of W is tangent to W, and has positive inner product with the gradient of $f/|f|$: we can later divide by this inner product to normalise it to 1, so that the projection of ξ will indeed be the unit tangent vector to S^1. For the second assertion, we seek a

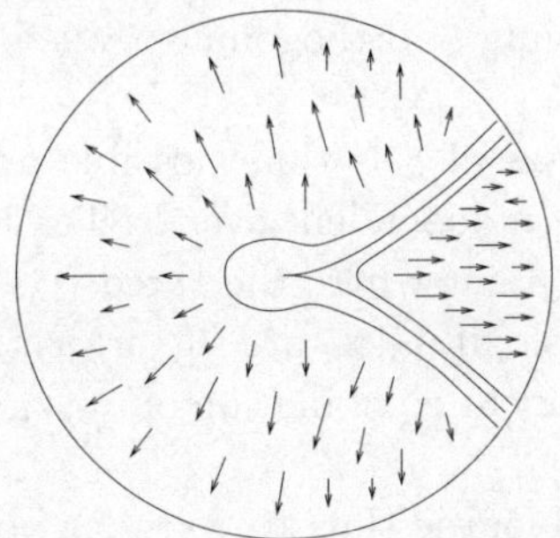

Fig. 6.2. Equivalence of fibrations

homeomorphism of $B_\epsilon \cap f^{-1}(D^*_\eta)$ onto W which carries f/η to $f/|f|$ and thus gives an isomorphism between the two fibrations. This will be obtained by integrating a vector field ξ'. We require that ξ' is tangent to the level sets of $f/|f|$ on $B_\epsilon - f^{-1}(0)$, and that both $|f|$ and $\|z\|$ increase strictly along the integral curves of ξ'. It will then follow that each integral curve of ξ' lies in a level set, crosses $f^{-1}(D_\eta)$ transversely, and goes 'outwards' till it crosses S_ϵ transversely, as illustrated in Figure 6.2. By Theorem 5.1.5 it suffices to show that each of ξ, ξ' can be constructed locally.

To express the conditions on the vector fields ξ, ξ', we need a mixture of real and complex notions. We identify the vector $(a + ib, c + id) \in \mathbb{C}^2$

with the vector $(a, b, c, d) \in \mathbb{R}^4$, and use only inner products in the real sense. If g is a real-valued function of $z_1 = x_1 + iy_1$ and $z_2 = x_2 + iy_2$, define

$$\nabla(g) := (\partial g/\partial x_1 + i\partial g/\partial y_1, \partial g/\partial x_2 + i\partial g/\partial y_2).$$

Thus if $f = g + ih$ is holomorphic, $\nabla(g) = i\nabla(h) = (\overline{\partial f/\partial z_1}, \overline{\partial f/\partial z_2})$; we also denote this by $\nabla(f)$. As the modulus-argument decomposition of f corresponds to the real-imaginary decomposition of $log\, f$, we will work with $\nabla log\, f$: write $\nabla(log\,|f|) = \mathbf{u}$; then $\nabla(arg\, f) = \mathbf{v} = -i\mathbf{u}$. Since W is a level of $\|\mathbf{z}\|^2$, its tangent space is the orthogonal complement of $\nabla\|\mathbf{z}\|^2 = 2\mathbf{z}$. The gradient of $f/|f|$ is proportional to $\mathbf{v}$.

Thus the conditions that ξ be tangent to W, and have positive inner product with the gradient of $f/|f|$, are equivalent respectively to having ξ orthogonal to $\mathbf{z}$ and having positive inner product with $\mathbf{v}$. The conditions that ξ' be tangent to the level sets of $f/|f|$, and that both $|f|$ and $\|z\|$ increase strictly along the integral curves of ξ', are equivalent respectively to having ξ' orthogonal to $\mathbf{v}$ and having positive inner product with $\mathbf{u}$ and $\mathbf{z}$.

The local construction of ξ is thus possible at any point where the vectors $\mathbf{v}, \mathbf{z}$ are linearly independent over $\mathbb{R}$. The local construction of ξ' offers no problem at a point where the three vectors $\mathbf{u}, \mathbf{v}, \mathbf{z}$ are linearly independent: for example, if $\mathbf{u_0}, \mathbf{z_0}$ are the projections of $\mathbf{u}, \mathbf{z}$ orthogonal to $\mathbf{v}$, there is a unique linear combination of $\mathbf{u_0}, \mathbf{z_0}$ having inner products 1 with each of $\mathbf{u}, \mathbf{z}$.

The crux of the problem is thus to consider points at which there is a non-trivial linear relation $a\mathbf{u} + b\mathbf{v} = c\mathbf{z}$ over $\mathbb{R}$. Since $\mathbf{v} = -i\mathbf{u}$, the relation becomes $(a - ib)\nabla log\, f = c\mathbf{z}$. If $a = b = 0$ we have $c\mathbf{z} = \mathbf{0}$, and since $\mathbf{z}$ is not zero, $c = 0$. But then the linear relation would be trivial. Thus we may divide by $a - ib$ and write the relation as $\nabla log\, f = \lambda\mathbf{z}$. The next Lemma 6.2.3 will show that in this situation we may suppose that λ is close to the positive real axis – to be specific, it is enough to insist that $|arg\,\lambda| < \frac{\pi}{2}$. Let us accept this for now and show how to conclude the proof.

For the first assertion, it suffices to observe that since λ has non-zero real part, the coefficient a in the above equation is non-zero, so $\mathbf{v}$ and $\mathbf{z}$ are linearly independent. For the second, observe first that, since $\mathbf{u}$ and $\mathbf{v}$ are independent, the linear relation is unique; we may write it as $a\mathbf{u} + b\mathbf{v} = \mathbf{z}$ where, by the lemma, $a > 0$. The projection of $\mathbf{z}$ orthogonal to $\mathbf{v}$ thus has the desired properties in some neighbourhood. □

Lemma 6.2.3 *For any holomorphic f defined near O and with $f(O) = 0$, we can find ϵ such that at any point $\mathbf{z}$ with $\|\mathbf{z}\| < \epsilon$ and at which there is a linear relation $\nabla log\, f = \lambda\mathbf{z}$, we have $\Re\lambda > 0$.*

Proof The proof of this lemma depends crucially on the following result. Its proof, while not more difficult than those involved in the rest of this chapter, demands a new range of concepts. We thus give a bare statement, and refer the reader to [132] for further details.

A subset $X \subset \mathbb{R}^n$ is said to be *semi-algebraic* if it is defined by a finite list of equations and inequalities $f_i(\mathbf{z}) = 0, g_j(\mathbf{z}) > 0, h_k(\mathbf{z}) \geq 0$, where all the f_i, g_j, h_k are polynomials. We say X is *semi-analytic* if, for any point $\mathbf{x} \in \mathbb{R}^n$ there exists $\epsilon > 0$ such that $X \cap \{\mathbf{z} \mid ||\mathbf{z} - \mathbf{x}|| < \epsilon\}$ is defined by equations and inequalities with all the f_i, g_j, h_k defined by power series in $\mathbf{z} - \mathbf{x}$ convergent in $\|\mathbf{z} - \mathbf{x}\| < \epsilon$.

Theorem 6.2.4 *(Curve Selection Theorem). Let $X \subset \mathbb{R}^n$ be semi-analytic; suppose there are points of X arbitrarily close to O. Then there is a real analytic curve $\mathbf{p} : [0, \eta) \to \mathbb{R}^n$ with $\mathbf{p}(0) = O$ and $\mathbf{p}(t) \in X - \{O\}$ for $t > 0$.*

We continue with the proof of Lemma 6.2.3. It follows from Theorem 6.2.4 that either there is a neighbourhood of O containing no points $\mathbf{z}$ with $f(\mathbf{z}) \neq 0$ at which there is a linear relation of the type in question – in which case the result is vacuously true – or there is a real analytic curve $\mathbf{p} : [0, \eta) \to \mathbb{R}^n$ with $\mathbf{p}(0) = O$, $f(\mathbf{p}(t)) \neq 0$ for $t \neq 0$, and $\nabla log\, f(\mathbf{p}(t)) = \lambda(t)\mathbf{p}(t)$ for $t > 0$.

We expand as (convergent) power series in t, in each case making explicit the first non-zero term:

$$\mathbf{p}(t) = \mathbf{a}t^\alpha + \cdots,$$

$$f(\mathbf{p}(t)) = bt^\beta + \cdots,$$

$$\nabla f(\mathbf{p}(t)) = \mathbf{c}t^\gamma + \cdots.$$

Observe that since the chosen analytic curve does not lie in $f^{-1}(0)$, f is not constant along it, so none of the above vanish identically in t.

We first infer from $\nabla \log f(\mathbf{p}(t)) = \lambda(t)\mathbf{p}(t)$ that

$$\nabla f(\mathbf{p}(t)) = \lambda(t)\mathbf{p}(t)\overline{f}(\mathbf{p}(t)),$$

so that

$$(\mathbf{c}t^{\gamma} + \cdots) = \lambda(t)(\bar{b}\mathbf{a}t^{\alpha+\beta} + \cdots),$$

and $\lambda(t)$ is a quotient of two convergent power series, so can itself be expanded as a series

$$\lambda(t) = \lambda_0 t^{\gamma-\alpha-\beta}(1 + \cdots),$$

where the initial exponent may be (and, in fact, is) negative. However, since t is positive real, we see that the argument of $\lambda(t)$ converges to that of λ_0 as $t \to 0$. The proof will be concluded by showing that λ_0 is a positive real number.

Substituting for λ in the preceding equation and taking leading terms gives $\mathbf{c} = \lambda_0 \bar{b}\mathbf{a}$. By the chain rule,

$$\frac{df(\mathbf{p}(t))}{dt} = \frac{\partial f}{\partial x}\frac{dx}{dt} + \frac{\partial f}{\partial y}\frac{dy}{dt},$$

where x, y denote coordinates in $\mathbb{C}^2$. The leading term on the right hand side is the (Hermitian) inner product of the leading terms $\alpha\mathbf{a}t^{\alpha-1}$ of $d\mathbf{p}(t)/dt$ and $\lambda_0\bar{b}\mathbf{a}t^{\gamma}$ of $\nabla f(\mathbf{p}(t))$, and so is $\alpha\lambda_0 b\|\mathbf{a}\|^2 t^{\alpha+\gamma-1}$. Since the coefficient is non-zero, this must equal the leading term $b\beta t^{\beta-1}$ on the left hand side. Hence indeed $\lambda_0 = \beta/(\alpha\|\mathbf{a}\|^2)$ is real and positive. □

The fibration of Theorem 6.2.1 fits well into general constructions; the fibration f_2 is crucial to a more detailed study of the knot (or link) K. We refer to either of the fibrations of Theorem 6.2.2 as the *Milnor fibration* of the curve $f = 0$, and to the fibre F of these fibrations as the *Milnor fibre*. An alternative proof of part of these fibration theorems will be given in Chapter 9.

The function $f/|f|$ is also defined on $S_\epsilon - K$. We may identify $N(K)$ with a product $K \times D_\eta$ in such a way that the projection on the second factor is given by f; hence the closure of the preimage of a point θ by $f/|f|$ is a smooth surface F_θ^o whose closure is a compact smooth surface F_θ with boundary K. Thus F_θ consists of a fibre of the Milnor fibration, extended by attaching a copy of $\partial F \times [0, 2\pi]$ to the boundary, so that the new boundary is the singularity link K. The construction of the monodromy extends to these surfaces, and it follows by continuity that the monodromy reduces to the identity on the boundary K.

It is important to observe that the sphere S_ϵ can be reconstructed from the data (F, h). Take the product $F \times [0, 2\pi]$; for each $y \in \partial F$, identify all the points (y, θ) together, and for each $x \in F$ identify $(x, 2\pi)$

with $(h(x), 0)$: the result is homeomorphic to S_ϵ, with $F \times \theta$ giving the surface F_θ.

In general, given a manifold M with boundary and a self-homeomorphism h of M which restricts to the identity on ∂M we can construct a new, closed manifold W by making the corresponding identifications on $M \times [0, 2\pi]$. The identification is the same as we have just had for constructing a fibration with monodromy h, but then we make a further identification of the boundary in all fibres.

We then say we have an *open book* of W, with *leaves* the images of $M \times \theta$ and *spine* the image of ∂M. In the above situation, we have an open book decomposition of the sphere S_ϵ, with leaves the F_θ and spine K.

6.3 First properties of the Milnor fibre

Let C be a curve defined by the reduced equation $f(x, y) = 0$. In this section we give the basic facts about the Milnor fibre F.

Proposition 6.3.1 *The Milnor fibre F is a compact, connected, oriented surface with r boundary components, where the curve C has r branches.*

Proof Since – except at the origin – the map f is a submersion of a 4-manifold to a 2-manifold (we count real dimensions here), its fibres are 2-manifolds. They acquire orientations from those of the source $\mathbb{C}^2$ and target $\mathbb{C}$; or equivalently from orientations of S_ϵ and of the normal bundle of F in it, given by construction. Also by construction, F is compact, and its boundary coincides with K, so the number of components is as stated.

It remains to establish connectedness. Suppose if possible F were not connected: choose a component F_1 with non-empty boundary. Since h is the identity on the boundary, it must take F_1 into itself, hence also preserve the complement F_2 of F_1 in F. Now we gave above a construction on (F, h) yielding S_ϵ. If we apply it here, we obtain a space with a component arising from F_1 and at least one other connected component arising from F_2. But S_ϵ is connected. This contradiction establishes the result. □

This result gives a complete description of the topology of the fibre F up to saying what its genus is. The rank of the first Betti number of F is known as the *Milnor number* of the singularity, and denoted μ. This is a very important invariant, and we will obtain several formulae for computing it. Meanwhile we observe that μ determines the genus g of F. For this is the same (by definition) as the genus of the closed surface $\hat{F}$ obtained from F by attaching a disc along each boundary component. The Euler characteristic $\chi(\hat{F}) = r + \chi(F)$, and in turn $\chi(F) = 1 - \mu$. Thus $1 - \mu + r = 2 - 2g$, so $g = \frac{1}{2}(\mu - r + 1)$.

In the case when C consists of just one branch, we can identify μ with an invariant we have already encountered.

First observe that the complement $S^3 - K$ of the knot has the same homotopy type as the complement (denoted M above) of an open tube surrounding K. Now M is fibred over S^1 with fibre F. The homotopy sequence [169] of the fibration yields, since $\pi_2(S^1)$ is trivial, an exact sequence of fundamental groups

$$1 \to \pi_1(F) \to \pi_1(M) \to \pi_1(S^1) \to 1.$$

Here we can identify $\pi_1(M)$ with the fundamental group $\pi_1(S^3 - K)$ of the knot, which was denoted by G in Section 5.5. The group $\pi_1(S^1)$ is infinite cyclic, and can be identified with the first homology group of S^1, or indeed with that of $S^3 - K$, and hence with G^{ab}, the group G 'made abelian'. It follows from the exact sequence that the kernel G' of the projection $G \to G^{ab}$ is isomorphic with $\pi_1(F)$.

Making this kernel abelian gives an isomorphism of $(G')^{ab}$ with $H_1(F)$, which is an abelian group whose torsion free rank we have agreed to denote by μ. On the other hand, we defined the Alexander polynomial $\Delta^K(t)$ of the knot as the characteristic polynomial of the map from $(G')^{ab}$ to itself induced by the monodromy. But the degree of this characteristic polynomial coincides with the rank of the module. Thus we have

Proposition 6.3.2 *If C just has one branch, then*

$$\mu = \deg \Delta^C(t) = e_{g-1}\overline{\beta}_g - \beta_g - m + 1.$$

Thus $\mu(C) = N(S(C)) + 1 = 2\delta(C)$.

Proof The second equality is given by Corollary 5.5.6, as is the relation with $N(S(B))$. The final relation comes from Corollary 4.3.7. □

We next consider the simplest case. The result here is worth deriving directly.

Lemma 6.3.3 *The singularity at a transverse intersection of two smooth curves has Milnor number $\mu = 1$.*

Proof Since the intersection is transverse, if the equations are $f = 0$, $g = 0$ then f and g have linearly independent differentials at the point in question, so may (by the Inverse Function Theorem) be taken as local coordinates. It thus suffices to consider the curve $f(x, y) \equiv xy = 0$.

The Milnor fibre is the set of points $(x, y) \in S_\epsilon$ such that $\frac{f(x,y)}{|f(x,y)|} = 1$, i.e. the product xy is real and positive. We can thus write $x = ae^{i\theta}, y = be^{-i\theta}$. Here a, b are real, with $a > 0, b > 0, a^2 + b^2 = \epsilon^2$; θ may be arbitrary. The Milnor fibre is thus homeomorphic to the product of a circle by an interval on the line. Hence its first Betti number $\mu = 1$. □

A singularity as in the lemma is said to be *non-degenerate* : this is equivalent to saying it has type A_1. We will show in Corollary 6.5.2 that the converse holds: all singularities with $\mu = 1$ are of this form, and will see in Proposition 6.5.4 that any singularity C can be regarded in a certain sense as made up of $\mu(C)$ non-degenerate singularities.

There is also a direct characterisation.

Lemma 6.3.4 *A singular point of f is non-degenerate if and only if the Hessian $H(f) \neq 0$ at the point.*

Proof Since, on a change F of coordinates, the Hessian is multiplied by the square of the Jacobian of F, both conditions are invariant by such changes. If f is as above, we may take $f = xy$ and then the Hessian is $-1 \neq 0$.

Conversely, suppose the Hessian non-zero. Then the equation has order 2. By Theorem 2.2.7, there exist coordinates in which C is given by $y^2 + x^k = 0$ for some $k \geq 2$ (or by $y^2 = 0$). Only if $k = 2$ is the Hessian non-zero.

We can also argue more directly. Let the terms of degree 2 in the Taylor expansion of f at O be $ax^2 + 2bxy + cy^2$. Then the Hessian at O is $ac - b^2 \neq 0$. Hence the quadratic terms factorise, with distinct factors. Take these factors as new coordinates. In the new system, the coefficients of x^2 and y^2 are 0; the coefficient of xy is not. Thus the Newton polygon has two sides. By Lemma 2.4.4, f factorises, with one factor equal to x plus higher terms, the other factor to y plus higher terms. □

6.4 Euler characteristics and fibrations

In order to obtain further formulae concerning the Milnor number, we need further recourse to topology. It turns out that all the formulae we require can be expressed in terms of the most basic of all topological invariants: the Euler characteristic. In this section, we summarise its basic properties, and go on to the results needed for the applications.

Let X be a compact triangulated topological space. For each i write α_i for the number (necessarily finite) of i-simplices. Then the Euler characteristic is $\chi(X) = \sum_i (-1)^i \alpha_i$. In fact all the spaces we consider are triangulable, though it is not convenient to exhibit triangulations. The topological invariance is established by showing that $\chi(X)$ is likewise equal to the alternating sum of the ranks of the homology groups of X. (Thus in the case of the Milnor fibre, the only non-zero homology groups are H_0, of rank 1, and H_1, of rank μ, so that $\chi(F) = 1 - \mu$, as already noted.)

The Euler characteristic is an invariant of homotopy type, so that if X is a cone, or more generally is contractible, $\chi(X)$ equals the value of χ for a single point, namely 1.

We have an additive property for unions (again all spaces are assumed compact and triangulable)

$$\chi(X \cup Y) = \chi(X) + \chi(Y) - \chi(X \cap Y).$$

There is also a multiplicative formula for products:

$$\chi(X \times Y) = \chi(X)\chi(Y).$$

It follows that χ is also multiplicative for fibrations $\pi \mid E \to B$ (with fibre denoted F). For by definition we can cover B by sets B_i above which the fibration is a product; by compactness, we need only a finite number N, say, of such sets. We proceed by induction on N. Write $B = B_1 \cup B_2$, where B_1 is the union of $N-1$ such sets, so the result holds here by induction, and the fibration is a product over B_2. Then

$$\begin{aligned}\chi(E) &= \chi(\pi^{-1}(B_1)) + \chi(\pi^{-1}(B_2)) - \chi(\pi^{-1}(B_1 \cap B_2)) \\ &= \chi(F)\chi(B_1) + \chi(F)\chi(B_2) - \chi(F)\chi(B_1 \cap B_2) = \chi(F)\chi(B).\end{aligned}$$

We now consider a map which has the properties of a fibration except at a finite set of points. For example, this is so if X is a compact complex analytic surface, S a connected curve, and $f : X \to S$ a proper map with only finitely many critical points. We allow S to have a boundary (e.g. $S = D_\eta$ above), and require that f has no critical points over the boundary; thus the restriction of f to $f^{-1}(\partial S) \to \partial S$ is a fibration. Likewise, we allow X to have a boundary, partitioned as $\partial_R X \cup \partial_S X$, where $\partial_S X = f^{-1}(\partial S)$ and the restriction of f to $\partial_R X$ is a submersion, and hence by Theorem 6.1.2 a fibration.

More generally, we may contemplate a family $\{f_u \mid u \in U\}$ of maps f_u each of which is a fibration except at a finite set of points. We regard the family as constituting a single map $g : X \times U \to S \times U$.

Our first concern is with the Euler characteristics of the fibres. If we remove from S the interiors of small discs surrounding the critical values of f, to obtain S', say, then over S' we have a fibration by Theorem 6.1.2; thus all the fibres are homeomorphic, and in particular have the same Euler characteristic.

For a singular fibre, we will assume that the fibre contains only one singular point: if there are more, we perform the same constructions and arguments as below for each one separately. Enclose the singular point in a small disc: we regard this as the disc B_ϵ of Section 6.2. In S we have a very small disc which we identify with D_η. We have already seen that the restriction of f to $S_\epsilon \cap f^{-1}(D_\eta)$ is a submersion; we can thus apply the fibration Theorem 6.1.2 to the restriction of f to the part of $f^{-1}(D_\eta)$ *outside* B_ϵ; in particular, all these fibres are homeomorphic.

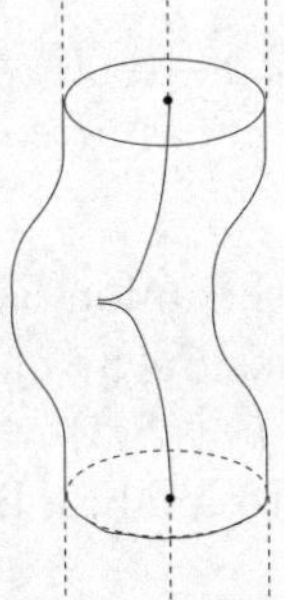

Fig. 6.3. The space $B_\epsilon \cap f^{-1}(D_\eta)$ used for gluing

The singular fibre of f is thus obtained, up to homeomorphism, from a non-singular one by removing the part in B_ϵ, viz. the Milnor fibre, and replacing it by $B_\epsilon \cap f^{-1}(0)$. But, as remarked in Lemma 5.2.1, this latter is homeomorphic to the cone on K. We now apply the additive property of the Euler characteristic to obtain the following.

Theorem 6.4.1 *In the above situation, for any fibre $F_s = f^{-1}(s)$ we define $\chi'(F_s)$ to be obtained from $\chi(F_s)$ by subtracting the sum of the Milnor numbers of the singular points on F_s. Then $\chi'(F_s)$ is independent of s.*

Proof First consider (as just before the Theorem) the case of a map f from X to S. For the non-singular fibres, all values of $\chi' = \chi$ are the same. For a singular fibre, we decompose it into the part F_i inside B_ϵ and the part F_o outside (both including their common boundary). A nearby non-singular fibre F' is likewise decomposed. Now $\chi(F_o) = \chi(F'_o)$, since they are homeomorphic; $\chi(F_i) = 1$, since it is a cone, hence contractible; and $\chi(F'_i) = 1 - \mu$, essentially by the definition of μ. Hence

$$\begin{aligned}\chi(F) - \mu &= (\chi(F_i) - \mu) + \chi(F_o) - \chi(F_i \cap F_o)\\ &= (\chi(F'_i)) + \chi(F'_o) - \chi(F'_i \cap F'_o) \;=\; \chi(F').\end{aligned}$$

For the case of a map g assembled from a family of maps f_t, we note that over a dense open subset of $S \times U$ we have a fibration as usual, so that all fibres are homeomorphic; by hypothesis, this open set intersects each level $S \times \{u\}$. The result follows from this, together with the above argument applied to f_u. □

Secondly, we return to the case of $f : X \to S$ and seek to calculate $\chi(X)$. The arguments are very similar to the above: we will prove

Theorem 6.4.2 *We have $\chi(X) = \chi(F)\chi(S) + \sum_P \mu_P$, where F denotes a non-singular fibre, and the sum is extended over all singular points P.*

Proof We decompose S into the union S_0 of small discs D_η surrounding the singular values and the closure S_1 of their complement. We have a fibration over S_1, so that $\chi(f^{-1}(S_1)) = \chi(F)\chi(S_1)$. By the additive property again, it will suffice to analyse by how much each $\chi(f^{-1}(D_\eta))$ differs from $\chi(F)$.

But over D_η we have again decomposed $F^{-1}(D_\eta)$ into an exterior part, which is fibred, and an interior part, which may be identified with the space $B_\epsilon \cap f^{-1}(D_\eta)$ of Section 6.2. The above difference may thus be calculated by looking at the 'interior' part, so we obtain the difference of χ of this space and χ of the Milnor fibre. But the space is contractible, as we saw in Lemma 5.2.1. The result follows by putting these remarks together. □

6.5 Further formulae for μ

The Milnor number is a key measure of the complexity of a singularity. Here we will give three further formulae for calculating it. We begin by obtaining the value of μ for a curve obtained as a union. Then, after some preparation, we establish that μ is equal to the intersection number of the polar curves $\partial f/\partial x = 0$ and $\partial f/\partial y = 0$. Finally we obtain a formula in terms of the sequence of multiplicities of infinitely near points.

We first apply Theorem 6.4.1 to calculate μ for the union of two curve-germs.

Theorem 6.5.1 *We have* $\mu(C \cup C') = \mu(C) + \mu(C') + 2C.C' - 1$. *Hence if* C *has irreducible components* $\{B_i \mid 1 \leq i \leq r\}$, *we have*

$$\mu(C) = \sum_i \mu(B_i) + 2\sum_{i<j}(B_i.B_j) - r + 1.$$

Proof Choose ϵ small enough so that the disc B_ϵ can be used for all the Milnor fibrations involved, with the disc D_η in the target. Let f, f' be equations for the germs C, C', and set $g_{t,u}(x,y) = (f(x,y)-t)(f'(x,y)-u)$. We choose δ so small that for $|t|, |u| < \delta$, each $g_{t,u}$ restricts to a submersion $\partial D_\eta \cap g_{t,u}^{-1}(B_\epsilon) \to B_\epsilon$. We thus have a family of fibrations, and can apply Theorem 6.4.1.

For $t = u = 0$ we have the defining equation $g_{0,0} = 0$ of $C \cup C'$, so the Euler characteristic of a non-singular fibre is $1 - \mu(C \cup C')$. For other values of (t, u), the zero locus of $g_{t,u}$ consists of the union of Milnor fibres F_t, F'_u of f and f', with respective Euler characteristics $1 - \mu(C), 1 - \mu(C')$. We claim that for most pairs (t, u), F_t and F'_u intersect transversely, and second, that the number of these intersections is $C.C'$.

Consider in fact small (non-zero) values of (t, u) such that the curves $f = t, f' = u$ have a non-transverse intersection. Since transversality fails, $\frac{\partial(f,f')}{\partial(x,y)} = 0$. This equation is not an identity, so defines a curve. We can choose arbitrarily small values (t, u) such that the intersection points $f = t, f' = u$ avoid this curve.

Since (t, u) is very small, the links given at any level by intersection with S_ϵ are isotopic. Hence the linking numbers are constant. But the linking number at $(t, u) = (0, 0)$ is the intersection number $C.C'$. Hence this is true for all small (t, u). But this linking number is equal to the intersection number of the surfaces F_t, F'_u bounding the given curves. Thus the number of intersections is as stated.

Thus $\chi(F_t \cup F'_u) = \chi(F_t) + \chi(F'_u) - C.C'$; whereas $F_t \cup F'_u$ has $C.C'$ singular points, each a union of two smooth curves meeting transversely, hence with $\mu = 1$. Hence by Theorem 6.4.1, the Euler characteristic of a nonsingular fibre is

$$\chi(F_t) + \chi(F'_u) - C.C' - C.C'.1 = \chi(F_t) + \chi(F_u) - 2C.C'.$$

The result follows. □

Corollary 6.5.2 *A curve singularity with $\mu = 1$ is the union of two smooth branches meeting transversely.*

Proof By Proposition 6.3.2, $\mu(C)$ is even if C is a single branch. Thus any curve with $\mu(C) = 1$ has at least 2 branches. Since $\mu(C \cup C') > \mu(C)$, each branch must have $\mu = 0$, hence be smooth. If B and B' are smooth and $B.B' = k$, $\mu(B \cup B') = 2k - 1$, so we arrive at $\mu = 1$ only if $k = 1$; and for more branches, μ must be greater. □

We next work towards an important formulae for μ as an intersection number.

A 1-parameter deformation $\{f_t\}$ of f is called a *morsification* if for (small) $t \neq 0$ all singularities of f_t are non-degenerate. The following picture illustrates (in the real case) the curves $f_t(x, y) = y^2 - (x^2 - t)^2 = 0$ for $t = -1, 0, 1$. For $t = 0$, f_t has an A_3 singularity at O; for $t = 1$ the curve has two A_1 singularities and f_1 also has a non-degenerate maximum at O (for $t = -1$ two of the singularities do not appear in the real picture).

We next show that any germ with isolated singularity admits a morsification.

Lemma 6.5.3 *The deformation $f + tx$ is a morsification of f* unless, *for some component C of the curve $\partial f/\partial y = 0$, $H(f)$ vanishes along C.*

Proof If it is not a morsification, there exist a sequence of numbers $t_i \to 0$ and a sequence of points $(x_i, y_i) \to (0, 0)$ such that (x_i, y_i) is a degenerate critical point of $f + t_i x$. Since it is a critical point, $0 = \partial(f + t_i x)/\partial y = \partial f/\partial y$. Passing to a subsequence if necessary, we may suppose that all these points lie on the same branch C of $\partial f/\partial y = 0$ at $(0, 0)$. Since these critical points are degenerate, it follows from Lemma 6.3.4 that $H(f)$ vanishes at them all. If the restriction of $H(f)$ to C did not vanish identically, it would have some order k in terms of a parameter on C, so there would be some neighbourhood of O on which it did not vanish (except at O itself). □

Non-trivial functions with vanishing Hessians do exist: see Exercise 6.7.12.

Proposition 6.5.4 *Any function with an isolated singularity admits a morsification.*

Proof Let f be the function: first suppose $H(f)$ does not vanish identically. Then not both $\partial f/\partial x$ and $\partial f/\partial y$ can vanish everywhere along the curve $H(f) = 0$ (else the singular point would not be isolated). Since $H(f) = 0$ has only finitely many branches, we can find constants α, β such that $\alpha\partial f/\partial x + \beta\partial f/\partial y$ vanishes along none of them. Then $f + t(\beta x - \alpha y)$ is a morsification.

If $H(f)$ vanishes identically, consider $g_t(x, y) = (1 + tx)f(x, y)$. A simple calculation yields

$$\begin{aligned} H(g_t) &= (1+tx)^2 H(f) + 2t(1+tx)\left(\frac{\partial f}{\partial x}\frac{\partial^2 f}{\partial y^2} - \frac{\partial f}{\partial y}\frac{\partial^2 f}{\partial x \partial y}\right) \\ &\quad - t^2\left(\frac{\partial f}{\partial y}\right)^2 . \end{aligned}$$

For this to vanish identically for all small t one needs, in particular, $\partial f/\partial y \equiv 0$. But in this case, $f(x, y) = \phi(x)$ depends only on x; since f has a critical point, $\phi'(0) = 0$; but now the critical point of f is non-isolated, contradicting our hypothesis.

Thus we have a morsification of g_{t_0} by adding a linear function $t\lambda$; combining these two facts gives a morsification of f, for example, $f(x, y) + t(1 + t_0 x)^{-1}\lambda$. □

Lemma 6.5.5 *Choose discs B_ϵ, D_η defining the Milnor fibration of f; let f_t be a 1-parameter deformation of $f_0 = f$. Then for small t, the sum of the Milnor numbers of f_t at points inside B_ϵ is equal to $\mu(f)$.*

Proof For a suitable small value of t we can regard f_t as defining a singular fibration with the same total space X as that of f. By Theorem 6.4.2, we have $\chi(X) = \chi(F)\chi(S) + \sum_P \mu_P$, where F denotes a non-singular fibre, and the sum is extended over all singular points P. Thus if we have two distinct fibrations with thc same base and homeomorphic fibres and total spaces, the values of $\sum_P \mu_P$ in both cases are equal. □

Theorem 6.5.6 *The value of μ is equal to the local intersection number of the polar curves $C_x : \partial f/\partial x = 0$ and $C_y : \partial f/\partial y = 0$.*

Proof Let f_t be a morsification of f. For small enough t, the link defined by the intersection of S_ϵ with the curves C_x and C_y is isotopic to the corresponding link obtained using f_t. Hence the corresponding linking numbers are the same. Thus the sum of the intersection numbers of C_x and C_y at points in B_ϵ is equal to the corresponding number obtained using f_t.

A point of intersection of C_x and C_y is a singular point of f. There is just one such point in B_ϵ: the origin. For f_t, however, we have μ such points. Each of these points has type A_1, or equivalently the function has non-degenerate Hessian. Thus its first partial derivatives give independent linear forms. It follows that at such a point, $C_{t,x}$ and $C_{t,y}$ are smooth and intersect transversely, so the intersection number is 1. The total intersection number is thus μ.

The theorem follows from these equalities. □

It follows from Theorems 6.5.4 and 6.5.6 that we can break up any isolated singularity into μ distinct ones, each with $\mu = 1$: μ is in this sense its multiplicity as a singularity.

Example 6.5.1 Let $f = x^a + y^b$. Then $\partial f/\partial x = ax^{a-1}$, so this polar curve consists of the y-axis, counted with multiplicity $a-1$; and similarly for $\partial f/\partial y$. It thus follows from Theorem 6.5.6 that $\mu(f) = (a-1)(b-1)$. The deformation $f_t(x, y) = x^a + y^b + t(x + y)$ is a morsification: the critical points of f_t are the $(a-1)(b-1)$ points where $0 = ax^{a-1} + t = by^{b-1} + t$.

In particular, taking $a = 2$, $b = k + 1$ we see that if f has type A_k, then $\mu(f) = k$. If C has type A_{2r-1}, then $C = B \cup B'$, where B and B' are smooth and $B.B' = r$. Thus by Theorem 6.5.1, $\mu(C) = \mu(B) + \mu(B') + 2B.B' - 1 = 0 + 0 + 2r - 1$, giving the same result.

Using Theorem 6.5.6, we may analyse the effect on μ of blowing-up, and hence obtain a formula for μ in terms of the tree of infinitely near points. First, we need a lemma.

Lemma 6.5.7 *Let f be a holomorphic function at O; let B be a branch of the curve defined by $\partial f/\partial x = 0$. Then the order of $pf + qy\frac{\partial f}{\partial y}$ along B is independent of p and q provided $p \geq 0, q \geq 0$ and $(p, q) \neq (0, 0)$.*

Proof Choose a (good) parametrisation of B, and let the least order terms in the power series expansions of y and $\partial f/\partial y$ (along B) in terms of the parameter be

$$y = at^\alpha + \cdots, \qquad \frac{\partial f}{\partial y} = bt^\beta + \cdots,$$

so that $a, b \neq 0$. Then along B we have

$$f = \int_0^t \frac{\partial f}{\partial t} dt = \int_0^t \frac{\partial f}{\partial y}\frac{\partial y}{\partial t} dt,$$

since $\partial f/\partial x$ vanishes along B. The lowest order term in the integrand is thus $ab\alpha t^{\alpha+\beta-1}$, so the lowest order term in f is $ab\frac{\alpha}{\alpha+\beta}t^{\alpha+\beta}$, and the lowest term in $pf + qy\partial f/\partial y$ is $ab\frac{p\alpha+q(\alpha+\beta)}{\alpha+\beta}t^{\alpha+\beta}$. Here, a and b are non-zero by hypothesis, and since α, β are positive integers the hypothesis shows that the coefficient is non-zero. Thus the order is $\alpha + \beta$ irrespective of the values of p, q. □

Theorem 6.5.8 *Suppose C has a single tangent and multiplicity m. Then the Milnor number of the strict transform is equal to*

$$\mu(C) - m(m-1).$$

Proof Consider the blow-up $(x, y) = (x_1y_1, y_1)$; if C has equation $f(x, y) = 0$, we write $f(x, y) = y_1^m f_1(x_1, y_1)$, so $C^{(1)}$ is given by $f_1(x_1, y_1) = 0$; the exceptional curve E is $y_1 = 0$. Differentiating with respect to x_1 and y_1 leads to

$$y_1\partial f/\partial x = \partial f/\partial x_1 = y_1^m \partial f_1/\partial x_1; \tag{6.1}$$

$$x_1\partial f/\partial x + \partial f/\partial y = \partial f/\partial y_1 = my_1^{m-1}f_1 + y_1^m\partial f_1/\partial y_1. \tag{6.2}$$

In particular, the strict transform of the polar curve $\partial f/\partial x = 0$ is given by $\partial f_1/\partial x_1 = 0$. We are concerned with intersection numbers of these curves with several others. It suffices to deal with each component separately (we recall from Section 4.5 that in general there are several components); choose a component B of $\partial f/\partial x = 0$, and parametrise it as $(\phi(t), \psi(t))$, and the corresponding component of $\partial f_1/\partial x_1 = 0$ as $(\frac{\phi(t)}{\psi(t)}, \psi(t))$. Then we can calculate intersection numbers as orders in t.

Applying Theorem 6.5.6 to C and its strict transform $C^{(1)}$ gives

$$\mu(C) = \sum_B \text{ord}_t(\partial f/\partial y), \quad \mu(C^{(1)}) = \sum_B \text{ord}_t(\partial f_1/\partial y_1).$$

From (6.2), since $\partial f/\partial x$ vanishes along B,

$$\text{ord}_t(\partial f/\partial y) = \text{ord}_t(\partial f/\partial y_1) = \text{ord}_t(my_1^{m-1}f_1 + y_1^m\partial f_1/\partial y_1).$$

It follows from Lemma 6.5.7 that

$$\text{ord}_t(mf_1 + y_1\partial f_1/\partial y_1) = \text{ord}_t(y_1\partial f_1/\partial y_1).$$

Collecting these results, we have

$$\mu(C) - \mu(C^{(1)}) = m \sum_B \operatorname{ord}_t(y_1) = m \sum_B B_1.E$$

is m times the intersection number of E with the strict transform of the polar curve $\partial f/\partial x = 0$. By Lemma 3.4.2, this intersection number is equal to the multiplicity of the polar curve, namely $m-1$. The conclusion follows. □

An alternative proof of this result for the case when C is a single branch was sketched in Exercise 4.7.12.

Theorem 6.5.9 *The Milnor number of a curve-germ C is given by*

$$\mu = \sum_P m_P(m_P - 1) - r + 1,$$

where r is the number of branches of C and the sum is extended over infinitely near points P where the multiplicity m_P of C is at least 2.

Proof If follows by induction from Theorem 6.5.8 that the result holds for the case of a single branch. Suppose that C is the union of B and B', with multiplicities m_P, m'_P, at infinitely near points P, and with s, s' branches respectively, and that the result holds for each of these. Then

$$\mu(B \cup B') = \mu(B) + \mu(B') + 2B.B' - 1,$$

by Theorem 6.5.1. By hypothesis,

$$\mu(B) = \sum_P m_P(m_P - 1) - s + 1,$$

and similarly for B', and by Lemma 4.4.2,

$$B.B' = \sum_P m_P n'_P.$$

Substituting these results in the right hand side of the equation gives

$$\begin{aligned}\mu(B \cup B') &= \sum_P \{m_P(m_P - 1) + n'_P(n'_P - 1) + 2m_P n'_P\} \\ &\qquad - s + 1 - s' + 1 - 1 \\ &= \sum_P (m_P + n'_P)(m_P + n'_P - 1) - (s + s') + 1,\end{aligned}$$

so the result holds also for $B \cup B'$. It thus follows in general by induction on the number of branches. □

Remark The *double point number* $\delta(C)$ of a curve singularity is sometimes used in place of μ. We met it for a single branch in Section 4.3, and saw in Proposition 6.3.2 that we then have $\mu(B) = 2\delta(B)$. In general we can define $\delta := \frac{1}{2}(\mu + r - 1)$, so by Theorem 6.5.1, $\delta(C \cup C') = \delta(C) + \delta(C') + 2C.C'$. Then Theorem 6.5.9 may be written as $\delta = \sum_P \binom{m_P}{2}$, where P runs through infinitely near points.

6.6 Notes

The fibration named in the title is due to Milnor (see [132]). This book gives a very clear account, and has been extremely influential. Although Milnor establishes his main results in arbitrary dimension, he also includes a chapter devoted to some special features of the case of curve singularities. The notation μ for the rank of the first Betti number of F is also taken from [132]. Our proofs of the fibration theorems follow Milnor's fairly closely, but our presentation diverges from his from that point on. However, most of the results in the chapter originate in Milnor's book. The rider Exercise 6.7.1 is taken from a paper of A'Campo [7].

The proof of Theorem 6.2.1 uses little more than the fact that f is a submersion, and the theorem generalises to give a fibration for any analytic function on a complex analytic set. Milnor's treatment of Theorem 6.2.2 applies for a function with an isolated singular point on a complex manifold of any dimension n, and he establishes (what is almost trivial when $n = 2$) that the fibre is homotopy equivalent to a bouquet of spheres S^{n-1}. The only formula for the number μ of such spheres that generalises to this case is Theorem 6.5.6 (the proof also generalises). However, further generalisations are possible, and perhaps the most general decomposition of a Milnor fibre as a bouquet (up to homotopy) is due to Tibăr [177].

A more detailed study of the Milnor fibration will be given in Chapter 10.

The importance of the Milnor number as invariant is illustrated by the fact that if $\{f_t(x, y) \mid 0 \leq t \leq 1\}$ is a family of functions, continuous in t, with $\mu(f_t)$ constant, then f_0 and f_1 are equisingular.

We have given a more extensive development than is usual of the Euler characteristic: this is partly also to anticipate Section 7.3. This allows us to set the Riemann-Hurwitz formula as Exercise 6.7.5.

The term morsification is in honour of Marston Morse [135], who was the first to use functions all of whose singularities are non-degenerate to define a decomposition of a manifold from which information about its topology (the 'Morse inequalities') can be obtained.

A surprising consequence of Theorem 6.5.6 was observed by Bernard Scott: since the intersection number of the polar curves $\partial f/\partial x = 0$ and $\partial f/\partial y = 0$ is so large, these curves are sometimes forced to have infinitesimally near points in common with each other which do not lie on the curve $f = 0$. Perhaps the first example is when $f(x, y) = y^3 + x^8$. Resolving C gives multiplicity sequence 3,3,2,1,1: thus O_3 is proximate to O_1 and O_4 to O_2. The polar curves P are of the form $0 = ay^2 + bx^7$ (plus higher terms), so have multiplicity sequence 2,2,2,1,1; O_4 is proximate to O_2. Since $C.P = 16$, the infinitely near points in common are just O_0, O_1 and O_2. But the intersection number of two polar curves is $\mu(C) = 14$; so any two polar curves have all the points $O_0, \ldots, O_4$ in common.

An important result, due independently to A'Campo [6] and Gusein-Zade [82], asserts that every plane curve singularity is equisingular to one defined over $\mathbb{R}$ and admitting a real morsification f_t with only 3 critical values – say $f_t(x, y) = a_t$ if $(x, y) \in \mathbb{R}^2$ is a local minimum of f_t, and takes the value b_t for all saddle type critical points and the value c_t for all local maxima. This can be used for direct geometrical construction of a model of the Milnor fibre. We will give an alternative direct construction in Chapter 9.

In finite characteristic, though there is no fibration in the topological sense, the techniques of algebraic geometry give a construction of a sheaf of vanishing cycles, and one may compare this with ℓ-adic or étale cohomology of the fibres. The conclusions are considerably more complicated here than in characteristic zero.

For a curve C we can define $\delta(C)$ as the sum $\sum_P \binom{m_P}{2}$, where P runs through infinitely near points. For a branch, this equals the number of gaps in the semigroup; for a union, we have $\delta(C \cup C') = \delta(C) + \delta(C') + C.C'$. We can now define $\mu(C) = 2\delta(C) + 1 - r$, where r is the number of branches of C.

For a function $f(x, y)$ we can define $\mu(f) = \dim_k \frac{k[x,y]}{J(f)}$, where $J(f)$ denotes the Jacobian ideal $\langle \frac{\partial f}{\partial x}, \frac{\partial f}{\partial y} \rangle$. This is finite only if $(0, 0)$ is isolated as a singularity of f.

However, an isolated singularity of a curve need not be isolated as a singularity of its defining equation. Example (all our examples are in characteristic 2): $f_1(x, y) = x^2 + y^3$: the singular locus of f_1 is the line $y = 0$, which meets each level set $f_1 = c$ in a single point, which is the only singular point on that curve.

Further, if $f = 0$ is an equation of C we need not have $\mu(f) = \mu(C)$. Examples: $f_1 = x^2 + y^3$, $f_2 = (1 + x)(x^2 + y^3)$, $f_3 = (1 + x^3)(x^2 + y^3)$ all define the same curve in a neighbourhood of the origin, but $\mu(f_1) = \infty$, $\mu(f_2) = 4$, $\mu(f_3) = 8$.

It is shown in [49] that if f does have an isolated singularity C, then $\mu(f)$ is the sum of the number $\mu(C)$ of vanishing cycles and a 'wild dimension' coming from the Swan character of the monodromy representation on the étale cohomology of the fibre.

6.7 Exercises

Exercise 6.7.1 Write down explicitly the conditions on the vector field $\xi = a_1\partial/\partial x_1 + b_1\partial/\partial y_1 + a_2\partial/\partial x_2 + b_2\partial/\partial y_2$, where $z_1 = x_1 + iy_1$, $z_2 = x_2 + iy_2$, and a_1 etc. are functions of (x_1, y_1, x_2, y_2), required in the proof of the fibration theorem for $\eta^{-1}f : B_\epsilon \cap f^{-1}(S_\eta) \to S^1$.

Show that if f is given by a power series with real coefficients, then if ξ is as above, so is ξ^* defined by $\xi^*(z_1, z_2) = -\overline{\xi(\overline{z_1}, \overline{z_2})}$. Deduce that $\Xi := \frac{1}{2}(\xi + \xi^*)$ also does.

Show that the monodromy map $h : F \to F$ (where F is the fibre over $1 \in S^1$) given by integrating Ξ satisfies $h^{-1}(z_1, z_2) = \overline{h(\overline{z_1}, \overline{z_2})}$. Deduce that h is conjugate to h^{-1} in the diffeomorphism group of F.

Exercise 6.7.2 Show that, for C irreducible, μ is given in terms of the Puiseux characteristic by $\mu = \sum_{q=1}^{g}(\beta_q - 1)(e_{q-1} - e_q)$.

Exercise 6.7.3 Calculate μ for the curves B_q^- defined in the proof of Lemma 4.3.1 in terms of the Puiseux characteristic of B.

Exercise 6.7.4 Find the singular points of $f(x, y) = x^3 + y^3 + 2axy + bx + cy$ if (a, b, c) is any of $(0, 1, 1)$, $(1, 0, 0)$, $(1, -1, -1)$ or $(3, 3, 3)$. Verify in each case that the sum of the Milnor numbers is 4.

Exercise 6.7.5 Use the arguments in the proof of Theorem 6.4.1 to establish the Riemann-Hurwitz formula, that if $\pi : S \to T$ is a holomorphic map of smooth closed algebraic curves, of degree d (i.e. such that for a general point $P \in T$, $\#\pi^{-1}(P) = d$), we have $\chi(S) = d\chi(T) - \sum_{P \in \Delta(\pi)}(d - r_P)$. Here $\Delta(\pi)$ denotes the image in T of the set of points at which the derivative of π vanishes, and r_P denotes $\#\pi^{-1}(P)$.

Exercise 6.7.6 Let C be the curve $x^d + y^d + z^d = 0$ in $P^2(\mathbb{C})$. Define $\pi : C \to P^1(\mathbb{C})$ by $\pi(x : y : z) = (x : y)$. Calculate $\chi(C)$ by using the formula in Exercise 6.7.5.

Exercise 6.7.7 Show that $\mu(D_k) = k$ for $k \geq 4$ and that $\mu(E_k) = k$ for $k = 6$, 7 and 8.

Exercise 6.7.8 Calculate μ for the curve given by $x^5 - x^2y^2 + y^5 = 0$.

Exercise 6.7.9 Calculate μ for curves with Puiseux characteristics (i) $(4; 10, 13)$, (ii) $(6; 8, 13)$ and (iii) $(6; 15, 22)$; first using Proposition 6.3.2, and then using Theorem 6.5.9.

Exercise 6.7.10 Show that any singularity C with multiplicity m has $\mu \geq (m-1)^2$. Deduce that if $\mu(C) = 2$, C has type A_2, and if $\mu(C) = 3$ then C has type A_3.

Exercise 6.7.11 Show that any singularity with $\mu < 9$ is simple. (Hint: use the preceding Exercise, and for the cases $m(C) = 3$ use the result of Exercise 4.7.3).

Exercise 6.7.12 Show that the following functions have identically vanishing Hessians: $y + \phi(x)$, $\sqrt{(a+bx)(c+dy)}$ and $(px + qy + r)^s$.

Exercise 6.7.13 Calculate μ for the curves listed in Exercise 4.7.3.

Exercise 6.7.14 Suppose the Newton polygon of f consists of a single edge, from $(a, 0)$ to $(0, b)$, that the highest common factor of a and b is h, and the polynomial $\phi(T)$ defined in the proof of Theorem 2.1.1 has h distinct roots.

Show that $f = 0$ has h branches, that each branch is equisingular to $x^{a/h} = y^{b/h}$, and that the intersection number of any two branches is ab/h^2. Hence calculate $\mu(f)$.

Exercise 6.7.15 For $f \in \mathbb{C}\{x, y\}$, the Newton polygon is called *commode* if it has a vertex on each coordinate axis (i.e. f is divisible by neither x nor y), and f is *Newton non-degenerate*, or NPND for short, if it is commode and, for each edge of the polygon, the corresponding polynomial $\phi(T)$ defined in the proof of Theorem 2.1.1 has distinct roots.

Suppose f is NPND. Let $f = \prod_{i=1}^{k} f_i$ be the factorisation of Lemma 2.4.4 corresponding to the edges of the Newton polygon. Using the calculation of Exercise 6.7.14 for $\mu(f_i)$, and the calculation of Exercise 4.7.6 for the intersection numbers, obtain a formula for $\mu(f)$.

Show that your answer can be reduced to the form $2A - L + 1$, where A is the area of the region below the Newton polygon and L the sum of the lengths cut by it on the axes.

Exercise 6.7.16 Suppose f is NPND. Show that the number r of branches of $f(x, y) = 0$ is one less than the number of points with integer coordinates (*lattice points*) lying on the Newton polygon.

Show that if N_a denotes the number of lattice points in the interior of the region bounded by the Newton polygon and the coordinate axes, and N_e the number of lattice points lying on the polygon (but not on the axes), $\mu(f) = 2N_a + N_e$ and $\delta(f) = N_a + N_e$.

Exercise 6.7.17 Show that for any singularity with multiplicity 3, there exist coordinates with respect to which it is NPND. (Hint: use the ideas in the proof of Theorem 2.2.7).

List the possible Newton polygons, and relate them to the types of triple points as listed in Exercise 4.7.3.

7

Projective curves and their duals

For curves in the projective plane, the most basic invariant is the degree of the defining equation. This gives a qualitative bound for the possible complexities of the curve, and also of its singularities. We will see that this may be turned into precise quantitative bounds.

First, however, we discuss two classical topics. The first is a formula for the genus of a curve in terms of its degree and its singularities. The second is a corresponding formula for its *class*: the number of tangents to the curve from a generic point. This can be reinterpreted. The tangents to a curve are lines in the plane, which can also be interpreted as points in a dual plane. The locus of these points, or rather its closure, is the dual curve, and the formulae yield relations between the original curve and its dual. The oldest version of these formulae consists of the celebrated Plücker equations.

Our proofs of these results use the technique involving the Euler characteristic which we developed in Theorems 6.4.1 and 6.4.2. A further refinement of this technique consists of the calculus of so-called constructible functions. We describe this, and give a further application, to Klein's formula concerning the singularities of a curve in the real projective plane.

These formulae do not exhaust the relations between the singularities of a curve and those of its dual, and we give a full discussion of the relation between the two.

We conclude with a survey of some of the known results concerning the possible configurations of singularities of curves of a given degree.

7.1 The genus of a singular curve

Let C be a reduced curve of degree d in the projective plane $P^2(\mathbb{C})$. The topology of C is determined by the singularities and the Euler characteristic. We begin by calculating $\chi(C)$, and deduce the genus of the curve obtained by resolving the singularities of C.

Since we are now studying curves C rather than just germs at a point, we introduce the notations $m_P(C)$ for the multiplicity of (the germ of) C at the point $P \in C$, $\mu_P(C)$ for its Milnor number, $r_P(C)$ for the number of branches and $\delta_P(C)$ for the double point number. We omit C from the notation if it is clear from the context. We also write $\mu_{\text{tot}}(C)$ for the sum $\sum_{P \in \text{Sing } C} \mu_P$ over the set Sing C of all singular points of C.

Theorem 7.1.1 *If C is a reduced curve of degree d in $P^2(\mathbb{C})$, then*

$$\chi(C) = 3d - d^2 + \mu_{\text{tot}}(C).$$

Proof We first show how this situation can be adapted to the context of Theorem 6.4.1. Fix a curve C, and choose a line L meeting C in d distinct points (such a line exists since C is assumed reduced). Take projective coordinates $(x : y : z)$ with L given by $z = 0$. Substitute $z = 1$ in the equation $\tilde{f}(x, y, z) = 0$ for C to obtain $f(x, y) = \tilde{f}(x, y, 1)$. The equation $f(x, y) = 0$ in the affine plane $\mathbb{C}^2$ determines a curve C', the intersection of C with $\mathbb{C}^2$.

The curve C' is non-compact, so we intersect it by a disc B_R of large radius R with centre O. We claim that for R large enough the curve C is transverse to the boundary sphere S_R. Indeed, by construction C is transverse to L, and S_R is the boundary of a tubular neighbourhood of L.

We consider the family $\{f_t\}$ obtained by varying all the coefficients in f: however, we shall shortly restrict to variations close to f itself. This gives curves $C_t \subset P^2(\mathbb{C})$ and $C'_t \subset \mathbb{C}^2$. For t small enough, all these are also transverse to S_R. Thus if we fix (as usual) a small disc $D_\eta \subset \mathbb{C}$, we have a family of maps giving fibrations over D_η.

These fibrations have fibres $C'_t \cap D_R$, but if we extend the fibres to C_t we still have a fibration, and there are no further singular points. This follows from Theorem 6.1.2: all we have done is to introduce a device to get round the fact that we do not have actual maps from $P^2(\mathbb{C})$ to $\mathbb{C}$ (what we are considering are, in fact, zeros of cross-sections of a bundle over $P^2(\mathbb{C})$ with fibre $\mathbb{C}$).

We can apply Theorem 6.4.1 to deduce that the numbers $\chi(C_t) - \mu_{\text{tot}}(C_t)$ do not depend on t. Although up to here we have only dealt with small deformations t, if we allow arbitrary deformations but exclude singular curves, the same arguments show that we have an actual fibration: all non-singular curves that meet L in d distinct points are homeomorphic. Since each singular curve C has arbitrarily close non-singular curves, we see that this constancy holds in general.

To obtain the result it remains to evaluate the constant value of $\chi(C) - \mu_{\text{tot}}(C)$. To do this, take C to be the union of d lines in general position. Each line is a copy of $P^1(\mathbb{C})$, hence is homeomorphic to S^2, with Euler characteristic 2. Any two lines meet just once: there are $\frac{1}{2}d(d-1)$ such intersections, all at distinct points, and all transverse so that by Lemma 6.3.3 each has $\mu = 1$. Thus $\chi(C) = 2d - \frac{1}{2}d(d-1)$, and $\chi(C) - \mu_{\text{tot}}(C) = 2d - \frac{1}{2}d(d-1) - \frac{1}{2}d(d-1) = 3d - d^2$. This completes the proof. □

An alternative to the final paragraph of the proof is to take C to be the curve $x^d + y^d + z^d = 0$ and use the calculation of $\chi(C)$ in Exercise 6.7.6.

Corollary 7.1.2 *A non-singular curve of degree d is connected, and has genus* $\frac{1}{2}(d-1)(d-2)$.

Proof Such a curve cannot have several components, else these would have positive degree, and would intersect each other, making singularities. It is thus a connected, compact orientable surface, of some genus g; then it has Euler characteristic $2 - 2g$. The result follows. □

For our second corollary, recall that we have defined the normalisation $\pi : \widetilde{C} \to C$ by resolving the singularities of C. As far as the topology is concerned, the map π is bijective except at singular points, and over a singular point of C with r branches, there are just r points of $\widetilde{C}$. Hence

Corollary 7.1.3 *The normalisation* $\widetilde{C}$ *of a reduced plane curve* C *of degree* d *has Euler characteristic* $3d - d^2 + \sum_{P \in \operatorname{Sing} C}(\mu_P + r_P - 1)$. *In particular, if* C *is irreducible,* $\widetilde{C}$ *is connected, of genus*

$$g = \tfrac{1}{2}(d-1)(d-2) - \sum_{P \in \operatorname{Sing} C} \tfrac{1}{2}(\mu_P + r_P - 1).$$

This formula may be written as $g = \binom{d-1}{2} - \sum_P \delta_P$, where the sum is extended over singular points P of C. It may also be combined with Theorem 6.5.9 to give $g = \binom{d-1}{2} - \sum_Q \binom{n_Q}{2}$, where Q runs through all the infinitely near points to all the (singular) points P of C.

We may use Corollary 7.1.3 to list possible configurations of singularities on curves of given degree. There are relatively few equisingularity classes of singularities with $\frac{1}{2}(\mu + r - 1) \leq 4$ and, according to Exercise 7.7.1 these are as follows:

$\frac{1}{2}(\mu + r - 1)$	Cases
1	A_1, A_2
2	A_3, A_4
3	A_5, A_6, D_4, D_5, E_6
4	$A_7, A_8, D_6, D_7, E_7, E_8$

For example, if C is an irreducible curve of degree 4, it follows from Corollary 7.1.3 that $\sum_P \frac{1}{2}(\mu_P + r_P - 1) \leq 3$. In this case, any collection of singularities satisfying the inequality appears as the set of singularities of some irreducible quartic curve (see Exercise 7.7.5). However, the corresponding statement fails to be true for curves of higher degrees: see e.g. Example 7.7.2.

Corollary 7.1.4 *If C, C' are curves of degrees d, d' with no common component,*

$$\mu_{\text{tot}}(C \cup C') = \mu_{\text{tot}}(C) + \mu_{\text{tot}}(C') - \#(C \cap C') + 2dd'.$$

This follows at once by applying the theorem to each of C, C' and $C \cup C'$.

7.2 The degree of the dual curve

Let C be an algebraic curve in $P^2(\mathbb{C})$. At each non-singular point $P \in C$ there is a tangent line L_P to C, which determines a point (which we denote by the same symbol) in the dual projective space $P^2(\mathbb{C})^\vee$. Consider the locus of these points as P varies: its closure is another algebraic curve $C^\vee$, called the *dual curve* of C. It will be convenient to regard C as defined by a reduced equation $f(x, y, z) = 0$, and also to suppose that no component of C is a straight line (for the dual of a line is just a point, not another curve). In this section we will calculate the degree of $C^\vee$. This leads to formulae relating the singularities of C and $C^\vee$.

This degree is the number of points at which $C^\vee$ meets a general line in $P^2(\mathbb{C})^\vee$. This line is dual to a point P_0 in $P^2(\mathbb{C})$, and we must count the number of tangents to C which pass through P_0. Let P_0 have coordinates $(x_0 : y_0 : z_0)$. The equation of the tangent at a smooth point P of C is $(\partial f/\partial x)_P x + (\partial f/\partial y)_P y + (\partial f/\partial z)_P z = 0$, so the condition for this tangent to pass through P_0 is $(\partial f/\partial x)_P x_0 + (\partial f/\partial y)_P y_0 + (\partial f/\partial z)_P z_0 = 0$, i.e. that P lies on the polar curve

$(\partial f/\partial x)x_0 + (\partial f/\partial y)y_0 + (\partial f/\partial z)z_0 = 0$ of P_0 with respect to C, which we denote for now by C_0. We have shown that the feet of the tangents from P_0 to C are the intersection points of C with C_0.

If C has degree d, then its polar curves all have degree $d-1$. By Bézout's theorem, the total intersection number of C and C_0 is $d(d-1)$, so if all these intersections are transverse there are $d(d-1)$ intersections, and hence tangents from P_0 to C, so the degree of $C^\vee$ is also equal to $d(d-1)$. To obtain a general formula, we must first consider possible non-transverse intersections at smooth points of C and then determine the effect of the singular points.

Take coordinates with P_0 at $(1,0,0)$ and the intersection point P_1 of C and C_0 at $(0,1,0)$; we may then use affine coordinates $z=1$ so that P_0 is at infinity on the x-axis, P_1 is at the origin and C_0 is given by $\partial f/\partial x = 0$. Write $f(x,y,1) = \sum_{r+s\le d} a_{r,s}x^r y^s$; then since C and C_0 pass through the origin, $a_{0,0} = a_{1,0} = 0$. Since we are supposing P_1 a smooth point of C, $a_{0,1} \neq 0$, so C has tangent $y=0$ at this point. Then C_0 fails to be transverse to it there if and only if it, too, has tangent $y=0$, hence if $a_{2,0}=0$. This is the condition that $y=0$ is an inflexional tangent (the intersection number is ≥ 3). Now – providing no component of C is a straight line – C has only a finite number of inflexional tangents: as we saw in Section 1.5, they are the tangents at the points of intersection of C with its Hessian curve. We can thus avoid failure of transversality by choosing P_0 to lie on none of these tangents.

Now consider the case when C is singular at P_1. Taking coordinates as before, we have

Lemma 7.2.1 *The intersection number of* $C : f=0$ *and* $C_0 : \partial f/\partial x = 0$ *at* O *is equal to the sum of the intersection numbers of* C_0 *with* $\partial f/\partial y = 0$ *and with* $y=0$.

Proof It follows from Lemma 6.5.7 that if B is a branch of C_0 at O, then the order of f along B is the sum of the orders of y and of $\partial f/\partial y$, so that the intersection number with B of C is the sum of the intersection numbers of $y=0$ and $\partial f/\partial y = 0$. The result thus follows by summing over branches B. □

The intersection number of $\partial f/\partial x = 0$ and $\partial f/\partial y = 0$ at P_1 is μ_{P_1}, by Theorem 6.5.6.

Write m for $m_{P_1}(C)$. Then $a_{m,0}=0$ if and only if $y=0$ is tangent to C at P_1; also if and only if $y=0$ is tangent to C_0 at P_1. Thus if

$a_{m,0} \neq 0$, the intersection number of C_0 with $y = 0$ at P_1 is equal to $m - 1$.

We thus insist that P_0 is chosen so that, as well as not lying on any inflexional tangent of C, it is also not on the tangent to any branch at a singular point of C. Then each singular point P accounts for an intersection multiplicity $\mu_P + m_P - 1$. The remaining intersections of C and C_0 each have multiplicity 1 and represent points of the dual curve. This completes the proof of

Theorem 7.2.2 *The degree of $C^\vee$ is given by*

$$d^\vee = d(d-1) - \sum_{P \in \mathrm{Sing}C} (\mu_P + m_P - 1).$$

Observe that direct elimination always gives an equation of degree $d(d-1)$ in the coordinates of a line, expressing the condition that the line meets C in two coincident points. The above argument shows that this equation factorises as a product of the equation of the dual curve and, for each singular point P of C, the equation of the line in $P^2(\mathbb{C})^\vee$ representing lines passing through $P \in P^2(\mathbb{C})$ raised to the $(\mu_P + m_P - 1)$th power.

We note the similarity of this result to Corollary 7.1.3. Combining the two we obtain

Corollary 7.2.3 *If C is an irreducible curve in $P^2(\mathbb{C})$ with degree d, genus g and class $d^\vee$, then*

$$d^\vee - 2g = 2(d-1) - \sum_{P \in \mathrm{Sing}C} (m_P(C) - r_P(C)).$$

Example 7.2.1 We have already determined the list of possible configurations of singularities of an irreducible curve C of degree 4 and genus 0. We can now give the degree of the dual in each case:

Class	Singularities
3	$3A_2$
4	$A_1 + 2A_2,\ A_2 + A_4,\ E_6$
5	$2A_1 + A_2,\ A_1 + A_4,\ A_2 + A_3,\ A_6,\ D_5$
6	$3A_1,\ A_1 + A_3,\ A_5,\ D_4$

Now suppose the curve C has 'ordinary singularities' only, which means in our present notation that both C and $C^\vee$ admit only singularities of types A_1 (ordinary nodes) and A_2 (ordinary cusps). Then if C has degree d and δ nodes and κ cusps, and genus g, while the corresponding numbers for $C^\vee$ are $d^\vee$, $\delta^\vee$ and $\kappa^\vee$ (the genus is necessarily the same), applying Corollary 7.1.3 and Theorem 7.2.2 to $C^\vee$ as well as to C, gives:

$$2g = (d-1)(d-2) - 2\delta - 2\kappa = (d^\vee - 1)(d^\vee - 2) - 2\delta^\vee - 2\kappa^\vee, \quad (7.1)$$

$$d^\vee = d(d-1) - 2\delta - 3\kappa, \quad d = d^\vee(d^\vee - 1) - 2\delta^\vee - 3\kappa^\vee. \quad (7.2)$$

Taking the second equality of (7.1) and subtracting twice the first and the second of equations (7.2) gives, after cancellation,

$$\kappa^\vee = 3d(d-2) - 6\delta - 8\kappa, \quad \kappa = 3d^\vee(d^\vee - 2) - 6\delta^\vee - 8\kappa^\vee. \quad (7.3)$$

Various other formulae may be obtained by combining these. For example,

$$d^\vee - 2g = 2(d-1) - \kappa, \quad d - 2g = 2(d^\vee - 1) - \kappa^\vee,$$

and so $\kappa - \kappa^\vee = 3(d - d^\vee)$.

Example 7.2.2 Let C be a nonsingular curve of degree 4; assume that its dual $C^\vee$ has ordinary singularities. Then we have $2g = 6$, $d^\vee = 12$, and hence

$$6 = 110 - 2\delta^\vee - 2\kappa^\vee, \quad 4 = 132 - 2\delta^\vee - 3\kappa^\vee.$$

Solving these equations gives $\delta^\vee = 28$, $\kappa^\vee = 24$.

These results may be formulated without explicit mention of the dual curve, since in the case of ordinary singularities a point of type A_1 on the dual gives a line in the original plane which is tangent to C at two distinct points: a *bitangent*, and conversely. Similarly a point of type A_2 on the dual corresponds to an inflexional tangent to C. Thus the formulae may be used to count bitangents and flexes: Example 7.2.2 shows that a general curve of degree 4 has 28 bitangents and 24 flexes. However, in the case of non-ordinary singularities further relations need to be taken into account, which we give in Section 7.4 below.

7.3 Constructible functions and Klein's equation

A useful extension of the technique of Euler characteristics is given by the calculus of constructible functions. We describe this fairly briefly, and show how it encapsulates some of the preceding arguments. Then we apply the technique to obtain a formula relating the singularities of a curve defined over $\mathbb{R}$ to those of its dual.

The term 'constructible' itself originally referred to subsets of $P^n(\mathbb{C})$ definable by polynomial equations and inequations. Working with real coefficients gives a much more flexible tool. We have already defined a subset of $\mathbb{R}^n$ to be semi-algebraic if it can be expressed as a finite union of sets of the form $\{f_i(z) = 0, g_j(z) > 0\}$, where all the f_i, g_j are polynomials. We can extend the definition to subsets of $P^n(\mathbb{R})$ or $P^n(\mathbb{C})$ (for example) either by using suitable embeddings of these in Euclidean space or (more simply) by taking sets whose intersection with each affine open set where some coordinate is $\neq 0$ is semi-algebraic in this sense. If X, Y are semi-algebraic sets, we say that a map $f : X \to Y$ is semi-algebraic if it is continuous and its graph is semi-algebraic: thus the pre-image of any semi-algebraic set is semi-algebraic.

If X is a semi-algebraic subset (which we usually suppose compact) of $\mathbb{R}^n$ (or $P^n(\mathbb{R})$, $P^n(\mathbb{C})$), a function ϕ on X is semi-algebraically *constructible* if

(i) it takes only finitely many values, all integers, and
(ii) for each $n \in \mathbb{Z}$, $\phi^{-1}(n)$ is semi-algebraic.

We write $\Omega(X)$ for the ring of constructible functions on X, where addition and multiplication are performed pointwise. Clearly each constructible function is an integer linear combination of characteristic functions 1_Z of constructible sets (where 1_Z is defined by $1_Z(x) = 1$ if $x \in Z$ and $1_Z(x) = 0$ if $x \notin Z$). Less trivially, if X is compact, each element of $\Omega(X)$ is an integer linear combination of characteristic functions 1_Z of constructible sets Z which may be supposed compact. For if Z is constructible, so is its closure $\overline{Z}$, and the difference $\overline{Z} \setminus Z$ has lower dimension, so the assertion follows by induction on dimension.

If Z is compact, $\chi(1_Z)$ is defined to be $\chi(Z)$: this is extended by linearity to define the Euler characteristic $\chi(\phi)$ for any constructible function ϕ. This may be written as $\chi(\phi) = \sum_n n\chi(\phi^{-1}(n))$. This is a special case (take Y to be a point) of a more general operation: if $f : X \to Y$ is a semi-algebraic map, we define $f_* : \Omega(X) \to \Omega(Y)$ by $f_*\psi(y) = \chi(\psi | f^{-1}(y))$. We also define $f^* : \Omega(Y) \to \Omega(X)$ by $f^*\phi = \phi \circ f$.

For the proof that these are well-defined, and for a full proof of the next result, see the references cited at the end of the chapter.

Theorem 7.3.1 *Let $f : X \to Y$ and $g : Y \to Z$ be semi-algebraic maps. Then*

(i) $(gf)^* = f^*g^*$,

(ii) $(gf)_* = g_*f_*$, *and in particular*

(iii) $\chi(f_*\phi) = \chi(\phi)$.

(iv) *(Pullback rule) Given* $\phi \in F_{PL}(Y')$, *and a pullback diagram*

$$\begin{array}{ccc} X' & \xrightarrow{f'} & Y' \\ \downarrow g' & & \downarrow g \\ X & \xrightarrow{f} & Y \end{array}$$

then $f^*g_*\phi = g'_*f'^*\phi$.

(v) *For* $f : X \to Y$ *and* $\phi \in F_{PL}(Y)$, $f_*f^*\phi = \phi.f_*1_X$.

Proof The assertion (i) is essentially trivial, since the map is defined by composition. However (ii) is definitely not trivial and is the main part of the Theorem. We will not even outline the proof here.

We turn to (iv), and begin by recalling that the *pullback* (X', g', f') of $X \xrightarrow{f} Y$ and $Y' \xrightarrow{g} Y$ is defined by letting X' be the subset of $X \times Y'$ consisting of the points (x, y) such that $f(x) = g(y)$, and g', f' its projections. For the proof it suffices to consider the case when $\phi = 1_Z$ for some compact subset Z of Y'. Evaluating at $x \in X$, where $f(x) = y$, we have

$$f^*g_*\phi(x) = g_*\phi(y) = \chi(g^{-1}(y) \cap Z),$$

$$g'_*f'^*\phi(x) = g'_*1_{f'^{-1}(Z)}(x) = \chi(g'^{-1}(x) \cap f'^{-1}(Z)).$$

But since the diagram is a pullback, $g^{-1}(y) \cap Z$ is homeomorphic to $g'^{-1}(x) \cap f'^{-1}(Z)$.

As to (v), we may suppose $\phi = 1_Z$ for some $Z \subseteq Y$. Then

$$f_*f^*\phi(x) = f_*(f^*1_Y)(x) = f_*(1_{f^{-1}(Y)})(x) = \chi(f^{-1}(Y) \cap f^{-1}(x)).$$

This equals 0 if $x \notin Y$ and $\chi(f^{-1}(x))$ if $x \in Y$; in either case it is equal to $(1_Y.f_*1)(x)$. □

By Theorem 6.2.1, if $f : (\mathbb{C}^2, \mathbf{0}) \to (\mathbb{C}, 0)$ is a holomorphic function-germ with an isolated singularity at O, f defines by restriction a fibration $D_\epsilon \cap f^{-1}B^*_\eta \to B^*_\eta$, for suitably small ϵ, η. The fibre F has $\chi(F) = 1 - \mu$. The singular fibre $F_0 = D_\epsilon \cap f^{-1}(0)$ is homeomorphic to a cone, so is contractible; $\chi(F_0) = 1$.

Using additivity, we may incorporate this local picture in a global one. Suppose X a smooth complex surface, with boundary ∂X, and that $f : X \to U$ (U open in $\mathbb{C}$) satisfies (i) f is proper, (ii) f has only isolated singularities, (iii) $f \mid \partial X$ is a fibration. We see from the above that each singular point contributes to the Euler characteristic of the fibre. Define the constructible function $[\mu]$ on X to take the value 0 except at singular points of f, where its value is equal to the corresponding Milnor number.

Proposition 7.3.2 *In the above situation, we have*

$$f_*(1_X - [\mu]) = \chi(F)1_U,$$

$$\chi(X) - \sum_P [\mu](P) = \chi(F)\chi(U).$$

Here the sum may be extended over all points $P \in X$, since $[\mu](P)$ vanishes if P is not a singular point of f.

Proof If we exclude the singular fibres we have a fibration, so all fibres are homeomorphic, to F say, and the first assertion follows from the definition of f_*. Near a singular point P, we can find a neighbourhood such that the map remains a fibration (i.e. locally trivial) outside the neighbourhood while inside we have the general fibre F_P and singular fibre F_0 a cone; thus $\chi(F_0) - \chi(F_P) = \mu$. The assertion thus follows by additivity.

The second assertion now follows by taking Euler characteristics on both sides. □

The first equation corresponds to Theorem 6.4.1; the second is a restatement of Theorem 6.4.2.

A small perturbation of f will still have the properties listed before the Theorem: thus it follows from the second formula that $\sum_P [\mu](P)$ is unaltered by such a deformation.

Given a smooth family $g = \{f_s\} : X \times S \to U \times S$ of functions f_s each satisfying the above conditions, the separate functions $[\mu]$ piece together to a single constructible function $[\mu]$ on $X \times S$, and thus we have

$$g_*(1_{X \times S} - [\mu]) = \chi(F)1_{U \times S}.$$

Moreover, since the argument is essentially local, the same holds if the product $U \times S$ is replaced by a fibre bundle over S with fibre U.

Now let $X = P^2(\mathbb{C})$ and S be the family of homogeneous polynomials f_s of fixed degree d which have only isolated singularities. The zero locus of f_s is a curve C_s whose singularities are those of f_s. Hence

$$\chi(C_s) - \mu_{\text{tot}}(C_s) = \chi(C)$$

is independent of s (here C denotes any smooth curve of degree d). We have thus recovered Theorem 7.1.1.

We next apply the theory of constructible functions to study duality. It is convenient to write $P, P^\vee$ for $P^2(\mathbb{C}), P^2(\mathbb{C})^\vee$ and to introduce the *incidence variety*, defined by

$$I := \{(x, H) \in P \times P^\vee \mid x \in H\}.$$

We denote the projections by $p : I \to P$ and $q : I \to P^\vee$. Then the dual of a constructible function $\phi : P \to \mathbb{Z}$ is defined as $\phi^\vee := q_* p^* \phi$. The same definitions may be made in the real case; here we write $\mathbb{R}P, \mathbb{R}P^\vee$ for $P^2(\mathbb{R}), P^2(\mathbb{R})^\vee$.

We begin by calculating the double dual $\phi^{\vee\vee}$. As part of the argument will be used again later, let us first suppose given a subvariety $j : Y \hookrightarrow P^\vee$ and discuss how to calculate $\chi(\phi^\vee|_Y) = \chi(j^*\phi^\vee)$. Write (Z, k, r) for the pullback of $I \xrightarrow{q} P^\vee \xleftarrow{j} Y$, so that $Z = (P \times Y) \cap I$.

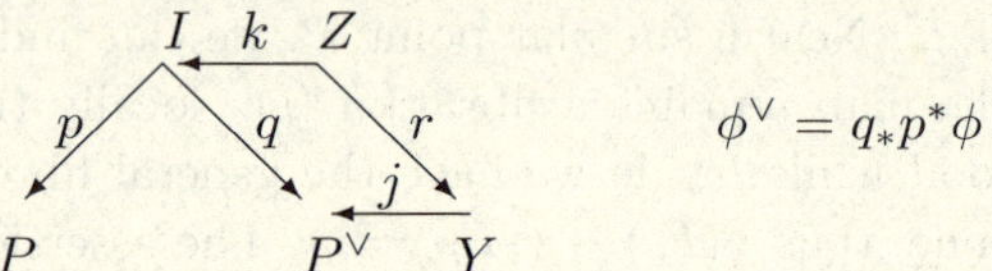

Then

$$\begin{aligned} \chi(j^*\phi^\vee) &= \chi(j^* q_* p^* \phi) = \chi(r_* k^* p^* \phi) = \chi(k^* p^* \phi) = \chi((pk)_* k^* p^* \phi) \\ &= \chi((pk)_* 1_Z . \phi), \end{aligned}$$

where we use in succession the definition of $\phi^\vee$, Theorem 7.3.1(iv), (iii), (iii) and (v). We need to calculate $(pk)_* 1_Z$, whose value at a point x is the Euler characteristic of the set $Y(x)$ of lines H in P which belong to Y and pass through x.

To evaluate $\phi^{\vee\vee}$ at a point x' we take Y to be the set of lines passing through x'. There are two cases for $Y(x)$: if $x = x'$ we have the set of all lines through a single point, which is isomorphic to $P^1(\mathbb{C})$ and has Euler characteristic 2, while for $x \neq x'$ there is just one line through two distinct points so the Euler characteristic is 1. Thus $(pk)_* 1_Z = 1_P + 1_{x'}$, and we obtain $\phi^{\vee\vee}(x') = \chi(\phi) + \phi(x')$, i.e.

Lemma 7.3.3 *For any constructible ϕ on $P^2(\mathbb{C})$, we have*

$$\phi^{\vee\vee} = \chi(\phi) 1_P + \phi.$$

For a semialgebraic function ϕ on $P^2(\mathbb{R})$, we have $\phi^{\vee\vee} = \chi(\phi) 1_{\mathbb{R}P} - \phi$.

The only difference in the real case is that the Euler characteristic takes different values for real projective spaces. However, a more interesting result is obtained by taking the dual in the complex sense and restricting to the real subspace.

Lemma 7.3.4 *With the above notation,* $\chi(\phi^\vee|_{\mathbb{R}P^\vee}) = \chi(\phi) - \chi(\phi|_{\mathbb{R}P})$.

Proof We apply the above method, taking $Y = \mathbb{R}P^\vee$. There are again two cases for $Y(x)$. If $x \in \mathbb{R}P$ is a real point, we have the set of real lines through x, which is isomorphic to $P^1(\mathbb{R})$; if not, any real line through x must pass also through the complex conjugate point $\overline{x}$ and hence coincide with the line joining the two. Thus $(pk)_*1_Z = 1_P - 1_{\mathbb{R}P}$. The result follows. □

Now consider a curve C in $P = P^2(\mathbb{C})$; we assume C irreducible and not a straight line, so $C^\vee$ shares the same properties. Write d for the degree of C and $d^\vee$ for the degree of $C^\vee$. Recall that the multiplicity $m_x(C)$ of C at a point x is the local intersection number of C at x with a general line through x. We define m^C by $m^C(x) := m_x(C)$. Then m^C takes the value 1 at smooth points, hence at all but finitely many points of C; hence m^C is a constructible function. It behaves better for duality than 1_C, which is closely related to it.

Proposition 7.3.5 *For any irreducible curve* C, $(m^C)^\vee = d.1_{P^\vee} - m^{C^\vee}$.

Proof We begin by recalling the proof of Theorem 7.2.2. Let H be any point of $C^\vee$; then a general line L_x through H in $P^\vee$ consists of all the lines through a point $x \in P$, where x is a general point of the line H. The intersections of L_x with $C^\vee$ are the tangents from x to C, where 'tangent' is suitably interpreted at the singular points of C. The feet of these tangents are the intersections of C with the polar curve C_x of x with respect to C.

To describe the effect of the singular points, we recall that by Lemma 7.2.1 the intersection numbers at a point z are related by

$$(C.C_x)_z = (C_y.C_x)_z + (L_1.C_x)_z,$$

and that $(C_y.C_x)_z$ is equal to the Milnor number $\mu_z(C)$ of C at z. Also $(L_1.C)_z = (L_1.C_x)_z + 1$; thus $(C.C_x)_z = \mu_z(C) + (L_1.C)_z - 1$.

Fix z and vary the line $xz = L_1$ through z. The formula just obtained shows that $(C.C_x)_z$ takes its least value when L_1 is not tangent to C

at z; this value being $\mu_z(C) + m^C(z) - 1$. Thus, as remarked earlier, one interpretation is that the dual curve contains the line dual to z counted $\mu_z(C) + m_z(C) - 1$ times; allowing for all these gives the result of Theorem 7.2.2.

Returning to our original problem, where now $H = L_1$, we see that from the intersection multiplicity $(C.C_x)_z = \mu_z(C) + (L_1.C)_z - 1$ we have to cancel $\mu_z(C) + m^C(z) - 1$, leaving $(H.C)_z - m^C(z)$ as the intersection number of L_x with $C^\vee$ corresponding to the point z of tangency of H with C. Summing over all points of $H \cap C$ gives

$$\begin{aligned} m^{C^\vee}(H) &= (L_x.C^\vee)_H = \sum_{z \in H \cap C} (H.C)_z - m^C(z) \\ &= H.C - \sum_{z \in H \cap C} m^C(z) = H.C - (m^C)^\vee(H) \\ &= d - (m^C)^\vee(H), \end{aligned}$$

which proves the proposition. □

There are several proofs of the next result.

Lemma 7.3.6 *For C a plane curve, $\chi(m^C) = 2d - d^\vee$.*

Proof If C has several irreducible components, each of d, $d^\vee$ and $\chi(m^C)$ is a sum of the values for the several components. We may thus suppose C irreducible.

By Corollary 7.2.3,

$$d^\vee - 2g(C) = 2(d-1) - \sum_{z \in \mathrm{Sing}\, C} (m_z(C) - r_z(C)).$$

Let $\rho : \widetilde{C} \to C$ denote the normalisation. Since $\widetilde{C}$ has $r_z(C)$ points lying over z, $\rho_* 1_{\widetilde{C}}(z) = r_z(C)$, so $m^C - \rho_* 1_{\widetilde{C}}$ is zero except at singular points z where it takes the value $m_z(C) - r_z(C)$. Taking Euler characteristics,

$$\chi(m^C) - \chi(\widetilde{C}) = \sum_z (m_z(C) - r_z(C)) = 2(d-1) - d^\vee + 2g(C).$$

Since $\chi(\tilde{C}) = 2 - 2g$, the result follows. □

We come to a celebrated result which is the only numerical relation between invariants which holds for real algebraic curves. Write $C_\mathbb{R}$ for the set of real points of C and $r_z(C_\mathbb{R})$ for the number of *real* branches of C at z.

Theorem 7.3.7 *(Klein's Equation) For any plane curve C,*

$$d - \sum_{z \in \mathrm{Sing}\, C_{\mathbb{R}}} (m_z(C) - r_z(C_{\mathbb{R}})) = d^{\vee} - \sum_{w \in \mathrm{Sing}\, C^{\vee}\mathbb{R}} (m_w(C^{\vee}) - r_w(C^{\vee}_{\mathbb{R}})). \tag{7.4}$$

Proof If we take $\phi = m^C$ in Lemma 7.3.4, we obtain $\chi((m^C)^{\vee}|_{P^{\vee}\mathbb{R}}) = \chi(m^C) - \chi(m^C|_{P\mathbb{R}})$. By Proposition 7.3.5 we may substitute $(m^C)^{\vee} = d.1_{P^{\vee}} - m^{C^{\vee}}$, and by Lemma 7.3.6, $\chi(m^C) = 2d - d^{\vee}$. Putting these together gives

$$d - \chi(m^{C^{\vee}}|_{P^{\vee}\mathbb{R}}) = 2d - d^{\vee} - \chi(m^C|_{P\mathbb{R}}).$$

We now evaluate: first consider $\chi(1_C|_{P\mathbb{R}}) = \chi(C_{\mathbb{R}})$. Since $C_{\mathbb{R}}$ can be regarded as a collection of circles with some points identified, a simple counting argument gives

$$\chi(C_{\mathbb{R}}) = -\sum_{z \in \mathrm{Sing}\, C_{\mathbb{R}}} (1 - r_z(C_{\mathbb{R}})).$$

We obtain $\chi(m^C|_{P\mathbb{R}})$ by adding a correction term for each point where m^C differs from 1, giving

$$\chi(m^C|_{P\mathbb{R}}) = \sum_{z \in \mathrm{Sing}\, C_{\mathbb{R}}} (m_z(C) - r_z(C_{\mathbb{R}})).$$

Substituting this formula, and the corresponding one for $C^{\vee}$, yields the result. □

Although Klein's equation is proved for an arbitrary curve, its force lies in studying curves defined over $\mathbb{R}$; i.e. by a polynomial equation with real coefficients.

We can illustrate its use by considering ordinary singularities. The contribution of a singular point z to the equation is $m_z(C) - r_z(C_{\mathbb{R}})$: the multiplicity minus the number of real branches. In the real case, there are two types of non-degenerate singularities: the union A_1^+, called a *crunode*, of two real smooth branches meeting transversely, typified by $xy = 0$, and the union A_1^-, called an *acnode*, of a smooth branch with non-real tangent and its complex conjugate, typified by $x^2 + y^2 = 0$. Each has multiplicity 2; we have 2 real branches, hence contribution 0, for a crunode, and 0 real branches, hence contribution 2, for an acnode. There is just one type of real cusp A_2: it has multiplicity 2 and 1 real branch, so contributes 0.

Example 7.3.1 It follows using (7.2) and (7.3) or by direct calculation that if C is a cubic curve with just one A_1 singular point, then its dual $C^\vee$ has degree 4 and its singularities are 3 cusps. If C is defined over $\mathbb{R}$ then if the node is a crunode, then $C^\vee$ has just 1 real cusp; if the node is an acnode then $C^\vee$ has 3 real cusps. This can be predicted from (7.4), which reads $3 - 0 = 4 - 1$ and $3 - 2 = 4 - 3.1$ in the respective cases. A cubic with 3 real cusps is illustrated in Figure 7.2.

Example 7.3.2 If C is a nonsingular quartic curve, then $C^\vee$ has degree 12, with 28 nodes and 24 cusps. It follows from (7.4) that twice the number of real acnodes plus the number of real cusps is 8. In particular, if there are no real cusps there are just 4 real acnodes (there may be as many as 24 real crunodes). Now a crunode of $C^\vee$ is a bitangent of C both of whose points of contact are real; for an acnode of $C^\vee$, the two points of contact are complex conjugates.

7.4 The singularities of the dual

If the singularities of C and $C^\vee$ are not ordinary, they may not be chosen independently: indeed a non-ordinary singularity of one forces a singularity of the other. In this section we will obtain a precise formulation of the relation. The key to this will be a calculation of exponents of contact of corresponding pro-branches. These considerations are local, so in this section we are concerned with germs, rather than with curves in the large.

Observe that if C admits a good affine parametrisation

$$(x, y) = (f(t), g(t)),$$

then the tangent at the point t is $g'(t)(x - f(t)) = f'(t)(y - g(t))$, so as a line in the dual projective plane it has coordinates $(g'(t) : -f'(t) : f'(t)g(t) - g'(t)f(t))$. Suppose the terms of least order in $f(t)$ and $g(t)$ respectively are f_0t^ϕ and g_0t^ψ, and that $\phi < \psi$ (so that the tangent at O is $y = 0$). Then we may take affine coordinates in the dual plane with $y^\vee = -1$, so the coordinates of the tangent line become $(\frac{g'(t)}{f'(t)}, g(t) - \frac{f(t)g'(t)}{f'(t)})$, giving a parametrisation of the dual curve $C^\vee$. Moreover, the leading terms of $\frac{g'(t)}{f'(t)}$ and $g(t) - \frac{f(t)g'(t)}{f'(t)}$ are $\frac{g_0\psi}{f_0\phi}t^{\psi-\phi}$ and $\frac{g_0(\phi-\psi)}{\phi}t^\psi$ respectively. Thus the leading terms in the blow-up C^1 of C are $(f_0t^\phi, \frac{g_0}{f_0}t^{\psi-\phi})$ while for the blow-up $C^{\vee 1}$ of the dual we obtain

$(\frac{g_0\psi}{f_0\phi}t^{\psi-\phi}, \frac{f_0(\phi-\psi)}{\psi}t^{\phi})$. The similarity between these two is our clue: our first main result will be that C^1 and $C^{\vee 1}$ are equisingular.

The above calculations suggest that both numbers ϕ and ψ will be important. The multiplicity of the branch is $m(\gamma) = \psi$. The number ϕ can be characterised as the intersection number of γ with its tangent line $x = 0$. In general, if C is a branch, or a union of branches with a common tangent, we define $i(C)$ to be the (local) intersection number of C with its tangent line. In the notation of the preceding paragraph, we have $m(\gamma) = \psi, i(\gamma) = \phi, m(\gamma^\vee) = \phi - \psi$ and $i(\gamma^\vee) = \phi$. Since all these numbers are additive for unions of branches with a common tangent, we have

Lemma 7.4.1 *Let C be a curve with a unique tangent. Then*

$$i(C) = i(C^\vee) = m(C) + m(C^\vee).$$

We will use the fact that the equisingularity class of a branch may be determined as follows. Using coordinates as usual, let the branch C yield pro-branches $\gamma_1, \dots, \gamma_m$; take the exponents of contact v_i of γ_m with γ_i for $1 \leq i < m$. This list contains $m - 1$ entries, so determines the multiplicity; the entries are of the form β_q/m (occurring $(e_{q-1} - e_q)$ times), so determine all the β_q. This follows from (iii) of Lemma 4.1.8.

We thus need to consider pro-branches of C^1 and $C^{\vee 1}$. Observe that we may also calculate using Puiseux series. Thus if we take $y = g(x)$, C^1 and $C^{\vee 1}$ admit respective parametrisations $(x, x^{-1}g(x))$ and $(g'(x), \frac{g(x)-xg'(x)}{g'(x)})$, with initial terms which reduce to $(x, g_0 x^{\psi-1})$ and $(g_0\psi x^{\psi-1}, \frac{1-\psi}{\psi}x)$.

First suppose that $\beta_1 > 2m$ (the parameters being calculated for the branch C itself). Then if C is given by $y = g(x)$ as above, we have $\psi \geq 2$ and the pro-branches of C^1 correspond in an obvious way to those of C. Since the initial exponents are the same, the same conclusion will go for $C^{\vee 1}$.

The case $\beta_1 < 2m$ is more complicated, since the multiplicity drops on blowing up. However we can trace what happens by looking at the leading term: again it suffices for now to consider C^1. After blowing up we need to solve for $x_1 = x$ in terms of $y_1 = y/x$ to obtain a standard parametrisation; then by Theorem 3.5.5 this produces the Puiseux characteristic $(\beta_1 - m; m, \beta_2 - \beta_1 + m, \dots, \beta_g - \beta_1 + m)$, where m is to be omitted if it is divisible by $\beta_1 - m$. All but $e_1 - 1$ of the exponents of contact obtained for pro-branches are equal to β_1/m for C; but to $\frac{m}{\beta_1 - m}$

for C^1 (note that in the case when m is divisible by $\beta_1 - m$ the latter is equal to e_1 so here there *are* only $e_1 - 1$ exponents of contact for C^1). Exclude these: the remaining e_1 pro-branches of C do correspond naturally to those of C^1, since multiplying x by ϵ with $\epsilon^{e_1} = 1$ does not affect the first term in y, hence also does not affect the first term of y_1.

We next give two lemmas to enable us to determine the exponent of contact of two pro-branches. We need these since for the calculation to follow we will not have close control of the parametrisation. Consider two pro-branches γ_1, γ_2 given by parametrisations (f_1, g_1) and (f_2, g_2) such that both f_1 and f_2 have the same leading term $f_0 t^\phi$. First suppose the leading terms $g_{0,1}t^{\psi_1}$ of g_1 and $g_{0,2}t^{\psi_2}$ of g_2 differ.

Lemma 7.4.2 *Suppose that either $\psi_1 \neq \psi_2$ or $\psi_1 = \psi_2, g_{0,1} \neq g_{0,2}$. Then the exponent of contact of γ_1 and γ_2 is*

$$\begin{array}{rl} \phi^{-1} \min(\psi_1, \psi_2) & \text{if this is } \geq 1, \\ \phi \max(\psi_1, \psi_2)^{-1} & \text{if this is } \geq 1, \\ 1, & \text{otherwise.} \end{array}$$

Proof In the first case, neither branch is tangent to the y-axis. Express y as a Puiseux series in x: since the leading terms differ, the exponent of contact is the lesser of the two exponents. The second case follows similarly by interchanging x and y. If neither applies, one branch is tangent to the x-axis and one to the y-axis: since the tangents are distinct, the exponent of contact is 1. □

If the leading terms coincide, the situation is more delicate. Suppose both g_1 and g_2 have leading term $g_0 t^\psi$. Write $\frac{f_2}{f_1} = 1 + h$, $\frac{g_2}{g_1} = 1 + k$, and let the leading terms of h, k be $r_1 t^\rho$ and $s_1 t^\sigma$. Then

Lemma 7.4.3 *For γ_1 and γ_2 as above, then provided either $\rho \neq \sigma$, or $\rho = \sigma$ and*

$$\operatorname{ord}(f_1'(g_2 - g_1) - g_1'(f_2 - f_1)) = \phi + \psi + \rho - 1,$$

the exponent of contact is $\frac{1}{\min(\phi,\psi)}(\max(\phi, \psi) + \min(\rho, \sigma))$.

Proof We may suppose without loss of generality that $\phi \leq \psi$. It is convenient to adjust the parametrisations so that the first coordinate is the same in both cases: this is achieved by a substitution of the form $t' = t(1 + \frac{r_1}{\phi} t^\rho + \cdots)$ in the parametrisation for γ_2. This will have the effect of multiplying the second coordinate g_2 by $(1 - \frac{r_1\psi}{\phi}t^\rho + \cdots)$, converting it to g_3, say. We then require the order of $g_3 - g_1$. The lowest term in $g_2 - g_1$ is $g_0 s_1 t^{\psi+\sigma}$ and the lowest term in $g_3 - g_2$ is $-\frac{g_0 r_1 \psi}{\phi} t^{\psi+\rho}$. The order is thus $\psi + \min(\rho, \sigma)$ unless these two terms cancel out, i.e. $\rho = \sigma$ and $\phi s_1 = \psi r_1$.

It remains only to calculate that the first terms occurring in the expansion of $(f_1'(g_2 - g_1) - g_1'(f_2 - f_1))$ are

$$\phi f_0 g_0 s_1 t^{\phi+\psi+\sigma-1} \text{ and } -\psi f_0 g_0 r_1 t^{\phi+\psi+\rho-1}$$

so that in the critical case $\rho = \sigma$ they only cancel if $\phi s_1 = \psi r_1$. □

The key step in the argument will be the following.

Proposition 7.4.4 *The exponent of contact of the blow-ups γ^1, γ'^1, say, is equal to the exponent of contact of the blow-ups $\gamma^{\vee 1}, \gamma'^{\vee 1}$ of the duals.*

Proof Suppose γ, γ' given by Puiseux series expansions $y = f(x), y = g(x)$ respectively. We have seen above that γ^1 and $\gamma^{\vee 1}$ are given respectively by $(x, x^{-1}f(x))$ and $(f'(x), \frac{f(x) - xf'(x)}{f'(x)})$, with respective initial terms $(x, f_0 x^{\phi-1})$ and $(f_0 \phi x^{\phi-1}, \frac{1-\phi}{\phi}x)$; corresponding results hold for γ'.

If $\phi \neq \psi$, the initial exponents for γ^1 and γ'^1 differ, so by Lemma 7.4.2 the exponent of contact is determined by these exponents. The same goes for the duals $\gamma^{\vee 1}$ and $\gamma'^{\vee 1}$ with the same exponents and hence the same exponent of contact.

Next suppose $\phi = \psi$ but $f_0 \neq g_0$. The argument is essentially unchanged, since the constant $\frac{1-\phi}{\phi}$ appearing before the leading term x is the same in both cases.

Otherwise f and g have the same leading term, so the initial terms for γ^1 and γ'^1 agree, as do those for $\gamma^{\vee 1}$ and $\gamma'^{\vee 1}$. Write $\rho = \text{ord}\,(\frac{g(t)}{f(t)} - 1)$. For each component in the above parametrisations the power series for γ and γ' or their duals or blow-ups differ (if at all) by terms beginning with t^ρ times the lowest term occurring. We are thus in the situation governed by Lemma 7.4.3. The desired result will thus follow if we show that both for comparing γ^1 and γ'^1 and for comparing $\gamma^{\vee 1}$ and $\gamma'^{\vee 1}$ the term corresponding to $f_1'(g_2 - g_1) - g_1'(f_2 - f_1)$ has the expected order $\phi + \rho - 1$, since the exponent of contact is then given by a rule involving only ϕ and ρ.

In the first case, the term simply reduces to $(g(x) - f(x))/x$, which clearly has the desired order. In the second, since the derivative of $\frac{f-xf'}{f'}$ is $\frac{ff''}{f'^2}$, it is

$$f''\left(\frac{g}{g'} - \frac{f}{f'}\right) + \frac{ff''}{f'^2}(g' - f') = \frac{f''}{f'^2 g'}\{f'(f'g - fg') + fg'(g' - f')\}.$$

Here, if we substitute $g = f + h$ in the term in braces, it becomes $f'^2 h + f h'^2$ where, since the first summand has order $3\phi + \rho - 2$ and the second has order $3\phi + 2\rho - 2$ the sum has the lesser order $3\phi + \rho - 2$. Hence the whole expression has order $\phi + \rho - 1$, as desired. □

Proposition 7.4.5

(i) *For any branch C, the blow-ups of C itself and of its dual are equisingular.*

(ii) *If C and C' are branches with the same tangent direction, the intersection numbers $C^1.C'^1 = C^{\vee 1}.C'^{\vee 1}$.*

Proof Since the equisingularity class is determined by the exponents of contact of distinct pro-branches, and by Proposition 7.4.4, the pro-branches of the given branches correspond in such a way that exponents of contact are preserved, assertion (i) follows.

Part (ii) follows similarly, since the intersection number is determined by the exponents of contact (we can either take their sum or apply Theorem 4.1.6 to their supremum). □

Since a point of $C^\vee$ corresponds to a point and a direction there on C, the final relation between the singularities of a curve and those of its dual is best stated in terms of the union C of the branches of C at a point P and having as tangent there a given line L.

Theorem 7.4.6 *There is an isomorphism between the trees of infinitely near points for C and $C^\vee$, which preserves the multiplicity of each branch at each infinitely near point (except at P itself), and is such that a point $O_s^\vee$ is proximate to $O_0^\vee$ if and only if O_s belongs to the tangent line L to C at O_0; and dually a point O_s is proximate to O_0 if and only if $O_s^\vee$ belongs to the tangent line $L^\vee$ to $C^\vee$ at $O_0^\vee$.*

Proof For a single branch B, (i) of Proposition 7.4.5 shows that the blown-up branches are equisingular, and this gives an isomorphism of the trees preserving multiplicities which trivially extends to an isomorphism between the trees of infinitely near points of B and $B^\vee$. When we have two branches, (ii) of the same proposition implies that the two trees have corresponding points in common, so the isomorphism extends to their union. Similarly for any number of branches.

For the second part, it is sufficient to consider the case of a single branch B. We write $m^\vee = mq + r$ with $q \in \mathbb{N}$ and $0 \leq r < m$. Then by Lemma 4.4.2, the intersection number $i = B.L$ – which equals $m + m^\vee$ by Lemma 7.4.1 – is the sum of the multiplicities (on B) of the common infinitely near points. There are no proximity relations among the infinitely near points on L, so these multiplicities must be m for $O_0, \ldots, O_q$ and r for O_{q+1}; otherwise (iii) of Proposition 3.5.1 would be violated. From what we have already seen, it follows that $O_1^\vee, \ldots, O_q^\vee$ have the same multiplicity m and that $O_{q+1}^\vee$ has multiplicity r. The same rule relating multiplicities and proximity relations shows that these, and no other infinitely near points are proximate to $O_0^\vee$. The same argument is valid with C and $C^\vee$ interchanged. □

Corollary 7.4.7 *If B is a branch with $i = 2m$ then B and $B^\vee$ are equisingular.*

For the isomorphism preserves all multiplicities, so the result follows from Theorem 5.5.9.

Example 7.4.1 Let B be a branch of type A_4. Then $m = 2$ and the possible values of i are 4 and 5. In the former case, the dual again has type A_4. In the latter, typified by $y^2 = x^5$, the dual has multiplicity sequence $3, 2, 1, 1$ and so type E_8.

Example 7.4.2 Let B be a branch of type E_6, hence with multiplicity sequence $3, 1, 1, 1$. The only possible value of m is 4, so the dual is smooth, having intersection number 3 with its tangent line. Such a point on a curve is called a *hyperflex*.

7.5 Singularities of curves of a given degree

In this section we present a survey, without proofs but with references, of known results on classification of singular curves.

For curves of low degree, it is possible to give an explicit list of types, indicating in each the set of singularities on the corresponding curves. In degree 2, this is trivial: we have the non-singular conics, and those that split as a pair of lines, meeting in an A_1 singularity (the case when the lines coincide is excluded, as we restrict to reduced curves).

In degree 3 we have the irreducible cubics, which may be non-singular, or have one node (A_1) or one cusp (A_2); a conic and a line which may be a chord, giving two A_1's, or a tangent, giving an A_3; or three distinct lines, giving 3 A_1's in general, but a D_4 if they concur. This, even in the real case, was known to Newton [144] who obtained the list from the affine classification.

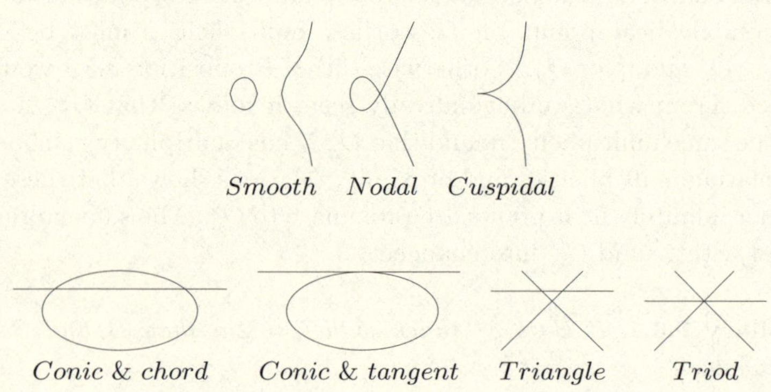

Fig. 7.1. Types of plane cubic curve

Explicit listing is also fairly easy in degree 4: except in the case of 4 concurrent lines, all the singularities are simple, and the sum of their Milnor numbers is 7 at most: cases were listed above. The list was essentially known to Cayley; an attractive and systematic account (using root systems) was given by Du Val [59] in 1934.

Listings are also available in degree 5 (Degtyarev [44]) and degree 6: see Urabe [181] and Degtyarev [43] for the case when at least one singularity is non-simple, and Yang [201] for curves with simple singularities only. A particularly interesting case is with simple singularities and $\mu_{tot} = 19$: Yang obtains 579 different configurations.

One is thus led to ask more qualitative questions. How can one bound the possible complexity that occurs? This may be interpreted in at least three ways: what is the maximum possible number of singular points? What is the largest possible value of μ at any critical point? What is the largest value of the sum μ_{tot} of Milnor numers at all critical points?

We can answer these questions. The maximum possible number of singular points is $\frac{1}{2}d(d-1)$, occurring for set of d lines in general position. For if C is irreducible of degree d, since its genus is non-negative it follows from Corollary 7.1.3 that $\sum_P \frac{1}{2}(\mu_P + r_P - 1) \le \frac{1}{2}(d-1)(d-2)$. Since each term $\frac{1}{2}(\mu_P + r_P - 1)$ with P singular is ≥ 1, the number of singular points is at most $\frac{1}{2}(d-1)(d-2)$. If C has r components C_i $(1 \le i \le r)$ with degrees d_i such that $\sum d_i = d$, then C_i has at most $\frac{1}{2}(d_i-1)(d_i-2)$ singular points, and $\#(C_i \cap C_j) \le d_i d_j$, so the maximum possible number of singular points of C is

$$\begin{aligned} \sum_i \tfrac{1}{2}(d_i-1)(d_i-2) + \sum_{i<j} d_i d_j &= \tfrac{1}{2}(\textstyle\sum d_i)^2 - \tfrac{3}{2}\textstyle\sum d_i + r \\ &= \tfrac{1}{2}d(d-3) + r, \end{aligned}$$

which is greatest when $r = d$ and all components are lines.

The largest possible value of μ at any critical point is $(d-1)^2$, occurring for a set of d concurrent lines; the same example gives the largest possible value of μ_{tot}. We can argue as in the preceding case. If C_i is irreducible of degree d_i then $\sum \mu_P \le \sum_P(\mu_P + r_P - 1) \le (d_i-1)(d_i-2)$. Now applying Corollary 7.1.4 inductively, adjoining one component at a time, gives $\mu_{\text{tot}}(C)$ equal to $\sum_i \mu_{\text{tot}}(C_i) + 2\sum_{i<j} C_i.C_j = d^2 - 3d + 2r$ diminished by $\sum_2^r \#(C_i \cap (\bigcup_{j<i} C_j))$. This last takes its smallest value when all the intersections occur at a single point, and is then $r-1$: now again the result is greatest when $r = d$. Alternatively, if $f(x,y,z) = 0$ is a curve meeting $z = 0$ in d distinct points, all curves in the 1-parameter family $f(x,y,tz) = 0$ are isomorphic to it until we reach $t = 0$, giving d concurrent lines, and a semicontinuity argument shows that the Milnor number cannot decrease on specialisation.

By Theorem 7.1.1, large values of $\mu_{\text{tot}}(C)$ correspond also to large values of $\chi(C)$. Now for C irreducible, we have $\chi(C) \le 2$, with equality only if C is rational with $r = 1$ at each singular point. Even in the reducible case, it seems that curves with $\chi(C) \ge 3$ are somewhat rare. In particular, it was shown by de Jong and Steenbrink [45], confirming a conjecture of Veys, that any curve C with $\chi(C) \ge 3$ has all components rational.

The examples giving the above answers are a little disappointing: one may also ask the same questions for irreducible curves. According to Corollary 7.1.3, for an irreducible curve C of degree d we have $\mu_{\text{tot}}(C) = d^2 - 3d + (2-2g) - \sum_P(r_P - 1)$. The maximum value of μ_{tot} is thus $d^2 - 3d + 2$, and is attained if $g = 0$ (the curve has a rational parametrisation) and each $r_P = 1$, so there is just one branch at each singular point. Equivalently, C itself is homeomorphic to $P^1(\mathbb{C})$, or equivalently, to S^2. Such curves C are called *rational cuspidal curves.* An example, in any degree d, is given by $x^d = y^{d-1}z$. A number of papers have been published on the enumeration of all such curves: e.g.

Flenner and Zaidenberg [71] analyse all cases where C has a singular point of multiplicity $d-2$. It is convenient to list the types of singularities: since each has just one branch, it suffices to give the multiplicity sequences (omitting multiplicities equal to 1). For $d \leq 6$, the complete list is known:

$$\begin{aligned}
d=3:&\quad 2.\\
d=4:&\quad 3,\ 222,\ 22+2,\ 2+2+2.\\
d=5:&\quad 4,\ 32+22,\ 3+222,\ 3+22+2,\ 222222,\ 2222+22,\\
&\quad 222+2+2+2,\ \text{and}\ 22+22+22.\\
d=6:&\quad 5,\ 42222,\ 4222+2,\ 422+22,\ 42+222,\ 3332,\ 333+2,\\
&\quad 332+3,\ 33+32.
\end{aligned}$$

We illustrate the quartic cases in Figure 7.2: the respective singularities, in our usual notation, are $3A_2$, A_2+A_4, A_6 and E_6. Pictures of all types

Fig. 7.2. Rational cuspidal quartic curves

of real rational quartic curves are given in my article [194].

One may also seek lists of examples where the value of μ is close to the maximum. The one explicit known result concerns a slightly different question, where the Milnor number is replaced by the *Tjurina number* τ, whose value at an isolated singularity O of $f(x,y)$ is defined to be $\dim_{\mathbb{C}} \frac{\mathcal{O}_{x,y}}{\langle f, f_x, f_y\rangle}$. Then [58] τ_{tot} achieves its maximum value $(d-1)^2$ only for a set of d concurrent lines; its next highest value is $d^2 - 3d + 3$, and moreover if $d \geq 6$ it takes the values $d^2 - 3d + 2$ and $d^2 - 3d + 3$ only for curves admitting a 1-parameter symmetry group; its next highest value being $d^2 - 4d + 7$. For a curve admitting the multiplicative group $\mathbb{C}^\times$ as symmetry group, the singularities are all weighted homogeneous and $\mu = \tau$; a curve admitting the additive group $\mathbb{C}^+$ is a union of r conics such as $\prod_1^r (y^2 - xz + a_i x^2)$ if $d = 2r$, and this union together with the common tangent $y = 0$ if $d = 2r + 1$. Since the intersection number between each pair of conics is 4, it follows from Theorem 6.5.1 that $\mu_{\text{tot}} = 4r^2 - 5r + 1$ or $4r^2 - r$ respectively: a higher value than in the $\mathbb{C}^\times$ cases.

Another interesting set of questions arises when we restrict attention to curves all of whose singularities are simple (in which case $\mu = \tau$). It was shown by Hirzebruch [90] that in this case that if $d = 2r$ then $\mu_{\text{tot}} \leq 3r^2 - 3r + 1$, and observed by Persson [148] that this bound is attained (if $r = m + 1$) by the curve

$$C_m : \; xy(x^{2m} + y^{2m} + z^{2m} - 2y^m z^m - 2z^m x^m - 2x^m y^m) = 0.$$

For $m \geq 3$, most of the singular points have complex coordinates. For $m = 2$, since

$$\begin{aligned} &x^4 + y^4 + z^4 - 2y^2 z^2 - 2z^2 x^2 - 2x^2 y^2 \\ &\quad = (x + y + z)(x + y - z)(x - y + z)(x - y - z), \end{aligned}$$

the curve is the union of 6 lines forming a complete quadrangle.

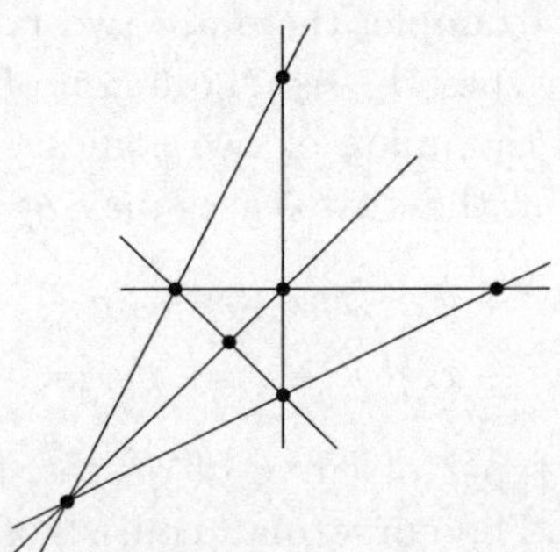

For the case when $d = 2r + 1$ is odd, an upper bound may be inferred since for any curve C, the union of C and a general tangent has degree $2r + 2$ and $\mu_{\text{tot}} = \mu_{\text{tot}}(C) + 2r + 2$. On the other hand, taking C_m and adding the line $z = 0$ (or removing $y = 0$) shows that the value $3r^2$ can be attained. For $r = 1, 2, 3$ we can do one better: if $d = 5$ four of the curves with 1-parameter symmetry group have $\mu_{\text{tot}} = 13$, and if $d = 7$ we have two examples: see Exercises 7.7.11 and 7.7.10.

In general requiring a curve to have a singularity imposes one condition on its equation, and in a certain sense requiring a simple singularity imposes μ conditions on the equation. So we may expect that a given collection of simple singularities to impose μ_{tot} conditions, and could hope that these define a smooth subvariety of this codimension of the variety of all (reduced) curves of degree d, and moreover that this subvariety is non-empty and connected; and that suitably formulated corresponding results hold also for non-simple singularities. However, these conditions are not always independent as is shown by the examples already given, so some restrictions are necessary. There is a growing literature of results showing that each of these expectations is fulfilled up to some point. The oldest such result, due to Shustin [167], holds for $\mu_{\text{tot}} \leq 4d - 5$; there are many recent improvements: see e.g. [80]. Among the results in this paper are that the variety is smooth of the expected dimension if $\sum(\tau_P + 1)^2 \leq (d+3)^2$; and that if $\max \tau_P \leq \frac{2}{5}d - 1$ and

$$\tfrac{25}{2}(\#A_1) + 18(\#A_2) + \tfrac{10}{9} \sum_{\tau_P \geq 3} (\tau_P + 2)^2 \leq d^2,$$

then the variety is connected or empty.

Although the singularities determine the genus in the case when there is only one component, there may be several different configurations for curves of a given degree with a given list of types of singularity: thus, in particular, the variety of curves with a given list of singularities need not be connected. For example, there are two types of curve of degree 4 with singularities of types A_5, A_1: the union of a nodal cubic and an inflexional tangent, or the union of two conics with a point of 3-point contact: the equations in these two cases may be taken respectively as

$$(x^3 + y^3 - xyz)(3x + 3y + z) = 0,$$
$$(y^2 - xz)(y^2 - xz + xy) = 0.$$

There are also two types of curve of degree 4 with just one singular point, of type A_5. The curve may consist of a smooth cubic with inflexional tangent, e.g.

$$z(yz^2 + x^3 + z^3) = 0,$$

or be irreducible, e.g.

$$(y^2 - xz)^2 + y^2z^2 + z^4.$$

It is harder to find examples with irreducible curves. The oldest such example is due to Zariski [202], who considered curves of degree 6 with 6 cusps (A_2 singularities). He showed that there were 2 families, distinguished from one another by the fact that in one case the 6 cusp points lie on a conic; in the other, they do not. In the former case, the equation may be written in the form $f(x,y,z)^3 = g(x,y,z)^2$ (where $f = 0$ gives the conic), and the fundamental group of the complement $P^2(\mathbb{C}) \setminus C$ is a free product $\mathbb{Z}_2 * \mathbb{Z}_3$. In the latter case, he later showed that the fundamental group is a cyclic group $\mathbb{Z}_6$ of order 6. These examples have stimulated many authors: equations have been studied, numerical invariants found to distinguish the cases, and many other examples of what are now called *Zariski pairs* given. See, for example, Artal [12].

7.6 Notes

Sections 7.1,7.2 The formulae of Theorem 7.1.1 and Theorem 7.2.2 have a long history. The identities (7.2) and (7.3) were first obtained (for curves with 'ordinary singularities') by Plücker [150], and (7.1) by A. Clebsch in 1864, the first paper to identify 'genus' as a crucial concept, following Riemann's work. Several authors, including Noether, contributed to the extension to arbitrary singularities: a detailed early account, with reference to other works, is given by Smith [168]. A traditional textbook treatment is in [165], who also give extensions of the formulae to space curves. Our presentation is somewhat closer to the elegant account of Milnor [132].

These formulae extend to numerous calculations in higher dimensions, of which we can only mention a handful. The extension of Plücker's formulae to space curves is classical. A wider study of enumerative formulae from the viewpoint of algebraic geometry was pioneered by Kleiman [100] among others; see also Piene [149], who develops formulae using the approach of polar varieties (generalising polar curves). There has also been a breakthrough in obtaining formulae from the viewpoint of singularity theory: see e.g. Rimányi [156] and Kazarjan [97].

The genus formula is also a special case of a general result, known as the *adjunction formula*, for curves on a smooth algebraic surface M: if K_M is the canonical class, $2g(C) + 2\delta_{\text{tot}}(C) = C.C + C.K_M + 2$: see e.g. [17] Section 2.11.

The total δ is also relevant for a different question. While curves of a given degree can be obtained from one another by deforming the equation, what is the condition that they can be related by deforming (in some sense) a parametrisation? If this is correctly interpreted, the answer is: yes, when the value of δ_{tot} is constant in the family.

Numerous papers have been devoted to the configuration of the 24 bitangents of a non-singular plane quartic curve from a wide variety of viewpoints: we mention only the relation to a root system of type E_6 with symmetries given by the corresponding Weyl group.

Section 7.3 The first reference for constructible functions in the spirit of this section is MacPherson [123]: in that paper only complex varieties were considered, and only the induced maps f_*. He gives a very brief sketch of proof for (ii) of Theorem 7.3.1. The fuller development of constructible functions in the context of Section 7.3 is due, I think, to Viro [184]. Following Viro, many authors prefer the notation $\int_X \phi d\chi$ for $\chi(\phi)$. We prefer to regard this as a special case of the operation $\pi_*\phi$. Our development is inspired by McCrory and Parusinski [124]. Part (iv) of Theorem 7.3.1 seems to be due to Ernström [67].

There is a large literature on characteristic classes for singular algebraic varieties, for which constructible functions have, from the first, been a key tool. A good survey of developments in this area is Brasselet [20].

It was Zeuthen's work [210] on the bitangents of a curve of degree 4 that motivated Klein [101] to discover his equation.

The application of constructible functions to prove Klein's equation is due to Viro [184]. Our treatment of Proposition 7.3.5, Lemma 7.3.6 and Theorem 7.3.7 are all inspired by this paper. A fuller account of the formula, with early references and a modern account of an early proof, is given in the paper [195] of the author. I regret that the important reference to the work [184] of Viro was overlooked in that article.

It was shown by Coolidge [40] that Klein's equation is the only relation additional to Plücker's equations that holds between the invariants in the real case.

Section 7.4 Part (i) of Theorem 7.4.6 is due to Zeuthen [211] and part (ii) to the author [193]: we follow the latter treatment.

Affine curves The study of singular curves in the affine plane $\mathbb{C}^2$ has flourished since the first draft of this chapter was written. One approach is to study polynomial maps $f : \mathbb{C}^n \to \mathbb{C}$: then there is a finite subset $B \subset \mathbb{C}$ such that f is topologically trivial over $\mathbb{C} \setminus B$. The set B includes the critical values of f but in general also others. In the case $n = 2$ the minimal set B can be characterised as the set of points x at which the Euler characteristic $\chi(f^{-1}(x))$ is smaller than its generic value.

In Section 5.5 we defined the Alexander polynomial starting from the abstract group, and this definition applies in other situations. Given a curve C in the projective plane $P^2(\mathbb{C})$, choose a line L meeting C in general position – i.e. in d distinct points, where C has degree d – and let G be the fundamental group of the complement $P^2(\mathbb{C}) \setminus (C \cup L)$: this is easily seen not to depend on L. Then G^{ab} is free abelian of rank r equal to the number of (global) branches of C, and its r variable Alexander polynomial was defined by Libgober [117] to be that of C.

Another range of questions arises from the study of the behaviour of f 'near infinity'. Here too it is possible to define a Milnor fibration, a Newton polygon, an Alexander polynomial with properties analogous to those holding in the neighbourhood of an isolated singular point.

We can also regard $\mathbb{C}^2$ as the complement of a line $z = 0$ 'at infinity' in the projective plane $P^2(\mathbb{C})$; then there is a homogeneous polynomial $F(x, y, z)$ of degree d with $f(x, y) = F(x, y, 1)$, or equivalently $f(\frac{x}{z}, \frac{y}{z}) = F(x, y, z)/z^d$. Thus the study of f at infinity can be regarded as the study of the meromorphic function F/G (where $G = z^d$) near a common zero of F and G. A number of the known results about polynomial functions on the plane have been generalised in this sense to results about meromorphic functions.

We refer to the survey article [83] for a partial guide to the literature in this area up to 2000.

Characteristic p If C is a reduced, irreducible projective plane curve of degree d, then its genus (i.e. that of its normalisation) is $g = \frac{1}{2}(d-1)(d-2) - \sum_{P \in \text{Sing } C} \delta_P$. This is the same as in characteristic 0. A proof may be found in [149].

The degree of the dual offers more difficulty. If we write $\widehat{C}$ for the closure in the incidence variety of the set of pairs $\{(x, L)\}$, where x is a smooth point of C and L the tangent to C at that point, then the projection onto the second factor gives a curve in the dual plane that is not necessarily reduced. I will write $C^\vee$ for the underlying reduced curve (note: this differs from the notation employed by Fulton [74]). Its multiplicity in the image divisor is the degree π of the field extension $[k(C) : k(C^\vee)]$. More precisely, we can express $\pi = \pi_s \pi_i$ as a product of separable and inseparable degrees. Geometrically (see e.g. [86]), the map $C \to C^\vee$ is generically π_s to 1 as a map of sets, while the intersection multiplicity of C with a tangent at a generic point is $\max(\pi_i, 2)$ (note that π_i is a power of p; if $p = 2$, we always have $\pi_i \geq 2$). It is known [197] that if $\pi_i = 1$ then $\pi = 1$ and the dual of $C^\vee$ is C again: such curves C are called reflexive.

The formula for the degree of the dual may now be written as $d(d-1) = \pi_i \pi_s d^\vee + \sum_P e_P(C)$, where again P runs over singular points of C. This result is given in [74]4.4.4. We can use the same proof as in characteristic 0: this leads to the expression of e_P as the local intersection number at P of C with a generic polar curve C_Q. This can be re-expressed more invariantly. To calculate the intersection number, we have a sum over branches B of C at P of $B.C_Q$; now take a parametrisation of B and an equation $a\partial f/\partial x + b\partial f/\partial y + c\partial f/\partial z$ of C_Q and substitute the former in the latter. For generic (a, b, c) we have the least of the orders of $\partial f/\partial x$, $\partial f/\partial y$ and $\partial f/\partial z$ as functions of t, i.e. the order of the ideal $J = \langle \partial f/\partial x, \partial f/\partial y, \partial f/\partial z\rangle$ in $\tilde{\mathcal{O}}$: this is Fulton's form. Recall that, as noted before, the codimension $\mu(f)$ of $\langle \partial f/\partial x, \partial f/\partial y\rangle$ in $\mathcal{O}_{x,y}$ is not the same as $\mu_O(C)$.

To see that e_P is an invariant of C, it suffices to observe first that, at a point where z is non-zero, the identity $df = x\partial f/\partial x + y\partial f/\partial y + z\partial f/\partial z$ shows that $\partial f/\partial z$ belongs to the ideal $J^+ = \langle f, \partial f/\partial x, \partial f/\partial y\rangle$, and hence that J and J^+ pull back to the same ideal in $\mathcal{O}$, where the image of f is 0. The codimension here is the Tjurina number τ_P: note that e_P is the codimension of the ideal $\tilde{J}$ generated by J in $\tilde{\mathcal{O}}$. We conjecture that the inequality $e_P \geq \tau_P$ always holds (it follows if $\tilde{J} \cap \mathcal{O} = J$).

A further formula is obtained by Piene [149]. For each point Q of the normalised curve $\tilde{C}$, corresponding to a branch of C at the image point P, the projection expressed in terms of local coordinates gives a parametrisation $(x(t), y(t))$ of the branch. The ideal $\tilde{I} = \langle dx/dt, dy/dt\rangle$ in the local ring $\mathcal{O}_Q$ is called the *ramification ideal.* Write κ_Q for its codimension, and κ_P for the sum of these numbers over all branches of C at P. Then $e_P = 2\delta_P + \kappa_P$.

7.7 Exercises

Exercise 7.7.1 Prove that the only singularities with $\frac{1}{2}(\mu + r - 1) \leq 4$ are those listed at the end of Section 7.1. (Compare Exercise 6.7.11.)

Exercise 7.7.2 Show that no plane curve of degree 5 can have 6 cusps. (Hint: calculate the class.)

Exercise 7.7.3 Let C be a curve of degree 6 with three singular points, all of type E_6. Show that C is irreducible, and calculate the genus of its normalisation. If $C^\vee$ has only node and cusp singularities, how many of each does it have?

Exercise 7.7.4 Enumerate the possible sets of singularities for curves of degree 4 in $P^2(\mathbb{C})$, one of whose components is a cubic. Hint: the cubic is smooth, nodal or cuspidal; in each case you need to describe the situation at each intersection point of the line and the cubic.

Exercise 7.7.5 For each collection of singularities such that $\sum \delta_P = 3$, give an equation for an irreducible quartic curve C in $P^2(\mathbb{C})$ whose singularities are of the given types. The following method can be used. If there are 3 singular points, choose coordinates with these points as $(1,0,0)$, $(0,1,0)$ and $(0,0,1)$: show that the transformation $\alpha(x,y,z) = (yz, zx, xy)$ satisfies $\alpha^2 = 1$ and takes C to a conic. If there are 2 singular points argue similarly with the transformation $\beta(x,y,z) = (xz, yz, y^2)$ and for A_5 and A_6 use $\gamma(x,y,z) = (y^2 - xz, yz, z^2)$. If there is a singular point of multiplicity 3, take it as $(0,0,1)$ and argue from first principles.

Exercise 7.7.6 Show that if $p : X \to Y$ is a fibration with fibre F, then for any constructible function ϕ on Y, $\chi(p^*\phi) = \chi(F)\chi(\phi)$.

Deduce that if ϕ is a constructible function on $P^2(\mathbb{C})$ you have $\chi(\phi^\vee) = 2\chi(\phi)$. Check that this is consistent with the calculation of $\phi^{\vee\vee}$ in Lemma 7.3.3 and with that of $(m^C)^\vee$ in Proposition 7.3.5.

What is the corresponding result for constructible functions on $P^2(\mathbb{R})$?

Exercise 7.7.7 Show that if C is a real quartic curve of class 4 with no triple point, a_C is the number of acnodes of C and c_C the number of real inflexions, then $2a_C + c_C = 2a_{C^\vee} + c_{C^\vee}$.

Exercise 7.7.8 Show that the singular point of the dual to a branch B with an A_6 singularity has Milnor number 6, 16 or 24.

Exercise 7.7.9 Show that if C is an irreducible curve of degree and class 4 and genus 0, the collection of singularities of C is one of the following: (i) two of type A_2 and one of type A_1; (ii) one of type A_4 and one of type A_2: (iii) one of type E_6.

Show that if C has type (ii), so has $C^\vee$. Give examples of curves C_1, C_2 of type (iii) with $C_1^\vee$ of type (i) and $C_2^\vee$ of type (iii).

Exercise 7.7.10 Show that a nodal cubic C has 3 flexes, and that the points of inflection are collinear. Let C' be the union of C, the 3 inflexional tangents and the line through the flexes. Find and describe the singularities of C', and hence show that $\mu_{\text{tot}}(C') = 28$.

Exercise 7.7.11 Show that if C is a quartic curve with 3 cusps, $C^\vee$ is a nodal cubic curve. Show also that the tangents at the cusps are concurrent. Determine the singularities of the curve C' which is the union of C and the cuspidal tangents, and hence show that $\mu_{\text{tot}}(C') = 28$.

Exercise 7.7.12 Show that if C is the curve defined by

$$(x^3 - y^2z + z^3)(x^3 - y^2z)z = 0,$$

then $\chi(C) = 2$ but C has a non-rational component. (7 is the least possible degree for such curves.) Determine all the singular points of C and describe their nature.

8

Combinatorics on a resolution tree

Although invariants for a curve with a single branch can be written in sequence, and calculated in terms of the Puiseux characteristic, for curves with several branches it is necessary to work on a tree. In this chapter we consider the dual graph of a tree produced from a plane by an arbitrary sequence of point blowings up. We will see that many invariants can be most conveniently expressed using the algebra of exceptional cycles on the surface T which is the result of the blowings up. This leads to many formulae; some of these complete the development of Chapter 4, others lead to a study of the 'topological zeta function'.

We also prepare for the discussion of the topology of the Milnor fibration in Chapters 9 and 10. Indeed, on the boundary we have an isomorphism $\partial T \to \partial S$, so ∂T includes the singularity link complement and allows a fairly explicit description of it, and of the Milnor fibration, which we will give in Chapter 9. Here we will introduce the invariants and notation in terms of which the later calculations will be expressed.

8.1 The homology of a blow-up

Let C be a curve defined near $O \in \mathbb{C}^2$, with branches B_j. We recall that by Theorem 3.4.4, C has a good resolution, which is a map $\pi : T \to S$, where S is a (small enough) disc neighbourhood of the point $O \in \mathbb{C}^2$. The map π gives an isomorphism $(T - \pi^{-1}(O)) \to (S - O)$, and the collection $\pi^{-1}(C)$ of curves has normal crossings. Moreover, π is constructed as a composite of maps $\pi_i : T_{i+1} \to T_i$ (with $T_0 = S$), each obtained by blowing up a single point O_i, which thus gives rise to an exceptional curve $E_i \subset T_{i+1}$. Thus $\pi^{-1}(C)$ is the union of the strict transforms of the curves E_i with those of the branches B_j of C. In this section we begin the study of the topology of T and of these curves lying in it.

We now take a more general viewpoint, and consider a map $\pi : T \to S$ obtained by an arbitrary sequence of blowings up $\pi_i : T_{i+1} \to T_i$ ($0 \leq i < N$) of points O_i (always assumed to lie in the respective preimage of $O \in \mathbb{C}^2$). We usually continue to denote the strict transform of the blow-up of O_i by E_i. However, when we need to specify the strict transforms lying in T_k, we will write $E_i^{(k)}$ or, for any curve C in S, $C^{(k)}$. It follows inductively from Proposition 3.4.3 that the E_i have normal crossings; in particular, two components may only intersect in one point. We recall that π is said to be a good resolution of C if $\pi^{-1}(C)$ has normal crossings.

As in Section 3.6, we define the dual graph Γ_R for the blow up π to have a vertex V_i corresponding to each curve E_i; and join two vertices if and only if the corresponding curves E_i intersect. We see inductively that Γ_R is always a tree. The augmented dual graph Γ_R^+ (for π and the curve C) has the vertices V_i and also vertices W_j corresponding to the transforms $B_j^{(N)}$ of the branches of C; again two vertices are joined if the corresponding curves intersect. It is customary to draw an arrow pointing to W_j along the edge which meets W_j, so these vertices are sometimes described as *arrowhead vertices.*

We next study the topology of the manifold T.

Lemma 8.1.1 *The only non-vanishing (reduced) homology group of T is $H_2(T)$, which is free abelian on N generators, which may be taken as the homology classes $[E_i]$ of the exceptional curves E_i ($0 \leq i < N$).*

Proof Each blowing up has the effect of replacing a point by a 2-sphere. The blow up $\pi : Y \to X$ of a point $x \in X$ was constructed by choosing a (small) disc D surrounding X, removing the interior of the disc D (giving M, say) and replacing it by the result Q of blowing up $x \in D$. Since Q is a smooth neighbourhood of the exceptional curve of the blow up, it is homotopy equivalent to a 2-sphere.

Since D is contractible, the maps $H_i(X) \to H_i(X, D)$ are isomorphisms, except when $i = 0$; and $H_0(X, D) = 0$. By excision, $H_i(X, D) \cong H_i(M, \partial D) \cong H_i(Y, Q)$. The map $H_3(Y, Q) \to H_2(Q)$ is zero since it factorises through $H_3(M, \partial Q) \to H_2(\partial Q)$, and $H_2(\partial Q) = H_2(\partial D) = 0$.

Thus the maps $H_i(Y) \to H_i(Y, Q)$ for $i \neq 0$ are isomorphisms except for those in the short exact sequence

$$H_2(Q) \to H_2(Y) \to H_2(Y, Q).$$

The first assertion of the lemma now follows by induction. The second also follows, since we obtain a basis for $H_2(Y)$ by adding the class of the exceptional curve which generates $H_2(Q)$ to a list of elements corresponding to generators of $H_2(Y, Q) \cong H_2(X, D) \cong H_2(X)$. □

We will shortly calculate the intersection numbers of the $[E_i]$.

In algebraic geometry, a *cycle* on an algebraic surface is a formal linear combination (with integer coefficients) of irreducible curves. When we study germs at O of curves on $S = \mathbb{C}^2$, we use 'cycle' to mean a linear combination of irreducible germs (or branches) at O. We associate a cycle with its local defining equation. The product fg is associated to the sum of the cycles $f = 0$ and $g = 0$. For any finite collection of cycles, we can choose a neighbourhood of O on which all the functions are defined. If C is defined by an equation $f = 0$ such that f has no repeated factors, we write $[C]$ for the corresponding cycle. Since intersection numbers behave additively for unions, they are also defined at the level of cycles.

In the blown up surface T, the cycle $f = 0$ in S has a *total transform* defined by $f \circ \pi$. This consists of two parts: a *strict transform*, which corresponds to the strict transform of the curve C, and an exceptional part, which is a linear combination of the curves E_i. We will use the term *exceptional cycle* for an element of $H_2(T)$, given as a linear combination of the $[E_i]$: the reader may prefer to think of this as a cycle in either the homological sense or the algebro-geometric. Note that this refines the set-theoretic notion of total transform defined in Section 3.2. We write $[\widehat{E}_i]$ for the class of the total transform of E_i.

We now describe the behaviour of a single blow up.

Lemma 8.1.2

(i) *Let $\pi : T' \to T$ be the map defined by blowing up the point $P \in T$, E the corresponding exceptional curve. Let C be a curve through P with local equation $f = 0$; let m be the multiplicity of C (or f) at P. Then if C' is the strict transform of C, we have $\pi^*[C] = [C'] + m[E]$.*

The intersection numbers in T and T' are related by

(ii) $[E].[E] = -1$,

(iii) $[E].\pi^*[F] = 0$,

(iv) $\pi^*[F].\pi^*[G] = [F].[G]$.

Proof If the blow-up is given by the local coordinate substitution $(x, y) = (x_1, x_1 y_1)$, then if m is the order of f, $f(x_1, x_1 y_1)$ is divisible by x_1^m, but by no higher power of x_1. Since $x_1 = 0$ is the local equation of E, the assertion follows.

For any curve F through P we have $[E].\pi^*[F] = \pi_*[E].[F]$ by Lemma 1.2.1. But this vanishes, since $\pi(E)$ is a point.

In particular, let L be a smooth curve through P in T, L' its strict transform. Then $\pi^*[L] = [L'] + [E]$ and so $0 = [E].\pi^*[L] = [E].[L'] + [E].[E]$, while by Lemma 3.4.1, $[E].[L'] = 1$. Hence $[E].[E] = -1$.

If G is a further divisor in S, then by the preceding result, $\pi^*[F].\pi^*[G] = [F'].\pi^*[G]$, and by Lemma 1.2.1 this is equal to $\pi_*[F'].[G]$, i.e. to $[F].[G]$. □

We return to the general situation, and recall the proximity matrix of Section 3.5. The same definition applies here.

$$p_{i,j} := \begin{cases} 1 & \text{if } i = j, \\ -1 & \text{if } i < j \text{ and } O_j \text{ is proximate to } O_i, \\ 0 & \text{otherwise.} \end{cases}$$

Again the proximity matrix P, and hence also its inverse Q, are (upper) unitriangular.

Lemma 8.1.3 *In T, we have $[E_i] = \sum_j p_{i,j}[\widehat{E}_j]$.*

Proof In terms of exceptional curves, O_j is proximate to O_i if and only if O_j lies on $E_i^{(j)}$, or equivalently, if $E_i^{(j+1)}$ and $E_j^{(j+1)}$ intersect (in T_{j+1}). Since the multiplicity of $E_i^{(j)}$ at O_j is necessarily 1 (by Lemma 3.4.1), it follows from (i) of Lemma 8.1.2 that $\pi_j^*[E_i^{(j)}] = [E_i^{(j+1)}] + [E_j^{(j+1)}]$ if O_j is proximate to O_i and $\pi_j^*[E_i^{(j)}] = [E_i^{(j+1)}]$ if not. Thus in both cases, $\pi_j^*[E_i^{(j)}] = [E_i^{(j+1)}] - p_{i,j}[E_j^{(j+1)}]$.

Write $[F_i^{(j)}]$ for the total transform $\pi_{j-1}^* \ldots \pi_{i+1}^*[E_i]$ of $[E_i]$ in T_j. We now show by induction on $j - i$ that, for $i < j$,

$$[F_i^{(j)}] + p_{i,i+1}[F_{i+1}^{(j)}] + \ldots + p_{i,j-1}[F_{j-1}^{(j)}] = [E_i^{(j)}].$$

This is a tautology if $j = i + 1$. If it holds for particular values of i and j then, applying π_j^* to both sides, we obtain

$$[F_i^{(j+1)}] + p_{i,i+1}[F_{i+1}^{(j+1)}] + \ldots + p_{i,j-1}[F_{j-1}^{(j+1)}] = \pi_j^*[E_i^{(j)}],$$

and the result follows for i and $j + 1$ since, as we have just shown,

$$p_{i,j}[F_j^{(j+1)}] + \pi_j^*[E_i^{(j)}] = [E_i^{(j+1)}].$$

Taking $j = k$ in the equation just obtained gives

$$[E_i] = [\widehat{E}_i] + p_{i,i+1}[\widehat{E}_{i+1}] + \cdots + p_{i,k}[\widehat{E}_k],$$

which, since $p_{i,i} = 1$ and $p_{i,j} = 0$ for $i > j$, proves the result. □

Corollary 8.1.4

(i) *The classes* $\{[\widehat{E}_i] \,|\, 1 \leq i \leq k\}$ *also form a basis of* $H_2(T)$.

(ii) *The proximity matrix* P *is the matrix of the change of basis between* $\{[E_i] \,|\, 1 \leq i \leq N\}$ *and* $\{[\widehat{E}_i] \,|\, 1 \leq i \leq N\}$.

(iii) *The classes* $[\widehat{E}_i]$ *are mutually orthogonal and each has square* -1.

Proof Assertion (i) follows since the $[E_i]$ are a basis and $[\widehat{E}_i] - [E_i]$ is a linear combination of the cycles $[E_j]$ with $j > i$; (ii) is a restatement of Lemma 8.1.3.

By Lemma 8.1.2 (ii), $[E_i^{(i+1)}]$ has self-intersection -1; by 8.1.2 (iii), it has zero intersection number with the total transform of $[E_j]$ for $j < i$; and by 8.1.2 (iv), we have the same intersection numbers for the total transforms $[\widehat{E}_i]$ and $[\widehat{E}_j]$: thus (iii) follows. □

In particular, the intersection matrix of $H_2(T)$ is negative definite. It is also unimodular, which we could have deduced from Poincaré duality of the 4-manifold T, or, better, of the closed manifold obtained from T by attaching a 4-disc to the 3-sphere ∂T.

Since P has determinant ± 1, its inverse matrix Q also has integer entries. We also write $[\epsilon_i]$ for the negative dual basis to $[E_i]$, in the sense that $[E_i].[\epsilon_j] = -\delta_{i,j}$. We have the immediate corollaries

Corollary 8.1.5 *We have*

(i) $[\widehat{E}_i] = \sum_j q_{i,j}[E_j]$,

(ii) $[\epsilon_i] = \sum_k q_{k,i}[\widehat{E}_k] = \sum_{j,k} q_{k,j}q_{k,i}[E_j]$; *in particular,* $[\epsilon_0] = [\widehat{E}_0]$,

(iii) $[E_i].[\widehat{E}_k] = \sum_j p_{i,j}[\widehat{E}_j].[\widehat{E}_k] = -\sum_j p_{i,j}\delta_{j,k} = -p_{i,k}$.

(iv) *The intersection matrix of the basis* $\{E_i\}$ *is* $-PP^t$.

We will see below that many formulae may be expressed using the proximity matrix, or more economically using cycles.

We can also calculate intersection multiplicities in a more direct manner. We see inductively from the definition of the notion of proximity that for $i > j$, E_i meets E_j if and only if O_i is proximate to O_j, but no O_k proximate to O_i is also proximate to O_j. As to the case $i = j$, we have

Lemma 8.1.6

(i) *If C is a curve with multiplicity m at P, and C' its strict transform when P is blown up, $[C'].[C'] = [C].[C] - m^2$.*

(ii) *If there are r_i points O_j proximate to O_i, then $[E_i].[E_i] = -(r_i+1)$.*

Proof We have $0 = [E].\pi^*[C] = [E].([C']+m[E]) = [E].[C'] - m$, using successively (iii), (i) and (ii) of Lemma 8.1.2. Now

$$\begin{aligned}[C].[C] &= \pi^*[C].\pi^*[C] = \pi^*[C].([C']+m[E]) = \pi^*[C].[C'] \\ &= ([C']+m[E]).[C'],\end{aligned}$$

using in turn (iv), (i), (iii) and (i) of Lemma 8.1.2; by the preceding relation, this reduces to $[C'].[C'] + m^2$.

Follows, since at its first appearance E_i has self-intersection number -1, and this is diminished by 1 each time a point on E_i is blown up. □

Assertion (i) of the lemma should be compared with Lemma 4.4.1. We will denote the number $-[E_i].[E_i]$ by a_i from now on.

Example 8.1.1 For a singularity of type $(4; 6, 7)$ we gave the proximity matrix in Example 3.6.1. It follows that the intersection matrix (with respect to the basis $[E_i]$) is

$$-PP^t = \begin{pmatrix} 3 & 0 & -1 & 0 & 0 \\ 0 & 2 & -1 & 0 & 0 \\ -1 & -1 & 3 & 0 & -1 \\ 0 & 0 & 0 & 2 & -1 \\ 0 & 0 & -1 & -1 & 1 \end{pmatrix}.$$

Observe that the diagonal elements do indeed count the proximity relations, and the non-zero off-diagonal elements correspond to the edges (V_0V_2, V_1V_2, V_2V_4 and V_3V_4) in the dual graph previously described.

We can deduce a convenient criterion for equisingularity from these results.

Theorem 8.1.7 *Two curves C and C' are equisingular if and only if there is an isomorphism of the dual trees $\Gamma_R(C) \to \Gamma_R(C')$ of minimal good resolutions which preserves the function a_i on vertices.*

Proof The vertices representing the exceptional curves which intersect the strict transform of C are precisely those with $a_i = 1$. A corresponding exceptional curve has self-intersection number -1, so there is a blow up $T \to T'$ producing this curve.

We now make an induction on the number of vertices in the tree, or equivalently, of blowings up required to produce the tree. At each stage of the procedure we have blowings up $T_r \to S$, $T'_r \to S$ and an isomorphism of the dual graphs respecting self-intersections of the exceptional curves. Choose a curve E_i with $a_i = 1$; we can then blow it down, and also blow down the corresponding curve in T'_r. The graph T_{r+1} is obtained from T_r by removing the vertex V_i; if this has valence 1, remove the edge V_iV_j ending at V_i; while if V_i has valence 2 and is adjacent to V_j and V_k replace the edges V_iV_j and V_iV_k by a single edge V_jV_k. According to Lemma 8.1.6 we must also diminish each of a_j and (in the valence 2 case) a_k by 1. We thus have a tree isomorphism at each stage of the process.

It follows that we can reverse the procedure and produce the trees $\Gamma_R(C)$ and $\Gamma_R(C')$ by sequences of blowings up whose centres correspond at each stage. Hence all the proximity relations are preserved. As this is so for trees resolving the separate branches, the criterion of Proposition 4.3.9 defining equisingularity is satisfied. □

Define the *fundamental cycle* to be $[Z] := \sum_i [\widehat{E}_i]$. We now give two results justifying this definition. Consider the differential 2-form $\omega := dx \wedge dy$ on $S = \mathbb{C}^2$. Then the resolution $\pi : T \to S$ induces the 2-form $\pi^*(\omega) = d(x \circ \pi) \wedge d(y \circ \pi)$ on T. At each point of T this can be expressed as a multiple $\phi(x', y')dx' \wedge dy'$ of the 2-form defined in terms of local coordinates x' and y'. The function $\phi(x', y')$ will vanish only along the exceptional curves E_i, and we write $\overline{\nu}_i$ for the order of vanishing along E_i.

Proposition 8.1.8 *We have*

$$[Z] = \sum_i \overline{\nu}_i [E_i], \qquad \overline{\nu}_j = 1 - \sum_{i<j} p_{i,j} \overline{\nu}_i$$

and, for each i,

$$[E_i].[E_i] = -2 - [Z].[E_i].$$

Proof We prove the final equation first. We have just seen that $[E_i].[E_i] = -\sum_j p_{i,j}^2$, while it follows from the definition of $[Z]$ that $[Z].[E_i] = -\sum_j p_{i,j}$. Thus $([E_i] + [Z]).[E_i] = -\sum_j p_{i,j}(p_{i,j} + 1)$. The jth term cancels if $p_{i,j}$ is 0 or -1; otherwise $i = j$ and we get $-1 - 1 = -2$.

We calculate the $\overline{\nu}_i$ by induction on the blowings up. Suppose at some stage we can express $\pi^*\omega$ as the product of ${x'}^r {y'}^s dx' \wedge dy'$ and a function non-vanishing at the origin, and perform a further blow up, setting $x' = x''y''$, $y' = y''$. Then substitution gives $(x''y'')^r {y''}^s (x''dy'' + y''dx'') \wedge dy''$, multiplied by a function non-vanishing at the origin. This reduces to $x''^r y''^{r+s+1} dx'' \wedge dy''$, which vanishes to order $r+s+1$ along the new exceptional curve $y''=0$. This order is the sum of 1 and the values (r and s) of the values of $\overline{\nu}_i$ at the infinitely near points to which the new point is proximate (if it is proximate only to one point, take $s=0$). Thus $\overline{\nu}_j = 1 - \sum_{i<j} p_{i,j}\overline{\nu}_i$, so $1 = \sum_i p_{i,j}\overline{\nu}_i$. We deduce that $\overline{\nu}_k = \sum_j q_{j,k}$, and so $[Z] = \sum_j [\widehat{E}_j] = \sum_{j,k} q_{j,k}[E_k] = \sum_k \overline{\nu}_k [E_k]$. □

8.2 The exceptional divisor of a curve

We recall from Section 3.5 that in the case when T is constructed as a good resolution of the curve C, the multiplicity $m_i(C)$ was defined as the multiplicity at O_i of the strict transform $C^{(i)}$ of C in T_i and that, by Lemma 3.5.3, we have $m_r(C) = q_{r,N-1}(C)$. We now generalise these results by defining and studying the exceptional cycle determined by C.

For any blow up $\pi : T \to S$, and any curve germ C in S, write $m_i(C)$ for the multiplicity at O_i of the strict transform $C^{(i)}$ of C; thus $m_i(C) = 0$ if the strict transform does not pass through O_i. Let us calculate the total transform $\pi^*[C]$ of $[C]$ in $T = T_N$. This certainly contains the strict transform $[C^{(N)}]$. Write $[C]_E := \pi^*[C] - [C^{(N)}]$ for the difference cycle.

Lemma 8.2.1 *We have* $[C]_E = \sum_i m_i(C)[\widehat{E}_i]$.

Proof We prove the result by induction. It then suffices to consider a single blow up $\pi_j : T_j \to T_{j-1}$. If $C^{(j-1)}$ does not pass through the centre O_{j-1} of the blow up, then $\pi_j^*[C^{(j-1)}] = [C^{(j)}]$; if it does, then by Lemma 8.1.2 we must add $[E_{j-1}]$ multiplied by the multiplicity of $C^{(j-1)}$ at O_{j-1}, which by definition is m_{j-1}. □

This leads to numerous formulae. First we have

Corollary 8.2.2 *For any curve C, we have*

(i) $m_i(C) = -[C]_E.[\widehat{E}_i]$. *In particular, C has multiplicity* $m_0(C) = -[C]_E.[\widehat{E}_0]$.

(ii) $[C]_E = \sum_{i,j} m_i(C) q_{i,j} [E_j]$.

Proof Follows from the lemma and the orthonormal property of the $[\widehat{E}_i]$.

Follows by substituting for $[\widehat{E}_i]$ from Corollary 8.1.5 (i) in the lemma. □

The proximity equations of Proposition 3.5.1 for the curve C do not hold in the present context, since T need not contain, together with a curve E_i corresponding to a given infinitely near point O_i of C, curves corresponding to all the infinitely near points of C proximate to it. We thus only have inequalities

$$m_i(C) \geq \sum \{m_j(C) \mid O_j \text{ proximate to } O_i\}. \tag{8.1}$$

Define the *defects*

$$\delta_i(C) := m_i(C) - \sum \{m_j(C) \mid O_j \text{ proximate to } O_i\} = \sum_j p_{i,j} m_j(C),$$

so the proximity inequalities just assert non-negativity of the defects.

It follows from the definition that $m_i(C) = \sum_j q_{i,j}\delta_j(C)$. The following is now immediate.

Corollary 8.2.3 *For any curve C, we have*

(i) $[C]_E = \sum_k \delta_k(C)[\epsilon_k] = \sum_{i,j,k} \delta_k(C) q_{i,k} q_{i,j} [E_j]$.

(ii) $[C]_E.[E_j] = (\sum_i m_i(C)[\widehat{E}_i]).[E_j] = -\sum_i p_{j,i} m_i(C) = -\delta_j(C)$.

Observe that $\delta_i(C) \geq 0$ also follows from (ii), since $\delta_i(C) = -[C]_E.[E_i]$, which is equal to $[C^{(N)}].[E_i]$ since $[E_i]$ is orthogonal to the total transform $\pi^*(C)$, and the intersection number of the curves $C^{(N)}$ and E_i – which have no component in common – is non-negative.

In Section 3.5 we introduced the notion of a curvette: recall that, for any infinitely near point O_i, a curvette ϵ_i is the image in S of a smooth curve in T meeting E_i transversely in a single point which lies on no other E_j. For example, a general line through O in S is a curvette ϵ_0. The strict transform of C is a curvette if and only if C is a single branch, and $\pi : T \to S$ provides a good resolution of it. We have a good resolution of a curve C with several branches if and only if the strict transform of C is a disjoint union of curvettes. We will say for short that 'T resolves C'.

The only failure of the proximity inequalities for a curvette ϵ_i is at the infinitely near point O_i itself, and we have $\delta_j(\epsilon_i) = 0$ if $i \neq j$ and 1 if $i = j$. Hence $[\epsilon_i]_E = [\epsilon_i]$, justifying our notation.

We define a partial ordering on the group H of exceptional cycles by:

$$\sum a_i[E_i] \geq 0 : \text{ if } a_i \geq 0 \text{ for each } i.$$

Define

$$\mathcal{E} := \{[D] \in H \mid (\forall i)\ [D].[E_i] \leq 0\}.$$

Equivalently, $[D] = \sum d_i[\widehat{E}_i] \in \mathcal{E}$ if and only if the sequence d_i satisfies the proximity inequalities. The letter $\mathcal{E}$ stands for *effective*. This is explained by

Proposition 8.2.4

(i) *We have $[x] \in \mathcal{E}$ if and only if there is a curve C such that $[C]_E = [x]$.*

(ii) *$\mathcal{E}$ is the free additive semigroup consisting of the non-negative linear combinations of the $[\epsilon_i]$.*

(iii) *For any $[x] \in \mathcal{E}$, $[x] \geq 0$.*

Proof The assertion that for any C, $[C]_E \in \mathcal{E}$ is just a reformulation of the proximity inequalities, as we have just seen. Conversely, if $[x] \in \mathcal{E}$, then for each i, $b_i := -[x].[E_i] \geq 0$. Thus $[x] = \sum_i b_i[\epsilon_i]$. Hence if C is a union of disjoint curvettes, with just b_i of them corresponding to E_i for each i, we have $[C]_E = [x]$.

Follows since the $[\epsilon_i]$ are the dual base to $[E_i]$. Finally, that $[C]_E \geq 0$ follows from (ii) of Corollary 8.2.2, since the terms $q_{i,j} \geq 0$. □

We can apply these ideas to the intersection number of two curves C and D in S.

Lemma 8.2.5 *Let C, D be curves in S and $\pi : T \to S$ as above. Then we have*

$$C.D = C^{(N)}.D^{(N)} - [C]_E.[D]_E.$$

If (and only if) the strict transforms $C^{(N)}$ and $D^{(N)}$ of C and D in T are disjoint, we have $C.D = -[C]_E.[D]_E$.

Proof Consider the effect of blowing up. By Lemma 4.4.1, we have $C.D = m_0(C)m_0(D) + C^{(1)}.D^{(1)}$. Applying this to the sequence of blowings up that defines T, we arrive at $C.D = \sum_i m_i(C)m_i(D) + C^{(N)}.D^{(N)}$.

By Lemma 8.2.1 applied to C and D, and the orthonormal property of the $[\widehat{E}_i]$, we have $[C]_E.[D]_E = -\sum_i m_i(C)m_i(D)$. The result follows.

We can also argue directly, using $C.D = \pi_*(\pi^*(C)).D = \pi^*(C).\pi^*(D)$. Now substitute the decompositions $\pi^*(C) = C^{(N)} + [C]_E$ and $\pi^*(D) = D^{(N)} + [D]_E$, and subtract the expressions $\pi^*(C).[D]_E$ and $[C]_E.\pi^*(D)$ which vanish, since exceptional curves are orthogonal to strict transforms. Cancelling out yields the result.

The final assertion is now immediate. □

It follows that, if the strict transforms of C and D are disjoint, $C.D = -[C]_E.[D]_E = \sum_i m_i(C)m_i(D)$. There is an analogous result for a single curve. By Theorem 6.5.9, if π resolves C, $\mu(C) = \sum_i m_i(C)(m_i(C) - 1) - r(C) + 1$, where $r(C)$ is the number of branches of C. We can rewrite these expressions in terms of cycles:

$$\sum_i m_i(C)(m_i(C) - 1) = -[C]_E.([C]_E - [Z]), \tag{8.2}$$

$$r(C) = \sum_i \delta_i(C) = -[C]_E.\left(\sum_i [E_i]\right). \tag{8.3}$$

Thus $\mu(C) - 1 = -[C]_E.([C]_E - [Z] + \sum_i [E_i])$.

By applying the above formulae to the curve C_f defined by $f = 0$, we obtain identities which we can rewrite in terms of the function f.

We have seen that if f is a function on S having order n at $O = O_0$, then at each point $P \in E_0$, $f \circ \pi_1$ is divisible by z^n, where $z = 0$ is a local equation for E_0. We say that f *vanishes to order at least n along E_0*, and extend this terminology to the vanishing of $f \circ \pi$ along the curves E_i in T. This is equivalent to saying that the coefficient of $[E_i]$ in $\pi^*[C_f]$ is at least n.

Write $[f]_E$ for $[C_f]_E$, and $m_j(f)$ for the order of vanishing of (the pullback of) a function f along the curve E_j. Then $m_j(f)$ is the coefficient of $[E_j]$ in the pullback $\pi^*[C_f]$, so $[f]_E = \sum_i m_i(f)[E_i]$. By Corollary (8.2.2) (iii), this gives

$$m_j(f) = [f]_E.[\epsilon_j] = \sum_{i,k} \delta_k(C_f) q_{i,k} q_{i,j}. \tag{8.4}$$

Let B be an irreducible curve. Then, as in Section 4.3, a parametrisation $\gamma : \mathbb{C} \to \mathbb{C}^2$ of B defines a ring homomorphism $\gamma^* : \mathcal{O}_O \to \mathbb{C}\{t\}$, and for each f we may consider the order $m_B(f)$ of $\gamma^*(f)$. This is independent of the choice of parametrisation, and equals the intersection number of B with the curve C_f. Hence $m_B(f) \geq \sum_i m_i(B)m_i(C_f)$, with equality if and only if the strict transforms $B^{(N)}$ and $C_f^{(N)}$ are disjoint.

We conclude this section by discussing the effect on all the above of an additional blowing up $\pi_N : T' = T_{N+1} \to T = T_N$. We use π_N^* to embed the group of cycles on T in that on T'. By Lemma 8.1.2, this respects intersection numbers. The proximity matrix P acquires an additional row and column; so do its inverse Q and the products P^tP and Q^tQ (but not PP^t and QQ^t).

Matters are clearer in terms of the group of exceptional cycles. Taking strict transforms embeds the group of such cycles on T in that of T'; this embedding preserves intersection numbers and (according to Lemma 8.2.1) respects the class $[C]_E$ of a curve C unless C passes through the centre O_N of the final blow-up. Thus the intersection number $[C]_E.[D]_E$ changes only if O_N lies on the strict transforms (in T) of both C and D.

8.3 Functions on the tree

In this section we study three numbers $m_i, M_i(C)$ and $\overline{\nu}_i$, each defined for each vertex V_i of the dual tree Γ_R of a blowing up $\pi : T \to S$. There are several ways of calculating these numbers: inductively on the successive blowings up, using properties of curvettes, and by a type of recurrence relation. The use of curvettes leads to explicit formulae when we work on the minimal resolution of an irreducible germ C; the recurrence relation allows us, in particular, to study how the quotients $m_i/M_i(C)$, $\nu_i/M_i(C)$ behave as we step along the tree, and thus leads on to basic properties of the so-called topological zeta function.

We define the number m_i as the multiplicity of a curvette ϵ_i. Thus by Corollary 8.2.2 (i), $m_i = -[\epsilon_i].[\widehat{E}_0]$. Since the $[\epsilon_i]$ are the negative dual base to the $[E_i]$, $[\widehat{E}_0] = \sum_i m_i[E_i]$; thus $m_i = q_{0,i}$. Since $[\epsilon_0] = [\widehat{E}_0]$, we have

$$[\epsilon_0] = [\widehat{E}_0] = \sum_i m_i[E_i] = \sum_{i,j} m_i p_{i,j}[\widehat{E}_j], \tag{8.5}$$

where the final equality arises by substituting for $[E_i]$ from Lemma 8.1.3.

If C is defined by $f = 0$ we write $M_i(C)$ for the order of vanishing of $\pi^*(f)$ along E_i; thus the $M_i(C)$ are the coefficients in $[C]_E = \sum_i M_i(C)[E_i]$. Recalling Lemma 8.2.1, we find

$$[C]_E = \sum_i M_i(C)[E_i] = \sum_{i,j} M_i(C)p_{i,j}[\widehat{E}_j] = \sum_i m_j(C)[\widehat{E}_j]. \tag{8.6}$$

Since, by Corollary 8.2.3, we have $[C]_E = \sum_{i,j,k} \delta_j(C) q_{k,j} q_{k,i} [E_i]$, we can also express $M_i(C)$ in terms of the proximity matrix by $M_i(C) = \sum_{j,k} \delta_j(C) q_{k,j} q_{k,i}$: thus if C has a single branch, we read the $M_i(C)$ off from the final row (or column) of $(PP^t)^{-1}$.

The number $\overline{\nu}_i$ was defined in Proposition 8.1.8, where the following formulae are given. We set $\nu_i := 1 + \overline{\nu}_i$.

$$[Z] = \sum_i \overline{\nu}_i [E_i] = \sum_{i,j} \overline{\nu}_i p_{i,j} [\widehat{E}_j] = \sum_j [\widehat{E}_j]. \tag{8.7}$$

Equating coefficients of $[\widehat{E}_j]$ in these formulae gives the following, describing the behaviour under blowing up.

$$m_i = \delta_{i,0} + \sum \{ m_j \mid O_i \text{ proximate to } O_j \}; \tag{8.8}$$

$$M_i(C) = m_i(C) + \sum \{ M_j(C) \mid O_i \text{ proximate to } O_j \}; \tag{8.9}$$

$$\overline{\nu}_i = 1 + \sum \{ \overline{\nu}_j \mid O_i \text{ proximate to } O_j \}. \tag{8.10}$$

In each case, the number of terms in the sum is either 1 or 2. The same relations also give rules for finding the numbers if additional blowings up are performed.

Several formulae can be regarded as 'recurrence relations' along Γ_R. We base them on the identities (8.5), (8.6) and (8.7).

For any cycle $[X] = \sum_j x_j [E_j]$, taking the intersection number with $[E_i]$ gives the identity $[X].[E_i] = x_i [E_i].[E_i] + \sum_{j \neq i} x_j [E_i].[E_j]$. Recall that for $i \neq j$, $[E_i].[E_j]$ equals 1 if the corresponding vertices V_i, V_j of Γ_R are adjacent, and 0 otherwise. The final term is thus the sum of the x_j for those vertices V_j adjacent to V_i. The self-intersection $[E_i].[E_i]$ is a negative integer which we will denote by $-a_i$. Thus the identity reduces to

$$a_i x_i = -[X].[E_i] + \sum \{ X_j \mid V_j \text{ adjacent to } V_i \}.$$

In the three cases mentioned, this gives

$$a_i m_i = \delta_{i,0} + \sum \{ m_j \mid V_j \text{ adjacent to } V_i \}; \tag{8.11}$$

$$a_i M_i(C) = \delta_i(C) + \sum \{ M_j(C) \mid V_j \text{ adjacent to } V_i \}; \tag{8.12}$$

and $a_i \overline{\nu}_i = [Z].[E_i] + \sum \{ \overline{\nu}_j \mid V_j \text{ adjacent to } V_i \}$. But $[Z].[E_i] = -\sum_j p_{i,j}$ $= 2 - a_i$, so this becomes $a_i(\overline{\nu}_i + 1) = 2 + \sum_j \overline{\nu}_j$ or, since $\nu_i = 1 + \overline{\nu}_i$,

$$a_i \nu_i = (2 - v_i) + \sum \{ \nu_j \mid V_j \text{ adjacent to } V_i \}, \tag{8.13}$$

where v_i denotes the valence of the vertex V_i in Γ_R.

Before proceeding, we note a simple application.

Proposition 8.3.1 *Two curves C and C' are equisingular if and only if there is an isomorphism of the dual trees $\Gamma_R(C) \to \Gamma_R(C')$ of minimal good resolutions which preserves the function ν_i, or the initial vertex V_0 and the function m_i, on vertices.*

Proof It follows from (8.13) or (8.11) respectively that the isomorphism also preserves the function a_i. The result now follows from Theorem 8.1.7. □

The relations (8.11), (8.12) and (8.13) can conveniently be used to calculate in turn the values of the coefficients m_i, $M_i(C)$ and ν_i at the various vertices of Γ_R. It will be convenient to extend the definitions to the vertices W_j of Γ_R^+ as follows:

$$m_j := 0, \quad M_j(C) := 1, \quad \nu_j := 1.$$

Then all the relations (8.11), (8.12) and (8.13) continue to hold at the vertices V_i (but not at the W_j) if the term $\delta_i(C)$ in (8.12) is deleted and v_i now denotes the valence in Γ_R^+.

Example 8.3.1 Consider again a germ with Puiseux characteristic $(8; 11)$. We have seen that the multiplicity sequence is $(8, 3, 3, 2, 1, 1)$; thus O_2 and O_3 are proximate to O_0, O_4 to O_2 and O_5 to O_3. Hence $a_0 = 4$, $a_1 = 2$, $a_2 = 3$, $a_3 = 3$, $a_4 = 2$ and $a_5 = 1$. The dual graph is a chain, with the vertices in the order $V_0V_3V_5V_4V_2V_1$; for Γ_R^+ we attach an edge V_5W.

Now $M_0(C)$ is the multiplicity, 8, of C. From the recurrence relation (8.12) we calculate successively $M_3(C) = 32$, $M_5(C) = 88$, $M_4(C) = 55$, $M_2(C) = 22$ and $M_1(C) = 11$. Similarly, $m_0 = 1$, as it is the multiplicity of a line, and now $m_3 = 3$, $m_5 = 8$, $m_4 = 5$, $m_2 = 2$ and $m_1 = 1$. Finally, $\overline{\nu}_0 = 1$ so $\nu_0 = 2$ and we find $\nu_3 = 7$, $\nu_5 = 19$, $\nu_4 = 12$, $\nu_2 = 5$ and $\nu_1 = 3$. We obtain the same numbers using successive blowing up (8.9) etc.: here, of course, we obtain the $M_i(C)$ in order of increasing i: 8, 11, 22, 32, 55, 88; as we also get from the proximity matrix. (Similarly, the sequence m_i is 1, 1, 2, 3, 5, 8 and the sequence ν_i is 2, 3, 5, 7, 12, 19.)

We can display this information on the resolution graph by marking the vertex V_i with $\binom{a_i}{m_i}\binom{M_i(C)}{\nu_i}$:

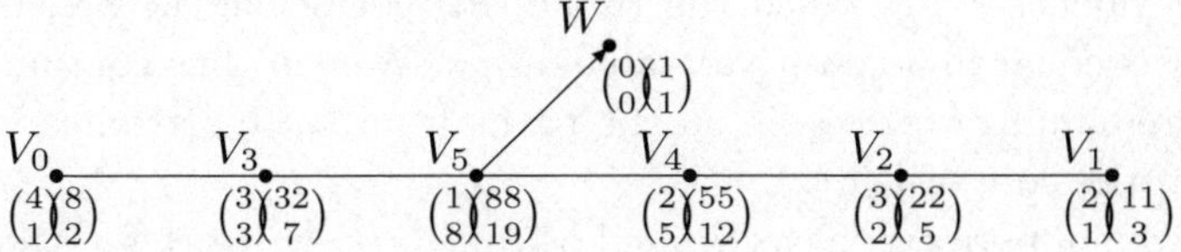

Example 8.3.2 We return again to a branch with Puiseux characteristic (4; 6, 7). We have seen in Example 3.6.1 that Γ_R^+ consists of a chain $V_0V_2V_4W$ with additional edges V_2V_1 and V_4V_3. From the proximity relations, $a_0 = a_2 = 3$, $a_1 = a_3 = 2$ and $a_4 = 1$. Using the blowing up relations we obtain successive values 4, 6, 12, 13, 26 for $M_i(C)$, 1, 1, 2, 2, 4 for m_i and 2, 3, 5, 6, 11 for ν. We can verify that these satisfy the recurrence relations in terms of the graph. We display the functions, with the same notation, as

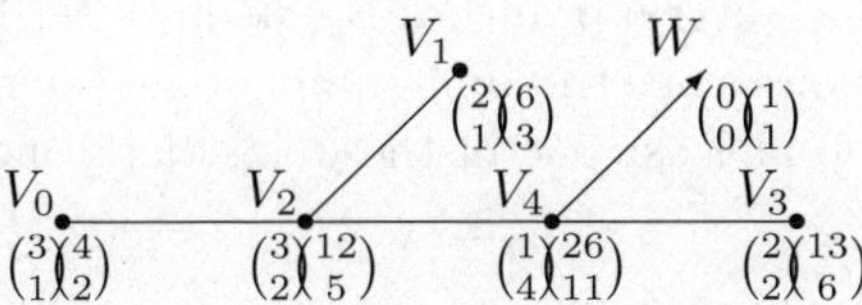

Now suppose C resolved by π, and study the behaviour of the ratio $M_i(C)/m_i$. If V_i has valence 1 in $\Gamma_R^+(C)$, $i \neq 0$, and the adjacent vertex is V_j, then $a_im_i = m_j$ and $a_iM_i(C) = M_j(C)$; in particular, $M_i(C)/m_i = M_j(C)/m_j$. If V_i has valence 2, and the adjacent vertices are V_j and V_k, then we obtain $M_i(C)/m_i = (M_j(C)+M_k(C))/(m_j+m_k)$. Since all the coefficients are positive, either $M_i(C)/m_i = M_j(C)/m_j = M_k(C)/m_k$ or (up to reordering of V_j and V_k) $M_j(C)/m_j < M_i(C)/m_i < M_k(C)/m_k$. This shows that the behaviour of the quotients $M_i(C)/m_i$ on the tree has a certain regularity.

Proposition 8.3.2 *The function $M_i(C)/m_i$ is strictly increasing along any path in Γ_R^+ from V_0 to an arrowhead vertex.*

For each edge not on such a path, it takes the same value at both ends of the edge.

Proof We prove the second assertion first. Choose a vertex V_i of valence 1, not V_0, and let V_iV_j be the edge at this vertex. By the above, the function takes the same values at V_i and V_j.

Now delete the vertex V_i and the edge V_iV_j from the graph, and decrease a_j by a_i. The relations (8.11) and (8.12) continue to hold. We can thus repeat the argument. Note that at no stage can we have a vertex

V_k of valence 1 for which the (modified) coefficient a_k vanishes, since $a_k m_k$ is equal to a (positive) value of m_r. We can thus continue till the only remaining vertices lie on the paths from V_0 to arrowhead vertices, i.e. on the core of $\Gamma_R(C)$.

Next suppose we have a single branch B; write γ for the core of $\Gamma_R(B)$. Now $a_0 m_0 = 1 + m_1$, $a_0 M_0(B) = M_1(B)$. Thus $M_0(B)/m_0 = M_1(B)/(1 + m_1) < M_1(B)/m_1$. We now proceed in steps along γ: let V_i, V_j and V_k arise consecutively in order. We have seen that for any other vertex V_r adjacent to V_j we have $M_r(B)/m_r = M_j(B)/m_j = \lambda$, say. Since $a_j M_j(B) - \lambda a_j m_j$ vanishes, so does $(M_i(B) - \lambda m_i) + (M_k(B) - \lambda m_k) + \sum(M_r(B) - \lambda m_r)$, and hence also $(M_i(B) - \lambda m_i) + (M_k(B) - \lambda m_k)$. Since $M_i(B) < \lambda m_i$, it follows that $M_k(B) > \lambda m_k$. Thus $M_i(B)/m_i$ is strictly increasing along γ, and hence is monotone increasing along any arc in $\Gamma_R^+(C)$ starting at V_0.

In general if C has several branches B_j, we have $M_i(C) = \sum_j M_i(B_j)$. The result now follows since the sum of monotone (increasing) sequences is monotone, and if at least one of these is strictly monotone, so is the sum. □

This result gives the further analysis promised in Section 4.4 of the relation between different ways of counting contact of branches: for if ϵ_i is a curvette at E_i it calculates the quotient $M_i(C)/m_i = \epsilon_i.C/m(\epsilon_i)$ in terms of the infinitely near point V_i.

Addendum 8.3.3 *For each branch B_j of C, the function $M_i(B_j)/m_i$ is strictly increasing along the geodesic $V_0 W_j$, and takes the same value at both ends of an edge not on this path.*

This follows by the same argument, since for $k \neq j$, $M_i(B_j)$ vanishes at $V_i = W_k$.

8.4 The topological zeta function

We next seek a similar analysis of the quotient M_i/ν_i. Here we must use the identity (8.13). The initial vertex V_0 now plays no special role, but the arrowhead vertices still do. We define the topological zeta function of the curve-germ by an explicit formula, and establish its basic properties using this analysis.

Let V_i be any vertex of Γ_R of valence 1. Then $M_i/\nu_i = a_i M_i / a_i \nu_i = M_j/(1 + \nu_j) < M_j/\nu_j$. As before, if V_i has valence 2, and the adjacent vertices are V_j and V_k, then either $M_i/\nu_i = M_j/\nu_j = M_k/\nu_k$ or (up to reordering of V_j and V_k) $M_j/\nu_j < M_i/\nu_i < M_k/\nu_k$. It also follows that

$M_j\nu_i - M_i\nu_j = M_i\nu_k - M_k\nu_i$, so the number $M_j\nu_i - M_i\nu_j$ defined on oriented edges V_iV_j is constant along a chain where intermediate vertices have valence 2.

For any edge V_iV_j of Γ_R^+, define $\alpha_{i,j} := \nu_j - \nu_i\frac{M_j}{M_i}$. For each i, write $L(i)$ for the set of vertices V_j of Γ_R^+ such that V_iV_j is an edge. Then by (8.13) and (8.12) we have, for each vertex V_i of Γ_R,

$$\begin{aligned}\sum_{j\in L(i)} \alpha_{i,j} &= \sum_{j\in L(i)} \nu_j - \nu_i\frac{\sum_{j\in L(i)} M_j}{M_i} = v_i - 2 + a_i\nu_i - \nu_i\frac{a_iM_i}{M_i}\\ &= v_i - 2. \end{aligned} \tag{8.14}$$

Lemma 8.4.1 *For any edge of $\Gamma_R^+(C)$, $|\alpha_{i,j}| \leq 1$. If equality holds, V_i has valence 1, and then $\alpha_{i,j} = -1$.*

Proof For any chain $V_1V_2\ldots V_k$ in Γ_R^+ such that V_1 has valence 1 and is not an arrowhead vertex and the intermediate vertices have valence 2, we have

$$M_{i+1}\nu_i - M_i\nu_{i+1} = M_2\nu_1 - M_1\nu_2 = a_1M_1(\nu_1) - M_1(a_1\nu_1 - 1) = M_1.$$

Thus $\alpha_{1,2} = -1$ and, since the other M_i exceed M_1 (since those vertices were created after V_1 in the blowing up process), the conclusion holds for the other edges in the chain.

First suppose C irreducible, and consider the minimal good resolution. The argument of the preceding paragraph covers all the edges not in the core. We number the vertices along the core consecutively from V_0 to W, and consider the edges in this order. If there is no rupture vertex, the result is already proved. Otherwise, if at some stage the next vertex V_i has valence > 2, it was created by blowing up the intersection point of two earlier exceptional curves E_j and E_{i-1}, so has neighbours V_j, V_{i+1} and V_{i+1}. We have shown that $0 < \alpha_{i,j} < 1$ and by induction $|\alpha_{i,i-1}| < 1$; thus $\alpha_{i,j} + \alpha_{i,i-1} < 2$. Since $M_i = M_{i-1} + M_j + m_i$ and $\nu_i = \nu_{i-1} + \nu_j$, we have

$$\alpha_{i,j} + \alpha_{i,i-1} = \nu_i - \nu_i\frac{M_{i-1} + M_j}{M_{i-1} + M_j + m_i} > 0.$$

Using the relation (8.14), we obtain $|\alpha_{i,i+1}| < 1$. If the next vertex V_i has valence 2, then $\alpha_{i,i+1} = \frac{M_{i-1}}{M_i}\alpha_{i-1,i}$, so as $M_i > M_{i-1}$, we again have $|\alpha_{i,i+1}| < 1$. It follows in either case that $|\alpha_{i+1,i}| < 1$.

Next we continue to suppose C irreducible, but consider an arbitrary resolution. It suffices to show that the property is not destroyed by a single blow-up. There are several cases. Blowing up a point of E_i lying on no other exceptional curve nor on the strict transform of C produces a new edge V_iV_I; Since V_I has valence 1, the result here was already proved above. Blowing up an intersection point $E_i \cap E_j$ subdivides the edge V_iV_j by inserting a vertex V_I. By (8.9) and (8.10) we have $M_I = M_i + M_j$ and $\nu_I = \nu_i + \nu_j$. Thus $M_I\nu_j - M_j\nu_I = M_i\nu_j - M_j\nu_i$; and we already know that $|M_i\nu_j - M_j\nu_i| \leq M_j < M_I$. Finally, if we blow up a point of $E_i \cap C$, we produce new edges V_iV_I and V_IV_C; we have $M_I = M_i + 1$ and $\nu_I = \nu_i + 1$, and $\nu_I M_i - \nu_i M_I = \nu_C M_I - \nu_I M_C = M_i - \nu_i$ which is positive and $< M_i < M_I$.

Finally consider C arbitrary, with branches B_j. For any vertex V_i we have $M_i(C) = \sum_j M_i(B_j)$. Hence

$$\alpha_{i,I}(C) = \nu_I - \nu_i \frac{M_I(C)}{M_i(C)} = \frac{\sum_j M_i(B_j)\alpha_{i,I}(B_j)}{\sum_j M_i(B_j)}.$$

The values of α for C are thus weighted means of those for the B_j. Since each of these lies between ± 1, so do the values for C. □

Corollary 8.4.2 *For any vertex V_i of Γ_R^+, either there is at most one adjacent vertex V_j such that $\alpha_{i,j} \leq 0$ or $v_i = 2$ and both values of $\alpha_{i,j}$ vanish.*

Proof By (8.14), $\sum_{j \in L(i)} \alpha_{i,j} = v_i - 2$. Since $|\alpha_{i,j}| \leq 1$ for each j, we could only have $\alpha_{i,j} \leq 0$ for two values of i if two of the values were zero and the rest equal to 1. But $\alpha_{i,j}$ does not take the value $+1$. □

An alternative way to formulate this conclusion is that either $\frac{\nu_j}{M_j} > \frac{\nu_i}{M_i}$ for all but at most one neighbour V_j of V_i or equality holds for both neighbours.

Corollary 8.4.3

(i) *The set of vertices V_i of Γ_R^+ at which the quotient $\frac{\nu_i}{M_i}$ attains its minimum forms a connected chain Y in the graph.*

(ii) *Intermediate vertices in Y have valence 2.*

(iii) *The quotient increases strictly along any path starting at and leaving Y.*

Proof Suppose the minimum value μ is attained at 2 distinct vertices of the tree; join them by a simple path Y. If the value of $\frac{\nu_j}{M_j}$ were not constant along Y, it would exceed μ somewhere; at a point where the quotient takes its highest value along Y, both neighbours would yield lower values, contradicting Corollary 8.4.2. Now (ii) also follows from that corollary.

Consider a simple path $V_0V_1\dots$ with $V_0 \in Y$, $V_1 \notin Y$. Then $\frac{\nu_0}{M_0} < \frac{\nu_1}{M_1}$, so by Corollary 8.4.2 we have $\frac{\nu_1}{M_1} < \frac{\nu_2}{M_2}$. We can continue along the path following the same argument. □

It follows from the proof of the lemma that in the case of C irreducible, the minimum is attained (exclusively) at the first rupture point.

We now define the *topological zeta function* of the curve C. Choose any resolution; introduce notation as above. Then set

$$Z_{\text{top}}(s) := \sum_i \frac{\chi(E_i^o)}{\nu_i + sM_i} + \sum_{i,j} \frac{1}{(\nu_i + sM_i)(\nu_j + sM_j)}, \tag{8.15}$$

where the first sum is extended over all vertices of Γ_R^+, E_i^o denotes the complement in E_i of the union of the other curves, χ is the Euler characteristic, and the second sum is extended over edges of Γ_R^+.

For any vertex W_j, the curve E_j is an open disc, E_j^o is the complement of the origin, so has $\chi(E_j^o) = 0$. For any other vertex V_i, E_i is homeomorphic to a 2-sphere, and we obtain E_i^o by making v_i punctures, so $\chi(E_i^o) = 2 - v_i$. We may thus rewrite the formula as

$$Z_{\text{top}}(s) := \sum_i \frac{2 - v_i}{\nu_i + sM_i} + \sum_{i,j} \frac{1}{(\nu_i + sM_i)(\nu_j + sM_j)}.$$

Lemma 8.4.4 *The function $Z_{\text{top}}(s)$ is independent of the choice of resolution.*

Proof It suffices to show that the result is not changed if we alter the resolution by a single blowing up. There are essentially two cases. If the point P to be blown up lies on a single exceptional curve E_i, the blowing up attaches to V_i a new edge V_iV_I with $M_I = M_i$ and $\nu_I = \nu_i + 1$. Since the valency of V_i is then decreased by 1, the effect on the zeta function is to add

$$-\frac{1}{\nu_i + sM_i} + \frac{1}{\nu_i + 1 + sM_i} + \frac{1}{(\nu_i + sM_i)(\nu_i + 1 + sM_i)} = 0.$$

If P is the point $E_i \cap E_j$ (where one of V_i and V_j may be an arrowhead vertex), we introduce a new vertex V_I in the edge V_iV_j; we have

$\nu_I = \nu_i + \nu_j$, $M_I = M_i + M_j$, and the effect is to add

$$\frac{1}{(\nu_i + sM_i)(\nu_I + sM_I)} + \frac{1}{(\nu_j + sM_j)(\nu_I + sM_I)} - \frac{1}{(\nu_i + sM_i)(\nu_j + sM_j)} = 0.$$

□

We now deduce from the foregoing some of the main properties of the topological zeta function.

Theorem 8.4.5

(i) *The function* $Z_{\text{top}}(s)$ *has at most one double pole.*

(ii) *Any pole is of the form* $s = -\frac{\nu_i}{M_i}$ *for some vertex* V_i *which is either an arrowhead vertex or a rupture point.*

(iii) *If* C *has just one branch, there is no double pole, and all the numbers in* (ii) *are poles.*

Proof Any double pole comes from an edge such that the factors $\nu_i + sM_i$ and $\nu_j + sM_j$ are proportional, i.e. $\nu_i M_j - \nu_j M_i = 0$, or equivalently $\alpha_{i,j} = 0$. By Corollary 8.4.3, such an edge belongs to Y. Since all the factors corresponding to vertices in Y are proportional, there is at most one double pole.

Write, for each edge $V_i V_j$ such that $\nu_i M_j - \nu_j M_i \neq 0$, or equivalently $\alpha_{i,j} \neq 0$,

$$\frac{1}{(\nu_i + sM_i)(\nu_j + sM_j)} = \frac{1}{M_i\nu_j - M_j\nu_i}\left(\frac{M_i}{\nu_i + sM_i} - \frac{M_j}{\nu_j + sM_j}\right).$$

The coefficient of $(\nu_i + sM_i)^{-1}$ in this expression is $\frac{M_i}{M_i\nu_j - M_j\nu_i} = \alpha_{i,j}^{-1}$. Thus the coefficient of $(\nu_i + sM_i)^{-1}$ in $Z_{\text{top}}(s)$ is $2 - v_i + \sum_{j\epsilon L(i)} \alpha_{i,j}^{-1}$.

Now if V_i is a vertex of Γ_R of valence 1, $\alpha_{i,j} = -1$, so the coefficient reduces to 0. If V_i has valence 2, then by (8.14) $\sum_{j\epsilon L(i)} \alpha_{i,j} = 0$, so the inverses of the two terms α also cancel.

Suppose V_i has valence 3; label the neighbouring vertices 1,2,3, and abbreviate $\alpha_{i,j}$ to α_j. We have the relation $\alpha_1 + \alpha_2 + \alpha_3 = 1$, and the coefficient of $(\nu_i + sM_i)^{-1}$ is

$$\alpha_1^{-1} + \alpha_2^{-1} + \alpha_3^{-1} - 1 = \frac{\alpha_2\alpha_2 + \alpha_3\alpha_1 + \alpha_1\alpha_2 - \alpha_1\alpha_2\alpha_3}{\alpha_1\alpha_2\alpha_3},$$

which is equal to

$$\frac{1-\alpha_1-\alpha_2-\alpha_3+\alpha_2\alpha_2+\alpha_3\alpha_1+\alpha_1\alpha_2-\alpha_1\alpha_2\alpha_3}{\alpha_1\alpha_2\alpha_3}$$
$$=\frac{(1-\alpha_1)(1-\alpha_2)(1-\alpha_3)}{\alpha_1\alpha_2\alpha_3}.$$

This does not vanish since the $\alpha_{i,j}$ do not take the value 1. □

We will see in Example 8.4.2 that the topological zeta function for an A_1 singularity has a double pole. Another example is given in Exercise 8.7.10.

Example 8.4.1 For a curve with Puiseux characteristic $(8; 11)$ we have described Γ_R^+ and calculated the values of $M_i(C)$ and ν_i in Example 8.3.1. We have 2 vertices (other than W) of valence 1, which contribute the terms $(2+8s)^{-1}$ and $(3+11s)^{-1}$ and 1 vertex of valence 3, which contributes $-(19+88s)^{-1}$ to the zeta function; and 6 edges, which contribute terms which we express in partial fractions as follows:

$$(2+8s)^{-1}(7+32s)^{-1} = 4(7+32s)^{-1} - (2+8s)^{-1},$$

$$(7+32s)^{-1}(19+88s)^{-1} = 11(19+88s)^{-1} - 4(7+32s)^{-1},$$

$$(19+88s)^{-1}(12+55s)^{-1} = 8(19+88s)^{-1} - 5(12+55s)^{-1},$$

$$(12+55s)^{-1}(5+22s)^{-1} = 5(12+55s)^{-1} - 2(5+22s)^{-1}$$

$$(5+22s)^{-1}(3+11s)^{-1} = 2(5+22s)^{-1} - (3+11s)^{-1},$$

and

$$(19+88s)^{-1}(1+s))^{-1} = \frac{1}{69}\left(88(19+88s)^{-1} - (1+s)^{-1}\right).$$

Ten of the fifteen terms cancel in pairs, and the four with denominator $19+88s$ can be combined to give the final conclusion

$$Z(s) = \frac{1}{69}\left(70\frac{19}{19+88s} - \frac{1}{1+s}\right).$$

For a curve with Puiseux characteristic $(4; 6, 7)$ we gave Γ_R^+, $M_i(C)$ and ν_i in Example 8.3.2. Thus the successive vertices $V_0, V_1, V_2, V_3, V_4, W$, with values of the v_i equal to 1,1,3,1,3,1, contribute $(2+4s)^{-1}$, $(3+6s)^{-1}$, $-(5+12s)^{-1}$, $(6+13s)^{-1}$, $-(11+26s)^{-1}$, $(1+s)^{-1}$; and from the edges

$$\begin{aligned}
V_0V_2: &\quad (2+4s)^{-1}(5+12s)^{-1} = -(2+4s)^{-1} + 3(5+12s)^{-1},\\
V_1V_2: &\quad (3+6s)^{-1}(5+12s)^{-1} = -(3+6s)^{-1} + 2(5+12s)^{-1},\\
V_2V_4: &\quad (5+12s)^{-1}(11+26s)^{-1} = 6(5+12s)^{-1} - 13(11+26s)^{-1},\\
V_3V_4: &\quad (6+13s)^{-1}(11+26s)^{-1} = -(6+13s)^{-1} + 2(11+26s)^{-1},\\
V_4W: &\quad (11+26s)^{-1}(1+s)^{-1} = \left(26(11+26s)^{-1} - (1+s)^{-1}\right)/15.
\end{aligned}$$

Summing these, the terms which do not cancel are $(5+12s)^{-1}$, with coefficient $-1+3+2+6=10$, $(11+26s)^{-1}$, with coefficient $-1+2+13+\frac{26}{15} = -\frac{154}{15}$, and $(1+s)^{-1}$ with coefficient $1-\frac{1}{15}=\frac{14}{15}$. Hence

$$Z(s) = \frac{10}{5+12s} - \frac{1}{15}\left(14\frac{11}{11+26s} + \frac{1}{1+s}\right).$$

Example 8.4.2 For a singularity of type A_{2k}, the graph Γ_R^+, with the vertex V_i marked with $\binom{\nu_i}{M_i}$, is

W

$V_0 \quad V_1 \quad V_2 \quad \cdots \quad V_{k-1} \quad V_{k+1} \quad V_k$

$\binom{2}{2} \quad \binom{3}{4} \quad \binom{4}{6} \quad \binom{k+1}{2k} \quad \binom{2k+3}{4k+2} \quad \binom{k+2}{2k+1}$

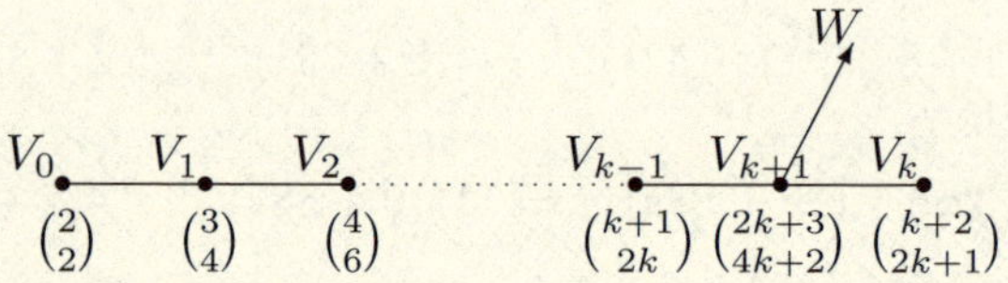

Contributions to $Z(s)$ are:

from V_0, $(2+2s)^{-1}$,
from V_{k+1}, $-((2k+1)+(4k+2)s)^{-1}$,
from V_k, $((k+2)+(2k+1)s)^{-1}$;
from the edge $V_{r-1}V_r$ $(1 \le r < k)$:

$$\begin{aligned}
&(r+1+2rs)^{-1}(r+2+(2r+2)s)^{-1} = \\
&\qquad (r+1)(r+2+(2r+2)s)^{-1} - r(r+1+2rs)^{-1},
\end{aligned}$$

from $V_{k-1}V_{k+1}$:

$$\begin{aligned}
&(2k+3+(4k+2)s)^{-1}(k+1+2ks)^{-1} = \\
&\qquad (2k+1)(2k+3+(4k+2)s)^{-1} - k(k+1+2ks)^{-1}
\end{aligned}$$

from $V_{k+1}V_k$:

$$\begin{aligned}
&(2k+3+(4k+2)s)^{-1}(k+2+(2k+1)s)^{-1} = \\
&\qquad 2(2k+3+(4k+2)s)^{-1} - (k+2+(2k+1)s)^{-1},
\end{aligned}$$

and from $V_{k+1}W$:

$$(2k+3+(4k+2)s)^{-1}(1+s)^{-1} = \frac{1}{2k-1}((4k+2)(2k+3+(4k+2)s)^{-1} - (1+s)^{-1}).$$

All terms cancel except the final $-\frac{1}{2k-1}(1+s)^{-1}$ and the terms involving $(2k+3+(4k+2)s)^{-1}$, whose coefficients are $(-1)+(2k+1)+(2)+\frac{4k+2}{2k-1}$. Adding up gives

$$Z(s) = \frac{1}{2k-1}\left(2k\frac{2k+3}{(2k+3+(4k+2)s)} - \frac{1}{1+s}\right).$$

For a singularity of type A_{2k-1}, the graph Γ_R^+ is

$$\underset{\binom{2}{2}}{V_0} - \underset{\binom{3}{4}}{V_1} - \underset{\binom{4}{6}}{V_2} \cdots \underset{\binom{k+1}{2k}}{V_{k-1}} \to W,\ W'$$

Contributions to $Z(s)$ are:

from V_0, $(2+2s)^{-1}$,
from V_{k-1}, $-((k+1)+2ks)^{-1}$;
from the edge $V_{r-1}V_r$ $(1 \le r < k)$ as before;
and from each of $V_{k-1}W$ and $V_{k-1}W'$:

$$(k+1+2ks)^{-1}(1+s)^{-1} = \frac{1}{k-1}(2k(k+1+2ks)^{-1} - (1+s)^{-1}).$$

Cancelling and collecting terms, this leaves

$$Z(s) = \frac{1}{k-1}\left(k+1\frac{k+1}{(k+1+2ks)} - \frac{2}{1+s}\right).$$

The case $k=1$ of A_1 is exceptional as V_0 then has valence 2; the only contributions are from V_0W and V_0W'; and $Z(s) = 2(2+2s)^{-1}(1+s)^{-1} = (1+s)^{-2}$.

8.5 Calculations for a single branch

Although the emphasis in this chapter is on functions on trees, in the case of a single branch more or less explicit calculations can be given. In particular, for a branch with $g=1$, the calculations involve some rather surprising properties of continued fractions. The general case can be investigated by putting together g continued fraction expansions suitably.

We begin with a curve consisting of a single branch with genus 1: say $x^{a+b} = y^a$, with a and b coprime. The resolution was described in Section 3.6. Suppose, as in (3.2), the steps in the Euclidean algorithm are:

$$\begin{array}{rcll} a & = & bq_1 + r_1 & (0 < r_1 < b) \\ b & = & r_1 q_2 + r_2 & (0 < r_2 < r_1) \\ & \cdots & & \\ r_{f-2} & = & r_{f-1} q_f; & \end{array}$$

write $s_k = \sum_{i=1}^{k} q_i$. Then O_0 has multiplicity a; the next q_1 points have multiplicity b and are proximate to O_0, as is the next, with multiplicity r_1. The next $q_2 - 1$ points also have multiplicity r_1, and are proximate to O_{q_1}. In general, O_{s_k} has multiplicity r_{k-1}; the next q_{k+1} points are proximate to it, each with multiplicity r_k, as is $O_{s_{k+1}+1}$, with multiplicity r_{k+1}; but the point following is proximate only to $O_{s_{k+1}}$. By Lemma 8.1.6 (ii), we have $a_0 = -[E_0]^2 = q_1 + 2$ and in general $a_{s_k} = -[E_{s_k}]^2 = q_{k+1} + 2$; for other values of i, $a_i = -[E_i]^2 = 2$. The pattern changes slightly at the end since $r_{f+1} = 0$; $a_{s_{f-1}} = q_f + 1$ and for the minimal good resolution the final vertex is O_{s_f} with $a_{s_f} = 1$.

The dual graph is a sequence of points on a line, in the order

$$\{0, s_1 + 1, \ldots, s_1 + q_2 = s_2, \ldots, \ldots s_2 + q_3 = s_3, \ldots, s_2 + 1, \\ q_1 = s_1, \ldots, 2, 1\}.$$

Starting from the left, we have the first group, then the third; the odd groups preceding the even ones which conclude with the fourth group, then the second. The point V_{s_f} joined to W is somewhere in the middle. Starting from the left, the sequence of values of a_i is: a $(q_1 + 2)$, followed by $(q_2 - 1)$ equal to 2, a $(q_3 + 2)$, then a string of $(q_4 - 1)$ 2's and so on; while starting from the right we have $(q_1 - 1)$ 2's, then a $(q_2 + 2)$, then $(q_3 - 1)$ 2's and so on; with the modification at the end as just described. Thus the sequence of quotients in the Euclidean algorithm determines the sequence of self-intersection numbers in the resolution on either side of the vertex attached to W, and either of these determines the sequence of quotients, but both the rule, and the 'mirror symmetry' between the two sides, appear somewhat mysterious. We now seek to explain this.

We can write the algorithm in continued fraction notation as

$$\frac{a}{b} = q_1 + \frac{1}{b/r_1} = q_1 + \frac{1}{q_2 + (1/(r_1/r_2))} = \cdots,$$

or, which will be more useful below, in matrix terms. Denote by $A(q)$

the matrix $\begin{pmatrix} q & 1 \\ 1 & 0 \end{pmatrix}$; then $\begin{pmatrix} a \\ b \end{pmatrix} = A(q_1)\begin{pmatrix} b \\ r_1 \end{pmatrix}$, and hence

$$\begin{pmatrix} a \\ b \end{pmatrix} = A(q_1)A(q_2)\cdots A(q_f)\begin{pmatrix} 1 \\ 0 \end{pmatrix}.$$

We also have a negative continued fraction expansion. First write $a = bQ_1 - R_1$ with $0 \le R_1 < b$ and then again continue as above until some R_i vanishes. Denote by $B(Q)$ the matrix $\begin{pmatrix} Q & -1 \\ 1 & 0 \end{pmatrix}$; then a precisely similar procedure leads to

$$\begin{pmatrix} a \\ b \end{pmatrix} = B(Q_1)B(Q_2)\cdots B(Q_F)\begin{pmatrix} 1 \\ 0 \end{pmatrix}.$$

We wish to compare expansions of these two types. Write

$$U_0 := \begin{pmatrix} 1 & 0 \\ 1 & -1 \end{pmatrix}, \quad U_1 := \begin{pmatrix} 1 & -1 \\ 0 & 1 \end{pmatrix}, \quad \text{and } V := \begin{pmatrix} 1 & 0 \\ 1 & 1 \end{pmatrix}.$$

Then $B(2) = VU_1V^{-1}$, so

$$\begin{aligned} B(2)^{k-1} &= VU_1^{k-1}V^{-1} = \begin{pmatrix} 1 & 0 \\ 1 & 1 \end{pmatrix}\begin{pmatrix} 1 & 1-k \\ 0 & 1 \end{pmatrix}\begin{pmatrix} 1 & 0 \\ -1 & 1 \end{pmatrix} \\ &= \begin{pmatrix} k & 1-k \\ k-1 & 2-k \end{pmatrix}. \end{aligned}$$

Now we observe the identities

$$U_0B(2)^{k-1} = \begin{pmatrix} k & 1-k \\ 1 & -1 \end{pmatrix} = A(k)U_1,$$

$$U_1B(k+2) = \begin{pmatrix} k+1 & -1 \\ 1 & 0 \end{pmatrix} = A(k)U_0.$$

Thus if we postmultiply the product $A(q_1)A(q_2)\cdots A(q_f)$ by U_s, for $s = 0$ or 1, the result is equal, according as $f + s$ is even or odd, to

$$U_0B(2)^{q_1-1}B(q_2+2)B(2)^{q_3-1}\cdots$$

or

$$U_1B(q_1+2)B(2)^{q_2-1}B(q_3+2)\cdots.$$

We have thus recovered the sequences obtained above for the self-intersection numbers.

It is now possible to calculate the numbers m_i, $M_i(C)$ and ν_i for each i either following the sequence of blowings up and using (8.8), (8.9) and

(8.10), or from induction along Γ_R using (8.11), (8.12) and (8.13). We leave the details as an exercise (Exercise 8.7.13).

We have observed that the coefficients of particular interest are those attached to rupture vertices of Γ_R^+. For the special case when the tree consists of the minimal good resolution of an irreducible curve C, we can calculate all these values in terms of the invariants of C. Let us denote by $AV_1, \ldots, AV_g$ the successive rupture points on the core of Γ_R^+, and by BV_q the point of valence 1 at the end of the chain attached to the core at AV_q.

We next identify corresponding curvettes: indeed, we already met C_q^- in (4.7).

Lemma 8.5.1 *Let C be an irreducible germ with parametrisation $x = t^m$, $y = \sum_{r=m}^{\infty} a_r t^r$. Write $(m; \beta_1, \ldots, \beta_g)$ for the Puiseux characteristic of C. Define curves by the parametrisations*

$$C_q^- : x = t^m, \quad y = \sum_{m \le r < \beta_q} a_r t^r,$$

$$C_q^0 : x = t^m, \quad y = \sum_{m \le r < \beta_q} a_r t^r + c t^{\beta_q},$$

where $c \neq a_{\beta_q}$. Then C_q^- and C_q^0 are curvettes corresponding to the respective points BV_q and AV_q in Γ_R^+.

These parametrisations are not good: a good parameter is $t^{e_{q-1}}$ in the first case, and t^{e_q} in the second.

Proof We recall the sequence of steps in the resolution of C. First there is a sequence of ordinary blowings up which do not decrease the multiplicity of C; then a battery of blowings up at satellite points corresponding to the Euclidean algorithm for finding the greatest common divisor e_1 of m and β_1 as described in Section 3.6; then the whole process is repeated g times. The vertex BV_q of valence 1 is introduced as the final non-satellite point before the qth battery, and the vertex AV_q of valence 3 as the final satellite point in the battery.

In the steps of the resolution of C, the coefficients telling us how the blow up behaves are identical for all of C, C_q^- and C_q^0 up to the end of the $(q-1)^{\text{st}}$ battery, and even up to the beginning of the qth. However C_q^- does not pass through the first satellite point of this battery; as its blow up is already nonsingular at the end of the $(q-1)^{\text{st}}$, it is indeed a curvette corresponding to the last non-satellite point BV_q. Similarly C_q^0

is resolved by the qth battery, so goes through the final satellite point AV_q, but – since we altered the coefficient of t^{β_q} – its strict transform meets the corresponding exceptional curve E_j in a different point to that in which the strict transform of C does. Thus indeed we have a curvette at AV_q. □

Theorem 8.5.2 *Let C be an irreducible germ, write $(m; \beta_1, \dots, \beta_g)$ for the Puiseux characteristic of C. Then the values of the invariants $M_k(C), m_k$ and ν_k at the rupture point AV_q are, respectively,*

$$M(C) = \frac{e_{q-1}}{e_q}\overline{\beta_q}, \qquad m = \frac{m}{e_q}, \qquad \nu = \frac{\beta_q + m}{e_q}.$$

Their values at the point BV_q of valence 1 are, respectively,

$$M(C) = \overline{\beta_q}, \qquad m = \frac{m}{e_{q-1}}, \qquad \nu = \left\lceil \frac{\beta_q + m}{e_{q-1}} \right\rceil.$$

Proof Suppose C has parametrisation $x = t^m, y = \sum_{r=m}^{\infty} a_r t^r$ as usual. By Lemma 8.5.1, the curves C_q^-, C_q^0 are curvettes at the points in question. Then C_q^- has multiplicity $\frac{m}{e_{q-1}}$ and C_q^0 has multiplicity $\frac{m}{e_q}$. These give the values of m_k for the corresponding vertices BV_q, AV_q.

Both C_q^- and C_q^0 have exponent of contact $\frac{\beta_q}{m}$ with C. Hence by Theorem 4.1.6, $C_q^0.C = \frac{m}{e_q}H(\beta_q) = \frac{e_{q-1}}{e_q}\overline{\beta_q}$ and $C_q^-.C = \frac{m}{e_{q-1}}H(\beta_q) = \overline{\beta_q}$, giving the values of $M_k(C)$ for the corresponding vertices AV_q, BV_q. Moreover, $\frac{\beta_q}{m}$ is also the exponent of contact of C_q^0 with any other curvette D_q^0 at the same point, so by Corollary 4.1.10, we have $C_q^0.D_q^0 = \frac{e_{q-1}\overline{\beta_q}}{e_q^2}$.

For the calculation of ν_i, first observe in general that since ϵ_i is irreducible, $\mu(\epsilon_i) = -[\epsilon_i].([\epsilon_i] - [Z])$ by (8.2), so

$$\overline{\nu}_i = -[\epsilon_i].[Z] = -[\epsilon_i].[\epsilon_i] + [\epsilon_i].([\epsilon_i] - [Z]) = -[\epsilon_i].[\epsilon_i] - \mu(\epsilon_i).$$

For the vertices of valence 3, since C_q^0 is irreducible with Puiseux characteristic $(\frac{m}{e_q}; \frac{\beta_1}{e_q}, \dots, \frac{\beta_q}{e_q})$, we have $\mu(C_q^0) = \frac{e_{q-1}}{e_q^2}\overline{\beta_q} - \frac{\beta_q}{e_q} - \frac{m}{e_q} + 1$. Thus

$$\overline{\nu}_k = \frac{e_{q-1}\overline{\beta_q}}{e_q^2} - \left(\frac{e_{q-1}}{e_q^2}\overline{\beta_q} - \frac{\beta_q}{e_q} - \frac{m}{e_q} + 1\right) = \frac{\beta_q}{e_q} + \frac{m}{e_q} - 1.$$

The same reasoning gives

$$\mu(C_q^-) = \mu(C_{q-1}^0) = \frac{e_{q-2}}{e_{q-1}^2}\overline{\beta_{q-1}} - \frac{\beta_{q-1}}{e_{q-1}} - \frac{m}{e_{q-1}} + 1.$$

However, matters are more complicated here since if D_q^- is another curvette at A_q, we do not immediately know the exponent of contact of C_q^- and D_q^-, nor their intersection number. We claim that the intersection number is

$$\left\lceil \frac{\overline{\beta_q}}{e_{q-1}} \right\rceil,$$

so the value of ν_k is

$$\left\lceil \frac{\overline{\beta_q}}{e_{q-1}} \right\rceil - \frac{e_{q-2}}{e_{q-1}^2}\overline{\beta_{q-1}} + \frac{\beta_{q-1}+m}{e_{q-1}} = \left\lceil \frac{X}{e_{q-1}} \right\rceil,$$

where

$$X = \left(\overline{\beta}_q - \frac{e_{q-2}}{e_{q-1}}\overline{\beta}_{q-1}\right) + \beta_{q-1} + m = (\beta_q - \beta_{q-1}) + \beta_{q-1} + m = \beta_q + m,$$

using (4.6).

To calculate the intersection number, write O_k for the infinitely near point corresponding to BV_q, and compare $m_j(C)$ with $m_j(C_q^-)$ for all the points O_j in the resolution with $j \leq k$. The proximity relations are the same in both cases, but the next point for C is proximate to O_{k-1} and the next point for C_q^- is not. Hence $\frac{m_j(C)}{m_j(C_q^-)} = \frac{m_0(C)}{m_0(C_q^-)} = e_{q-1}$ for all $j < k$, while $m_k(C_q^-) = m_{k-1}(C_q^-) = 1$ but $e_{q-1} = m_{k-1}(C) > m_k(C)$. Thus

$$\begin{aligned}
\overline{\beta_q} = C.C_q^- &= \sum_0^k m_j(C)m_j(C_q^-) \\
&= e_{q-1}\sum_0^k m_j(C_q^-)m_j(C_q^-) \\
&\quad + (m_k(C) - e_{q-1}m_k(C_q^-))m_k(C_q^-) \\
&= e_{q-1}C_q^-.C_q^- + (m_k(C) - e_{q-1}),
\end{aligned}$$

and so $C_q^-.C_q^- = \frac{\overline{\beta_q}}{e_{q-1}} + \eta$, where $\eta = \frac{e_{q-1}-m_k(C)}{e_{q-1}}$, so $0 < \eta < 1$. Our claim follows. □

Example 8.5.1 Apply Theorem 8.5.2 to a curve with Puiseux characteristic $(8; 11)$. The vertex AV_1 of the theorem is V_5; at it we have $M(C) = \frac{8}{1}11 = 88$, $m = \frac{8}{1} = 8$ and $\nu = \frac{11+8}{1} = 19$. The vertex BV_1 of the theorem is V_1, and there we have $M(C) = 11$, $m = \frac{8}{8} = 1$ and $\nu = \lceil\frac{11+8}{3}\rceil = 3$.

For a curve with Puiseux characteristic $(4; 6, 7)$ we have $m = \beta_0 = 4$, $\beta_1 = 6$, $\beta_2 = 7$, $e_0 = 4$, $e_1 = 2$, $e_2 = 1$ and $\overline{\beta}_1 = 6$, $\overline{\beta}_2 = 13$. Thus for

$AV_1 = V_2$ we have $M(C) = \frac{4}{2}6 = 12$, $m = \frac{4}{2} = 2$ and $\nu = \frac{6+4}{2} = 5$; for $AV_2 = V_4$ we have $M(C) = \frac{2}{1}13 = 26$, $m = \frac{4}{1} = 4$ and $\nu = \frac{7+4}{1} = 11$. For $BV_1 = V_1$ we have $M(C) = 6$, $m = \frac{4}{4} = 1$ and $\nu = \lceil\frac{6+4}{4}\rceil = 3$; for $BV_2 = V_3$ we have $M(C) = 13$, $m = \frac{4}{2} = 2$ and $\nu = \lceil\frac{7+4}{2}\rceil = 6$. These agree with the values already obtained in Examples 8.3.1 and 8.3.2.

In the calculation of monodromy in Chapter 10 an important role will be played by the function on edges of $\Gamma_R(C)$ defined by letting M_E be the highest common factor of the values of $M_i(C)$ at the two vertices at the ends of E. If C has just one branch, we can determine these values explicitly. We retain the notation of Theorem 8.5.2, and also write $AV_0 = V_0$ for the initial vertex.

Proposition 8.5.3 *For each edge of $\Gamma_R(C)$ between BV_q and AV_q we have $M_E = \overline{\beta}_q$. For each edge between AV_q and AV_{q+1} we have $M_E = e_q$.*

Proof For convenience, number the vertices from BV_q to AV_q consecutively as $V_q^0, V_q^1, \ldots, V_q^r$. By the Theorem, the value of M_q^0 is equal to $\overline{\beta}_q$. By (8.12), $M_q^1 = a_q^0 M_q^0$. Hence the value of M_E for the edge $V_q^0 V_q^1$ is equal to M_q^0. Now if $1 \leq i < r$, since $a_q^i M_q^i = M_q^{i-1} + M_q^{i+1}$, the values of M_E for the edges $V_q^{i-1} V_q^i$ and $V_q^i V_q^{i+1}$ agree. The assertion now follows by induction on i.

The same argument shows that M_E is constant along each chain of edges such that each interior point of the chain has valence 2, and also, since the value of M_i at AV_0 is the multiplicity of C, hence equal to e_0, that the constant value along $AV_0 AV_1$ is e_0. Similarly, arguing on $\Gamma_R^+(C)$ and using the fact that the value of M_i at the arrowhead vertex is 1, we see that for any edges between AV_g and W the value of M_E is $1 = e_g$.

In the case when $g = 1$ the result is now established. We may thus make an induction on g. Consider the curvette C_q^- of Lemma 8.5.1. The minimal resolution of C_q^- agrees with that of C up to the infinitely near point AV_{q-1}, but beyond that has no rupture points; so the assertion holds for C_q^- by inductive hypothesis. In view of the recurrence relation (8.12), the values for $M_i(C)$ and hence also $M_E(C)$ on the common part of the resolution graphs $\Gamma_R^+(C) \cap \Gamma_R^+(C_q^-)$, including the first edge following AV_{q-1}, are obtained from $M_i(C_q^-)$ and $M_E(C_q^-)$ by multiplying by the constant factor e_{q-1}. Since the values of M_E along $AV_{q-1}AV_q$ are equal to 1 for C_q^-, those for C are equal to e_{q-1}. □

8.6 Notes

Section 8.1, **Section 8.2** Some ideas for the approach, e.g. of Lemma 8.1.3, were taken from Lejeune [115], and go back to Lipman [119]. This was developed for a more general context: the study of rational (or general normal) surface singularities, and algebra on the exceptional cycles in a resolution. This study was originated by Artin: see e.g. [13]. Another paper on this topic from which we have borrowed ideas is Tosun's thesis: see [179] and [113]. A further useful source is Casas' text [35].

Negative definiteness conforms to a well known general result of Mumford. [136].

Section 8.3 Proposition 8.3.2 is given by Lê, Michel and Weber in [111]; they attribute the result to Zariski [208]. The approach also borrows from Veys' work.

Section 8.4 The topological zeta function of an algebraic variety was introduced by Denef and Loeser [50] by analogy with Igusa's p-adic zeta function, which is defined by an integral of the form $\int_{K^n} |f(x)|^s |dx|$, where K is a p-adic field and $f : K^n \to K$ is a polynomial function; and calculated using a resolution. There are, by now, several versions (e.g. local, global) and a generalisation defined using motivic integration: see e.g. [51].

Our treatment is borrowed from several papers by Veys: see for example [183], where he determines the poles of the zeta function for the case of curves. Several papers in this area are motivated by the conjecture of Denef and Loeser that for each pole z of the local zeta function, $e^{2\pi iz}$ is an eigenvalue of monodromy. For the case of curve singularities, this was established by Loeser [120], and follows from the calculation of the monodromy (see e.g. Chapter 10) together with cancellation arguments. It was also shown by Loeser (loc. cit.) that if z is a double pole of the zeta function, $e^{2\pi iz}$ corresponds to a Jordan block of size 2 of the monodromy.

Section 8.5 Duality for the sequences k_i is due to Hirzebruch [91]: his context was somewhat different, as he had periodic continued fractions and invariants $D = \sum k_i$, $D^* = \sum k_i^*$. See Ebeling and Wall [61] for a fuller account, and the relation to Arnold's strange duality, later interpreted as mirror symmetry.

8.7 Exercises

Exercise 8.7.1 Compute the proximity matrix and the intersection matrix for a resolution of the curve $y^3 = x^7$.

Exercise 8.7.2 Show that the entries in the matrices Q and Q^tQ are non-negative, and increase from left to right along the rows in the sense that if O_k is proximate to O_j, then the (i,k) entry is not less than the (i,j) entry.

Exercise 8.7.3 Write $\mathbf{m}$ for the (column) vector whose entries are m_i; similarly for $\mathbf{a}, \mathbf{m}(C), \mathbf{M}(C), \delta(C), \overline{\nu}, \nu, \mathbf{v}$; write $\mathbf{U}$ for the vector whose entries are all 1 and $\delta_{\mathbf{0}}$ for the vector with first entry 1 and the rest 0. Express (8.5)–(8.13) and the relevant identities of Proposition 8.1.8 and Corollary 8.2.3 in matrix terms.

Exercise 8.7.4 Compute the sequence of multiplicities of infinitely near points and the intersection matrix for the curve $x^3 = y^5$. Describe the resolution tree. Calculate the $M_i(C)$ using the recurrence relation (8.12). Check that your answer agrees with the final row (or column) of $(PP^t)^{-1}$.

Exercise 8.7.5 For each of the following curves, find a resolution tree and calculate the values of m_i, M_i and ν_i at all vertices of the tree: (i) $y^{11} = x^{19}$, (ii) $x = t^4$, $y = t^{10} + t^{11}$, (iii) $(y^3 - x^5)(y^4 - x^7) = 0$. Use each of the methods: using the known values at the initial vertex and the recurrence relations (8.11), (8.12) and (8.13); and resolving the curve in stages and using the relations (8.8), (8.9) and (8.10).

Exercise 8.7.6 (a) Check that the values of m_i, M_i and ν_i at the vertices of valence 1 or 3 obtained in (ii) of the preceding exercise agree with those given by Theorem 8.5.2.

Verify, using your calculations of m_i, M_i and ν_i in (iii) of the preceding exercise satisfy the monotonicity results obtained in Proposition 8.3.2 and Corollary 8.4.3.

Exercise 8.7.7 Verify that if the relations (8.11), (8.12) and (8.13) hold on a graph, and we blow up a point, the relations continue to hold at all points. (Hint: if the point blown up lies on E_i, then a_i is increased by 1.)

Exercise 8.7.8 Let C be an irreducible curve; form a good resolution, with dual tree Γ_R. Let V_k be a vertex in the core of Γ_R, and let X be the component of $\Gamma_R \setminus \{V_k\}$ containing the initial vertex V_0. Show that the determinant D_k of the submatrix of the intersection matrix whose rows and columns correspond to the vertices in X is equal, up to sign, with $m_k M_k(C)/m(C)$.

Exercise 8.7.9 For each of the examples in Exercise 8.7.5, deduce an expression for the topological zeta function, and check that cancellation does occur.

Exercise 8.7.10 Calculate the topological zeta function for the curve $(y^2 - x^3)(y^3 - x^2) = 0$.

Exercise 8.7.11 Calculate the topological zeta functions for
(i) $(y^2 - x^3)(y^3 - x^4) = 0$,
(ii) $(y^2 - x^3)(y^4 - x^3) = 0$.

Exercise 8.7.12 Show that the topological zeta function for any curve singularity satisfies $Z_{\text{top}}(0) = 1$. Hint: use induction on the blow-ups required for the resolution.

Exercise 8.7.13 Consider a monomial curve $x^a = y^b$ with a and b coprime. With the notation of (3.2) for the euclidean algorithm, and writing P_k/Q_k for the kth convergent in the continued fraction expansion of a/b, establish the following, where K_k denotes M_{s_k-1}.

(i) We have $K_k = \begin{cases} Q_k a &= P_k b - r_k \quad (k \text{ even}) \\ Q_k a - r_k &= P_k b \quad (k \text{ odd}) \end{cases}$.

(ii) For $0 \le j \le q_i$, we have $M_{s_k+i-1} = K_{k-1} + i(K_k + r_k)$.

Verify that these satisfy both the inductive formula (8.9) and the relation (8.12) on the tree.

Exercise 8.7.14 With the notation of the preceding exercise, calculate the corresponding numbers ν_i, and hence the topological zeta function.

9

Decomposition of the link complement and the Milnor fibre

In this chapter we begin the deeper study of the topology attached to the Milnor fibration. One key problem is to obtain an understanding of the monodromy. A major tool for this is a canonical decomposition of the Milnor fibre. Because the decomposition is intrinsic, it gives a better picture of the topology than we attained in Chapter 5, particularly when the curve has several branches. We discuss the decomposition theorems in this chapter, leaving the application to monodromy to Chapter 10. Although we present an introductory account of these matters, we will necessarily assume a higher level of mathematical sophistication than was the case in earlier chapters.

We may use the carousel of Section 5.3 or the resolution tree of Section 3.6 to obtain a decomposition. We will see directly that the same is obtained from each approach, but this fact is underpinned by major theorems of great generality. Although we do not need these results, we describe them to set our discussions in a wider context. We thus begin with a section stating the general decomposition theorems in 2- and 3-dimensional topology which underlie the constructions.

We now explain what we mean by 'decomposition'. A decomposition of a connected manifold M is effected by cutting along submanifolds of codimension 1. If T is a connected submanifold which separates M into two pieces, then if M_1, M_2 are the closures of the two complementary regions, the result of cutting is defined to be the disjoint union of M_1 and M_2. In general we must be a little more circumspect: a convenient formal definition is to take the metric completion of $M - T$, but it may be easier to picture the construction which takes a tubular neighbourhood of T in M – thus if T is 2-sided in M (the only situation we need) N is isomorphic to the product $T \times I$ – and then delete the interior of N from M. Provided T does not meet the boundary ∂M, the result of cutting has boundary consisting of ∂M and two copies of T. We recall that the closed complement of a link L in M was similarly defined (in Section 6.2) by removing from M the interior of a tubular neighbourhood of L, but here the tubular neighbourhood is homeomorphic to $L \times D^2$ and its boundary to the union $L \times S^1$ of tori.

A direct construction of the canonical decomposition of the closed complement of the singularity link in the case of plane curve singularities is given in Section 9.2 via the carousel of Section 5.3.

A regular neighbourhood of the union of the curves involved in the resolution of a singularity gives a 4-dimensional manifold with an easily described structure ('plumbing') whose boundary gives the singularity link. This gives an alternative approach to the decomposition which we give in Section 9.3: it leads to an explicit model for the Milnor fibre and the monodromy which we will use in Chapter 10.

After a preliminary discussion of the structure of Seifert fibre spaces, we proceed to introduce the notation of Eisenbud and Neumann [65] which gives a precise description of the structure of a decomposition. Finally we seek to evaluate the parameters that appear in this decomposition. This can be given in full in the irreducible case; in general, many of the relevant numerical invariants are those previously studied in Chapter 8.

A completely different approach to the decomposition was given by Lê, Michel and Weber [112] using polar curves and the polar discriminant; we will sketch it very briefly in Section 9.9.

9.1 Canonical decomposition theorems

In this section, by '3-manifold' we will always understand compact oriented 3-dimensional manifold; our surfaces also will be compact and orientable. The current overall understanding of the topology of 3-manifolds is summarised by the idea of Thurston (see [176]) that any 3-manifold should have a canonical decomposition into pieces each of which admits a geometric structure. The concept of 'geometric' here may be defined in terms of differential geometry; it implies (in the unbounded case) that the universal cover admits a transitive Lie group of automorphisms. A full account of the 8 types of structure that may appear is given in [161]. Although this geometrisation conjecture is not yet known in full, all the details are available in the case where the manifold is the closed complement of a link in S^3.

The first step in the programme consists in decomposing a 3-manifold as a connected sum using embedded spheres. The existence of a decomposition into primes was established by Kneser [102]; a uniqueness statement and further refinements were added by Milnor in [129]. A 3-manifold M such that every embedded copy of the 2-sphere S^2 bounds an embedded 3-disc is called *irreducible*. If M is not irreducible, we cut along such a 2-sphere: the result has two boundary components homeomorphic to S^2, to each of which we attach a 3-disc. If the 2-sphere separates M this gives two 3-manifolds, of which M is the connected sum.

Theorem 9.1.1 *Any 3-manifold may be expressed as a connected sum of 3-manifolds each of which is either irreducible or homeomorphic to* $S^2 \times S^1$. *The summands are unique up to homeomorphism.*

Observe that the 3-manifolds with which we shall be concerned are the closed complements of links in a 3-sphere. An embedded 2-sphere which does not bound a disc splits such a link into two separate pieces; in particular, the linking numbers of knots with one from each piece vanish. As we have seen, this does not occur for algebraic links, so the sum decomposition is necessarily trivial in this case.

The irreducible 3-manifolds M fall into two main types, according as the fundamental group is finite or not. In the former case, M is either homeomorphic to a disc D^3 or has no boundary and has universal cover homotopy equivalent to S^3. Thurston's geometrisation conjecture claims in this case not only that the cover is homeomorphic to S^3 but that M is homeomorphic to the quotient of S^3 by a subgroup of the orthogonal group SO_3. This part of the conjecture remains open.

If the fundamental group of an irreducible manifold is infinite, then it is torsion-free. This is, in particular, the case for link complements.

An embedded (connected) surface T in a 3-manifold M is *boundary-parallel* if there is an embedding of $T \times I$ in M with $T \times \{0\}$ mapped to T and $T \times \{1\}$ onto a component of ∂M. The closed surface T in M is *incompressible* if either T is homeomorphic to S^2 and does *not* bound an embedded disc or T has infinite fundamental group and the map $\pi_1(T) \to \pi_1(M)$ induced by inclusion is injective.

The next stage of the decomposition is invoked if some component N of ∂M is compressible, i.e. the induced map $\pi_1(N) \to \pi_1(M)$ fails to be injective. Then according to the Loop Theorem of Papakyriakopoulos, there is an embedding $(D^2, S^1) \hookrightarrow (M, N)$ such that the circle defines a non-trivial element of $\pi_1(N)$. We then cut M along the image of the disc. It follows from the main results of [92] and [93] that this procedure can be repeated until each component of ∂M is incompressible, and moreover that the result is essentially unique. We abstain from detailed discussion since in our examples all boundary components are indeed incompressible, so this step is superfluous.

The next step involves decomposing by cutting along embedded tori and annuli. A satisfactory result here – which provided the essential foundation for the whole idea – is due (independently) to Jaco and Shalen [92] and Johannsen [93], so is called the JSJ decomposition. We will give only the special case when all components of ∂M are tori, as this suffices for our needs and the statement in this case is simpler (in particular, there is no need for annuli).

To state the result we need to define a Seifert fibred space [162]. This is a 3-manifold X, with an effective fixed-point free action of S^1. The quotient space is then a surface S and we write $p : X \to S$ for the quotient map. It can be proved that it is enough to suppose that we have a map p with each fibre homeomorphic to S^1. We will further analyse the structure of Seifert fibre spaces in Section 9.7.

A 3-manifold X is called *atoroidal* if every incompressible embedded torus in the interior of X is boundary-parallel. It can be shown that this is equivalent to the condition that every subgroup of $\pi_1(M)$ isomorphic to $\mathbb{Z} \oplus \mathbb{Z}$ is conjugate to a subgroup of the fundamental group of some boundary component.

Theorem 9.1.2 *Let M be an irreducible 3-manifold such that either ∂M is empty and $\pi_1(M)$ is infinite, or M has non-empty boundary and each component of ∂M is an incompressible torus. Then there exists a collection $\{T_i\}$ of disjoint incompressible embedded tori in X such that if one cuts X along all these tori, each remaining piece is either atoroidal or a Seifert fibre space. Moreover a minimal collection of tori with this property is unique up to isotopy.*

The main respect in which the programme is incomplete is that it is not established in general that an atoroidal irreducible manifold admits a geometric structure. However we will see below that the pieces which arise in the decomposition of an algebraic link complement are Seifert fibred, and for these the geometry is well understood.

The other standard decomposition arises in the study of self-maps of compact surfaces. The basic idea is due to Thurston. If $h : F \to F$ is a self-map of a compact surface, one is interested in studying the dynamics of iterations of h, particularly in the case when h is a homeomorphism. It turns out that the results are the same for isotopic homeomorphisms. Since each self-homeomorphism of a compact surface is isotopic to a diffeomorphism, unique up to diffeotopy, we need not bother here about differentiability.

As with the theorems above, one can decompose (F, h) into pieces of one of two basic types. One type consists of homeomorphisms of finite order: some finite iterate of h is the identity.

A homeomorphism of the other type (for which, as we will not use it, we suppress full details) is called pseudo-Anosov. The picture here (see e.g. [36]) is that F admits measured foliations $\mathcal{F}^s$ and $\mathcal{F}^u$, each with isolated singularities, and transverse to each other, such that for some $\lambda > 1$ we have $h^*\mathcal{F}^s = \lambda^{-1}\mathcal{F}^s$, $h^*\mathcal{F}^u = \lambda\mathcal{F}^u$. Thus h preserves the leaves of each foliation, but contracts distances transverse to the stable foliation $\mathcal{F}^s$ and expands those transverse to the unstable foliation $\mathcal{F}^u$.

Theorem 9.1.3 *Let $h : F \to F$ be a self-diffeomorphism of a compact surface F. Then there exist a diffeomorphism h' isotopic to h and a family C, invariant under h', of disjoint simple closed curves on F, such that if cutting along C produces a diffeomorphism k of a surface G, then for each component of G either (i) the restriction of k to (the union of the images under powers of k of) that component is isotopic to a diffeomorphism of finite order or (ii) this restriction is pseudo-Anosov. If C is taken minimal, it is unique up to isotopy.*

For proofs see the books [36] (for closed surfaces only) or [70] (in general).

Now given a diffeomorphism h of a surface F we can regard h as the monodromy of a fibration over a circle as follows (compare Section 6.1). Take $F \times [0, 1]$, and identify, for each $x \in F$, the points $(x, 1)$ and $(h(x), 0)$. This defines a 3-manifold X, and the projection of $F \times [0, 1]$ on $[0, 1]$, composed with the map $t \mapsto e^{2\pi it}$ of $[0, 1]$ onto S^1, defines a map $X \to S^1$ which we can see is a fibration, with monodromy h. Observe that if h is replaced by an isotopic map the result is essentially the same. For if $A : F \times [0, 1] \to F \times [0, 1]$ is a level preserving map with $A(x, 0) = h(x)$ and $A(x, 1) = h'(x)$ for each $x \in F$, A induces a diffeomorphism between the two quotient spaces.

Remark 9.1.4 In the case when h has finite order n, we can define an action of S^1 on X as follows. The above construction of X is equivalent to defining X as the quotient of $F \times \mathbb{R}$ by the action of $\mathbb{Z}$ given by $r.(x, t) = (h(x), t - 1)$. This commutes with the action of $\mathbb{R}$ by $u.(x, t) = (x, nu + t)$, which thus passes to an action on X in which the subgroup $\mathbb{Z}$ of $\mathbb{R}$ acts trivially, so we obtain an action of the quotient $\mathbb{R}/\mathbb{Z} \cong S^1$.

This is fixed point free – indeed only if $nu \in \mathbb{Z}$ can the action of u have a fixed point – so we have a Seifert fibre space.

This illustrates the following result, which relates the two decomposition theorems.

Proposition 9.1.5 *If $p : X \to S^1$ is a fibration with monodromy $h : F \to F$, and C is a collection of curves as in Theorem 9.1.3, then the image T of $C \times [0, 1]$ under the quotient map $F \times [0, 1] \to X$ is the minimal collection of tori in Theorem 9.1.2. A piece of (F, h) where h is periodic gives a Seifert fibre space piece of X; if h is pseudo-Anosov we have an atoroidal piece of X.*

This result follows from the proofs of the preceding theorems.

To conclude this section, remark that in decomposing a manifold X, if a given collection of tori gives a decomposition (into Seifert fibre spaces and atoroidal pieces) which is not minimal, there is (at least) one torus that can be deleted without destroying this property. We can recognise this if either there is a piece homeomorphic to $S^1 \times S^1 \times I$, when either of the two parallel tori may be deleted; or if we have a torus on either side of which we have a Seifert fibre space and the Seifert fibres on the two sides are homologous in the torus, since then their union is again a Seifert fibre space.

9.2 The complement of an algebraic link

In this section we exhibit a torus decomposition of the closed complement of an algebraic link, and show that all the pieces into which it is decomposed are Seifert fibre spaces. It will follow that for the corresponding decomposition of the Milnor fibre, the monodromy has finite order in each piece.

We saw in Chapter 5 that a singular point of an algebraic plane curve determines a link in a 3-sphere, unique up to isotopy, whose number of components is equal to the number of branches of the curve at the point; we call such links *algebraic links*. The closed complement M is obtained from S^3 by deleting the interior of a tubular neighbourhood of the link, so is a compact manifold whose boundary consists of a union of tori, one surrounding each component of the link. We saw in Chapter 6 that M is a compact manifold fibred over S^1; the fibre is called the Milnor fibre F. The theorems of the preceding section are thus applicable.

To obtain an explicit description of the decompositions, we need a direct construction, rather than a reference to a general result. In our case, such a construction is at hand – much of it was already described in Chapter 5 – we can refer either to the carousel of Section 5.3 or to its more precise description as a cable in Section 5.4. Here we take the former, and consider first the case of a single branch.

We recall that starting with a Puiseux parametrisation $(t^m, a(t))$, where $a(t) = \sum_n^\infty c_r t^r$, we were led to consider the knot K given by $x = e^{im\theta}, y = \epsilon^{-1} a(\epsilon^{1/m} e^{i\theta})$ in the deformed unit sphere. The isotopy class is independent of ϵ provided this is small enough. As before, we seek to visualise this by drawing the picture for a fixed value of θ and then fitting the pictures together as θ increases from 0 to 2π.

We approximate K by the series of knots K_k given by replacing the power series $a(t)$ by the series $a_k(t) = \sum_n^k c_r t^r$ obtained from $a(t)$ by truncating at the t^k term. If $\beta_{q-1} \le k < \beta_q$ then y is unaltered if we multiply t by $e^{2\pi i e_{q-1}/m}$, so takes just m/e_{q-1} values for a given value of x. Further, as we saw in the proof of Proposition 5.3.1, if ϵ is small enough, the truncation error $|\, \epsilon^{-1} a(\epsilon^{1/m} e^{i\theta}) - \epsilon^{-1} a_k(\epsilon^{1/m} e^{i\theta}) \,|$ is small compared to the distances apart of these points. We may thus choose η_k which is also small compared to these distances, but dominates the remainder terms after truncation, and define the torus U_k as the union of the circles $x = e^{im\theta}, |y - \epsilon^{-1} a_k(\epsilon^{1/m} e^{i\theta})| = \eta_k$ as θ varies.

In the case $k + 1 < \beta_q$, the disc $|y - \epsilon^{-1} a(\epsilon^{1/m} e^{i\theta})| \le \eta_k$ in the slice $x = e^{im\theta}$ meets K_{k+1} in just one point and so meets the torus U_{k+1} in a disc. The region between U_k and U_{k+1} thus intersects the slice in the union of m/e_{q-1} annuli. Fitting these together as θ varies shows that this whole region is homeomorphic to the product of a torus and an interval: the two tori are parallel. Thus if we obtain a torus decomposition using these two, it will not be minimal: we may always omit U_{k+1}.

If, however, $k + 1 = \beta_q$, the above disc contains e_{q-1}/e_q points of K_{k+1}, equally spaced round a circle. As θ increases, we meet the same slice after an increase of $2\pi/m$, but do not return to the same disc of U_k till after $2\pi/e_{q-1}$, at which point the final term $\epsilon^{\frac{k+1}{m}-1} a_{\beta_q} e^{i\beta_q\theta}$ has been multiplied by $e^{2\pi i \beta_q / e_{q-1}}$. Thus the region between U_k and U_{k+1} is obtained up to homeomorphism as follows. Let Δ_q be obtained by removing from a large disc with centre O the interiors of e_{q-1}/e_q disjoint smaller discs, arranged uniformly round a circle with centre O, and all contained in the large disc. Take the product of Δ with the interval $[0, 2\pi/e_{q-1}]$, and identify the ends after rotating Δ through an angle $2\pi\beta_q/e_{q-1}$. It follows from Remark 9.1.4 that this yields a Seifert fibre space.

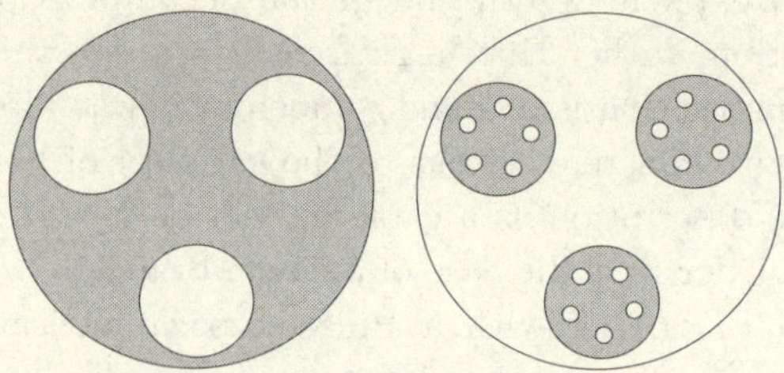

Fig. 9.1. Pieces of the decomposition

We do not need to continue iterating this procedure indefinitely: as soon as $k \geq \beta_g$, each of the discs we construct contains just one point of the final knot K, so U_k itself may be taken as the boundary of the neighbourhood N of K and its exterior as the closed complement M.

In Section 5.4 we defined the sequence of knots

$$K^q = K_{\beta_q} \text{ by } x = e^{im\theta}, \quad y = \sum_{r=1}^{\beta_q} \epsilon_r e^{ir\theta},$$

(where $e^{ie_q\theta}$ is a good parameter) lying on the tori T^{q-1}:

$$x = e^{im\theta}, \quad \left| y - \sum_{r=1}^{\beta_q - 1} \epsilon_r e^{ir\theta} \right| = \epsilon_q.$$

These, although similar to the above, will not do for our present purpose since they are not nested. We may, however, conveniently define $U^q = U_{\beta_q}$, and see that these do indeed give a minimal torus decomposition of M.

We now turn to the case of links with several components, which can be treated in the same way (compare Proposition 5.3.2). We refer to the situation described above as Case Ai. Choose parametrisations $(t^m, a_j(t))$ for B_j with $a_j(t) = \sum_n^\infty c_{j,r} t^r$: note that we take the same value of m for all j, even if this does not give good parametrisations for the branches. Suppose we have already constructed tori U_{k-1}; there will be one for each set J of branches such that the $c_{j,r}$ take the same value for all $r < k$. The next diagram depends on the coefficients $c_{j,k}$ $(j \in J)$ and not on higher terms.

Proceeding as above gives a torus $U_{j,k}$ for each $j \in J$, or rather, for each group with a common value of $c_{j,k}$. These are all exterior to each other, and each lying inside U_{k-1}. More precisely, each of the slice discs of U_{k-1} meets them in non-nested circles. The region interior to U_{k-1} and exterior to each of the others is a bundle over S^1 with fibre this plane region. For if we write the coordinates of the points in question as

$$x = e^{im\theta}, \quad y = \epsilon^{-1}\{a_k(\epsilon^{1/m}e^{i\theta}) + \xi\epsilon^{\frac{k}{m}}e^{ki\theta}\},$$

and recall that the torus U_{k-1} is the union of the circles

$$x = e^{im\theta}, \quad |y - \epsilon^{-1}a_{k-1}(\epsilon^{1/m}e^{i\theta})| = \eta_{k-1}$$

as θ varies, we see that the point is in U_{k-1} if $|\xi| = \epsilon\eta_{k-1}$, and in $U_{j,k}$ if $|\xi - c_{j,k}| = \epsilon^{1+\frac{k}{m}}\eta_k$. Thus projecting on ξ or taking the value of θ defines a bundle structure.

There are now two essentially distinct cases according as whether or not k is a characteristic exponent for at least one of the branches. First suppose (Case A) that it is. There is thus at most one $j \in J$ for which $c_{j,k} = 0$, and so for which k is not a characteristic exponent. In Case Ai we had a bundle over S^1 with fibre Δ the disc with circular holes punched out forming a cycle. The modification to be made here is that we have a cycle of e_{q-1}/e_q circular holes for each group of $j \in J$ with $c_{j,k} \neq 0$, all these being disjoint; if there is also a j with $c_{j,k} = 0$ we must punch out a further hole in the centre. Then the monodromy is constructed as before.

Now suppose (Case B) k is not a characteristic exponent for any of the branches. Then each (group of) $j \in J$ contributes a single hole to be punched out of the disc, and the monodromy (of this picture) is trivial.

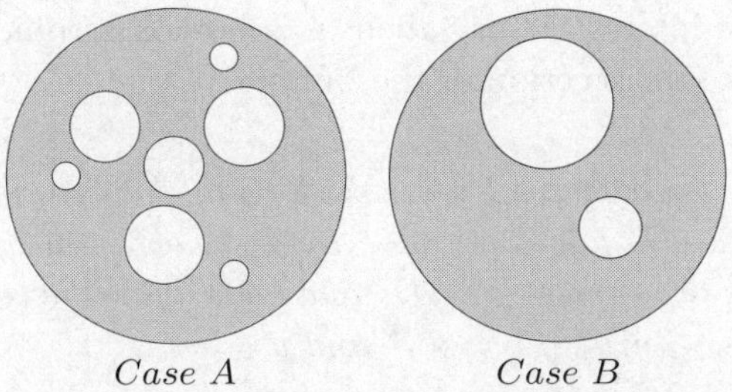

Fig. 9.2. The two cases of the decomposition

9.3 Resolution and plumbing

In this section we will show how to use a good resolution of a curve singularity to give a rather explicit construction of a 4-dimensional manifold whose boundary includes the singularity link complement. This construction yields directly a (non-minimal) torus decomposition of the link complement. Moreover, it gives explicit models for the Milnor fibre and the monodromy.

We recall that by Theorem 3.4.4 the curve singularity has a good resolution, which is a map $\pi : T \to S$ (with S a neighbourhood of $O \in \mathbb{C}^2$) such that the collection $\pi^{-1}(C)$ of curves has normal crossings. We adopt our usual notation and write E_i for the exceptional curves, B_j for the branches of C, and B_j^* (rather than $B_j^{(k)}$) for their strict transforms in T; write Γ_R for the dual graph of the resolution and Γ_R^+ for the augmented dual graph. Each curve E_i is isomorphic to a projective line $P^1(\mathbb{C})$ and homeomorphic to a 2-sphere. It has a tubular neighbourhood N_i which is diffeomorphic to a bundle ξ_i with fibre a disc D^2.

We interrupt the exposition to recall how such bundles are constructed. For any fibre bundle with fibre a circle S^1 or disc D^2 whose fibres may be consistently oriented, the structure group of the bundle may be taken as the circle group S^1 itself. It is shown in standard topology texts (see e.g. [169]) that such bundles over a base space X are classified by a cohomology class $c_1(\xi) \in H^2(X)$ known as the Chern class.

In particular, if X is a closed oriented surface, evaluating this class on the fundamental class $[X]$ determines an integer $c(\xi)$. If Y is a bundle over X with fibre D^2, $c(\xi)$ is equal to the self-intersection of X embedded in Y by the zero (or, indeed, any) cross-section of the bundle. If $c(\xi) = 0$, the bundle is a product. Otherwise a fibre gives an element of finite order $|c(\xi)|$ in the homology $H_1(\partial Y)$.

In the case of interest to us, when X is homeomorphic to a 2-sphere, there is a direct construction of the bundle.

Lemma 9.3.1 *An oriented D^2-bundle Y with characteristic class $c(\xi) = -a$ over a surface E homeomorphic to S^2 is obtained up to diffeomorphism by taking two copies of $\mathbb{C} \times D^2$ and identifying the copies of $\mathbb{C}^* \times D^2$ by letting $(u, x) \sim (v, y)$ if $v = u^{-1}$ and $y = xu^a$.*

Proof The bundle becomes trivial if we delete a point from E. We may regard E as a projective line, with coordinates $(u : v)$; thus over the subset $v \neq 0$ we may write u for u/v and have a copy of $\mathbb{C} \times D^2$ with coordinates (u, x); similarly over the subset $u \neq 0$ with (v, y). On the region of overlap we have $v = u^{-1}$, and must have $y = y(x, u)$; but may suppose y obtained from x by a rotation depending on u; say $y = x\phi(u)$. Up to homotopy, ϕ is a map from S^1 to S^1, so may be replaced by a power u^b for some $b \in \mathbb{Z}$.

The homology group $H_1(Y)$ is obtained from $H_1(\mathbb{C}^* \times (D^2 - \{0\}))$ by killing the homology classes which vanish on the two sides of the torus. Denoting homology bases by $\{U, X\}$ and $\{V, Y\}$ corresponding to the parametrisations, we have to kill X and $Y = X + bU$. We thus have a single generator U and relation $bU = 0$, hence a cyclic group of infinite order if $b = 0$ and order $|b|$ otherwise.

Hence $b = \pm a$. It remains only to check that we have the correct sign. In the special case of the normal bundle of $P^1(\mathbb{C})$ in $P^2(\mathbb{C})$ the self-intersection in P^2 is $+1$ and here we have $(u, x) = (u : x : 1) = (1 : u^{-1}x : u^{-1})$ corresponding to the case $a = -1$. □

It can be shown that for the neighbourhood of a holomorphic curve E in a complex 2-manifold with negative self-intersection number, this model is correct up to biholomorphic equivalence.

For the normal bundle ξ_i of E_i, the invariant $c(\xi_i)$ is equal to the self-intersection of E_i in N_i. We saw in Chapter 8 that this is negative, and denoted it by $-a_i$. Provided the neighbourhood S is suitably chosen (see e.g. Lemma 5.2.1) the curves B_j^* will be discs meeting the boundary ∂T transversely, and so themselves having neighbourhoods homeomorphic to $B_j^* \times D^2$.

Since $\pi^{-1}(C)$ has normal crossings, at each singular point there are just two components meeting transversely. It is thus possible to choose tubular neighbourhoods of these components such that, in a neighbourhood of the point of intersection, there is a chart taking the neighbourhood to $\{(x, y) \in \mathbb{C}^2 \mid |x| < 2\epsilon, |y| < 2\epsilon\}$ with the components meeting it in the coordinate axes, their tubular neighbourhoods being given by $|x| < 2\epsilon$, $|y| \leq \epsilon$ and $|x| \leq \epsilon$, $|y| < 2\epsilon$ respectively, with the projections of these on the components being the obvious ones.

The union of all these tubular neighbourhoods gives a neighbourhood N of $\pi^{-1}(C)$. We see that this can be synthesised (up to diffeomorphism) in the following way. Take the disjoint union of disc bundles N_i over E_i (with Euler classes $-a_i$) and $B_j \times D^2$ over B_j. For each intersection point of two components – say E_i and E_j – proceed as follows. Choose embeddings ϕ_i, ϕ_j of D^2 in E_i, E_j respectively; these will be taken small enough so as not to overlap any other such choice. The chosen tubular neighbourhood, which is a disc bundle over E_i, pulls back by ϕ_i to a trivial bundle, and we pick a trivialisation

$$\begin{array}{ccc} D^2 \times D^2 & \xrightarrow{\psi_i} & N_i \\ \downarrow & & \downarrow \\ D^2 & \xrightarrow{\phi_i} & E_i \end{array}$$

where the left hand vertical map denotes projection on the first factor. Now identify the two copies of $D^2 \times D^2$ with an interchange of factors. This construction is known as *plumbing.*

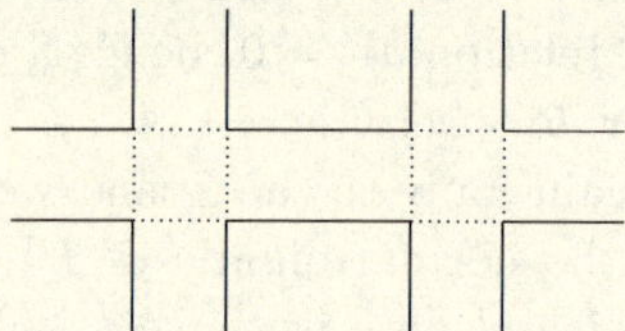

Fig. 9.3. Plumbing

The zero locus of $f \circ \pi$ in T (or N) is the union $\pi^{-1}(C)$ of the curves E_i and B_j. However, although, since f has no repeated factor, $f \circ \pi$ vanishes simply along the curves B_j, it will in general vanish to higher order along the E_i. As in Chapter 8 we write $M_i(C)$ for the order of vanishing, but as the curve C will be fixed throughout this section, we abbreviate this to M_i. Thus if the curve E_i is given in local coordinates at some point as $u = 0$, we have $f \circ \pi = u^{M_i}\phi$ for some function ϕ not vanishing at the point. Similarly, at a point of $E_i \cap E_j$, we can write $f \circ \pi = u^{M_i}v^{M_j}\phi$ with ϕ a function not vanishing at the point. We can include the curves B_i among the E_i provided we set $M_i = 1$ for them.

It will be convenient to write E_i^o for the result of removing the interiors of the plumbing discs from E_i, so that ∂N is the union of circle bundles over the E_i^o (or B_j^o), which intersect along certain tori. If the valence of the vertex V_i of the augmented resolution tree corresponding to E_i is v_i, there are just v_i such plumbing discs to remove. We will also sometimes also write f for $\pi \circ f : T \to \mathbb{C}$.

We have used plumbing to construct a model for the neighbourhood N. We next enhance the model to describe the link complement and incorporate the Milnor fibre and the monodromy. Since π is a resolution, it induces an isomorphism of $T - E$ onto $S - \{O\}$, and hence of $T - \pi^{-1}(C)$ onto $S - C$. It will thus suffice to work in T. The key point is the construction of suitable vector fields.

A tubular neighbourhood of E_i may be defined by an inequality $\rho_i \leq \epsilon$, where ρ_i denotes the squared distance from E_i in some metric. This general definition allows some flexibility in application: we can take more care in the choice of ρ_i, but this is not required for the proof that the model has the desired properties.

Proposition 9.3.2 *There exist a neighbourhood U of $\pi^{-1}(C)$, and a vector field ξ, defined on U, such that the following hold on $U \setminus \pi^{-1}(C)$:*

(i) $\xi(\Im \log(f)) = 0$, *so* $\arg f$ *is constant along integral curves of* ξ;

(ii) $\xi(\Re \log(f)) > 0$, *so* $|f|$ *increases along integral curves of* ξ;

(iii) $\xi(\rho_i) > 0$ *where* ρ_i *is defined, so* ρ_i *increases along integral curves of* ξ;

(iv) $\xi(||z||^2 \circ \pi) > 0$, *so* $||z||^2$ *increases along integral curves of* ξ.

Proof Each of the conditions (i)–(iv) defines a convex subset of the set of vectors at each point. It follows from Theorem 5.1.5 that it is sufficient to show that each point of U has a neighbourhood on which we may construct a suitable vector field ξ. Indeed, it suffices to consider points of $\pi^{-1}(C)$, since by compactness we may take U as the union of a finite number of neighbourhoods of such points.

First consider a point P of E_i lying in no other E_j or B_j. Then there are local complex coordinates (u, v) at P such that E_i is given by $u = 0$ and P is $(0, 0)$. Since the order of vanishing of f is M_i, we may write $f(u, v) = u^{M_i}\phi(u, v)$ where, since P lies on no other E_j, $\phi(0, 0) \neq 0$. But then there exists a local holomorphic function $\psi(u, v)$ such that $\psi^{M_i} = \phi$, and we can take $u\psi(u, v)$ as local coordinate in place of u. We may thus set $f(u, v) = u^{M_i}$.

Write $p = \Re u$, $q = \Im u$ and $r = \Re v$, $s = \Im v$ so that $u = p + iq$ and $v = r + is$. We claim that the vector field $\xi := p\partial/\partial p + q\partial/\partial q$ satisfies all the above conditions in some neighbourhood of P. Off E_i we may compute

$$\xi(\log f) = M_i\xi(\log u) = M_i u^{-1}\xi(u) = M_i u^{-1}(p + iq) = M_i.$$

Since this is real, (i) holds and since it is also positive, so does (ii).

Now express the distance function ρ_i in local coordinates. Since ρ_i vanishes when $p = q = 0$, by Lemma 1.4.3 we can write $\rho_i(p, q, r, s) = p\sigma(p, q, r, s) + q\tau(p, q, r, s)$ for some smooth functions σ and τ. Since also $\rho_i \geq 0$, the partial derivatives $\partial\rho_i/\partial p$ and $\partial\rho_i/\partial p$ must also vanish when $p = q = 0$. Since $\partial\rho_i/\partial p = \sigma + p\partial\sigma/\partial p + q\partial\tau/\partial p$, it follows that $\sigma(0, 0, r, s) = 0$, and similarly for τ. We may now apply Lemma 1.4.3 again to write $\sigma = pA + qB_1$ and $\tau = pB_2 + rC$, so (writing $2B = B_1 + B_2$) $\rho_i = Ap^2 + 2Bpq + Cq^2$.

Since ρ_i is a distance function (in some coordinates), its restriction to a slice where r and s are constant is comparable to the metric $p^2 + q^2$.

Thus along $p = q = 0$ we must have $A > 0$, $C > 0$ and $AC > B^2$. Then

$$\xi(\rho_i) = 2(Ap^2 + 2Bpq + Cq^2) + R, \qquad \text{where}$$

$$R = (\partial A/\partial p)p^3 + (\partial A/\partial q + 2\partial B/\partial p)p^2 q + (2\partial B/\partial q + \partial C/\partial p)pq^2 + (\partial C/\partial q)q^3$$

consists of terms of higher order in p and q. Since the quadratic form $Ap^2 + 2Bpq + Cq^2$ is positive definite along $p = q = 0$ it follows that $R/(Ap^2+2Bpq+Cq^2)$ tends to 0 as $(p, q) \to (0, 0)$. In particular, $\xi(\rho_i) > 0$ in some neighbourhood of the origin, except along $p = q = 0$. This proves (iii).

To study $||z^2||$, it is convenient to suppose coordinates (x, y) chosen in $\mathbb{C}^2$ so that neither $x = 0$ nor $y = 0$ is tangent to C. Then x and y each vanish to first order on E_0 with a single additional point, and vanish to constant order m_i along E_i for $i \neq 0$, since each of $x = 0$ and $y = 0$ is a curvette ϵ_0. Write $\pi^*(x) = u^{m_i}\phi(u, v)$ in the above local coordinates, so $\pi^*|x|^2 = (p^2 + q^2)^{m_i}|\phi(u, v)|^2$. Set $\psi(u, v) - |\phi(u, v)|^2$. Applying ξ to $\pi^*|x|^2$ thus gives

$$(p^2 + q^2)^{m_i}\{2m_i\psi(u, v) + p\partial\psi/\partial p + q\partial\psi/\partial q\}.$$

Here the expression in braces is a smooth function, and takes the value $2m_i\psi(0, v) = 2m_i|\phi(0, v)|^2$ when $u = 0$. Provided $\phi(0, 0) \neq 0$, this is strictly positive. Now repeat the calculation with y in place of x. We have just seen that the values corresponding to $\phi(0, 0)$ do not vanish for both x and y together. Thus $\xi(\pi^*(|x|^2 + |y|^2)) > 0$ at P and indeed (iv) holds.

Next we must consider a point $P \in E_i \cap E_j$. We can find local complex coordinates (u, v) at P such that E_i is given by $u = 0$ and E_j by $v = 0$. Arguing as above, we may also suppose that $f(u, v) = u^{M_i}v^{M_j}$. Define p, q, r, s as before. We claim that the vector field $\xi := p\partial/\partial p + q\partial/\partial q + r\partial/\partial r + s\partial/\partial s$ satisfies the above conditions in some neighbourhood of P.

First, $\xi(\log f) = M_i + M_j$, which is real and positive, so (i) and (ii) are satisfied. Next, we may again write $\rho_i = Ap^2 + 2Bpq + Cq^2$, where along $p = q = 0$ we have $A > 0$, $C > 0$ and $AC > B^2$. Calculating $\xi(\rho_i)$ gives the same formula as above, with the additional terms $A'p^2+2B'pq+C'q^2$, where $A' = r\frac{\partial A}{\partial r} + s\frac{\partial A}{\partial s}$ and similarly for B' and C'. Since A', B' and C' all vanish at P, they are all small in a neighbourhood of P, so these extra terms are small in comparison with $Ap^2 + 2Bpq + Cq^2$. It thus follows that indeed $\xi(\rho_i) > 0$ near P, so (iii) holds.

As to (iv), write $\pi^*(x) = u^{m_i}v^{m_j}\phi(u,v)$, so

$$\pi^*|x|^2 = (p^2+q^2)^{m_i}(r^2+s^2)^{m_j}|\phi(u,v)|^2.$$

Set $\psi(u,v) = |\phi(u,v)|^2$. Evaluating $\xi(\pi^*|x|^2)$ gives

$$(p^2+q^2)^{m_i}(r^2+s^2)^{m_j}\{2(m_i+m_j)\psi(u,v) + p\partial\psi/\partial p \\ +q\partial\psi/\partial q + r\partial\psi/\partial r + s\partial\psi/\partial s\}.$$

The argument now concludes as before.

As to B_j it is sufficient to consider a neighbourhood of the point P where it meets one of the E_i. We use the same calculations as when $P = E_i \cap E_j$, but must set $M_j = 1$ and $m_j = 0$: x and y do not vanish along B_j except at P. Then all the above arguments continue to apply. □

Since π gives a diffeomorphism of $T \setminus \pi^{-1}(O)$ onto $S \setminus \{O\}$, we can push down the vector field ξ just constructed to give a vector field in S. It will, however, be more convenient to work in T, and we will abuse notation to identify the two.

Choose ϵ small enough so that the preimage N^s in T of the entire ball B_ϵ is contained in the neighbourhood U of the Proposition. In Section 6.2 we also studied $B_\epsilon \cap f^{-1}(D_\eta)$, where η is sufficiently small compared to ϵ, and now write N^m for its preimage. We have now also described a plumbing model N^p, which is the union of tubular neighbourhoods N_i of E_i defined by $\rho_i \leq \epsilon_i$ (for small enough ϵ_i) and N_j of B_j^* defined by $|f| \leq \eta$ and $||z||^2 \leq \eta$.

Corollary 9.3.3 *There are homeomorphisms* $\partial N^p \to \partial N^m \to \partial N^s$, *which agree with the identity near where these manifolds intersect* C, *and are compatible with the function defined by* $f/|f|$ *on the complement of this intersection.*

The function defines a fibration on these complements; the three fibrations are equivalent.

Proof Each point of U lies on an integral curve $t \mapsto \gamma(t)$ of ξ, which tends to $\bigcup E_i$ as $t \to -\infty$, and along which all of $|f|$, $||z||^2$, and any relevant ρ_i increase with t. Since $\xi(\log|f|)$ is bounded below, if the integral curve does not lie in $C^{(N)}$, it cannot converge as $t \to \infty$ and leaves any neighbourhood of $C^{(N)}$.

We first deduce that each such curve crosses the boundary ∂N^s : $||z||^2 = \epsilon$ at a unique point. Second, if $|f| > \eta$ at this point, it must have previously crossed the boundary $\partial N^m : |f| = \eta$ at a unique point.

Thirdly, since the curve begins on $\bigcup E_i$ and reaches ∂N^s, it must have crossed ∂N^p somewhere; since each ρ_i increases along the curve, it can only cross the boundary from inside to outside, so crosses at a unique point. We now see that following the integral curves defines the desired homeomorphisms.

Since $\arg f$ is constant along integral curves of f, the homeomorphisms preserve the levels of the function $f/|f|$. It remains only to prove that one of our three maps is a fibration: the other two will then also be. Now since $\xi(\Im \log(f)) = 0$ but $\xi(||z||^2 \circ \pi) > 0$, the tangent spaces to levels of $\Im \log(f)$ (or equivalently, of $f/|f|$) and of $||z||^2 \circ \pi$ are transverse, so the unit tangent vector to S^1 can locally be lifted to a tangent vector to S_ϵ. Thus by Theorem 6.1.2, we have a fibration of ∂N^s. □

Observe that we have given an alternative proof of Theorem 6.2.1 and Theorem 6.2.2.

The plumbing model exhibits the Thurston decomposition in a particularly clear form. For the study of the monodromy it will be convenient to have, in addition to pieces corresponding to the vertices of gG_R^+, 'transition regions' at the frontiers corresponding to the edges. For example, if one model is defined locally by $|\rho_i| \leq 2\epsilon$ and $|\rho_j| \leq 2\epsilon$, then we may take a new model given by $|\rho_i| \leq \epsilon$ and $|\rho_j| \leq \epsilon$, but regard the region in between as a transition region.

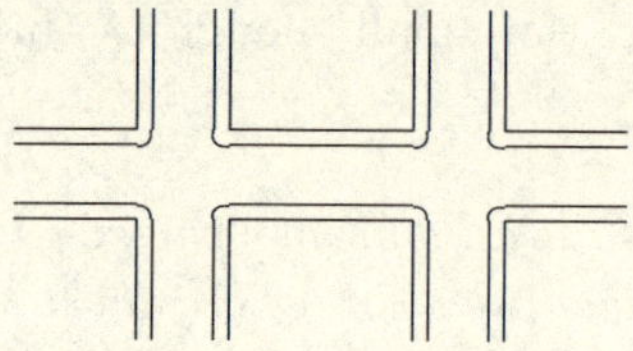

Fig. 9.4. Plumbing with rounded corners

It is more natural, however, to round the corners and take the transition region as smooth. Away from the intersection points $E_i \cap E_j$, we can take $|f|$ itself to play the role of ρ_i. We thus return to the model N^p, but now have a more explicit description. As in Section 5.2 we can replace $||z||^2$ by $|x|^2$ (where (x, y) are the coordinates in $\mathbb{C}^2$) for this purpose.

Proposition 9.3.4 *We can express N^p as a union as follows:*

For each vertex V_i of Γ_R^+, we have a region X_i:

if V is not an arrowhead vertex, $X_i = D^2 \times E_i^o$, with coordinates (u, v) where $u \in \mathbb{C}$ and $v \in E_i^o$, $f(u, v) = u^{M_i}$ and $|f| \leq \eta$;

if V is an arrowhead vertex, $X_i \subset \mathbb{C} \times B_j^o$, with local complex coordinates (u, v) and $f(u, v) = u$, $x = v^{m(B_j)}$; we have $|x|^2 \leq \epsilon$ and $|f| \leq \eta$.

For each edge $E = V_iV_j$ of Γ_R^+ we have a region Y_E in $\mathbb{C} \times \mathbb{C}$, with local complex coordinates (u_i, u_j) and $f(u_i, u_j) = u_i^{M_i} u_j^{M_j}$, $|u_i|, |u_j| \leq \alpha$ and $|f| \leq \eta$; η is small compared to α (in particular, $\eta < \alpha^{M_i+M_j}$).

The boundary of N^p is the union of the parts where $|f| = \eta$ and, for arrowhead regions, where $|x|^2 = \epsilon$. The regions have disjoint interiors, and each interior boundary $|u_j| = \alpha$ of Y_E is identified to an interior boundary component $v \in S_{i,j}$ in X_i by a homeomorphism which is the product of a homeomorphism of the circles $|u_j| = \alpha$ and $v \in S_{i,j}$ and the map $u = \alpha^{-M_j/M_i} u_i$.

Proof We have given the arguments for most of this above, but need to choose product structures in the regions (i) so that the identification maps are as stated. The projection f being fixed, all we need is to choose a retraction on E_i agreeing with the given map at the edge of the transition region. This will be a submersion near E_i, and necessarily transverse to f. □

This allows us to give an explicit construction of a monodromy map. We define a 'transition parameter' t in the region Y_E by

$$t := \frac{M_i \log |u/\alpha|}{M_j \log |v/\alpha| + M_i \log |u/\alpha|}. \tag{9.1}$$

Thus $t = 0$ along $|u| = \alpha$; $t = 1$ along $|v| = \alpha$ and, since $|f| = |u|^{M_i} |v|^{M_j} \leq \eta$ and $\eta < \alpha^{M_i+M_j}$, $0 \leq t \leq 1$ throughout. Equivalently,

$$\log |u|^{M_i} = (\log \alpha)\{(1-t)M_i - tM_j\} + t \log |f|,$$

and

$$\log |v|^{M_j} = -\log \alpha\{(1-t)M_i - tM_j\} + (1-t) \log |f|.$$

Define p, q, r, s to be real and imaginary parts as before.

Lemma 9.3.5 *We can take the monodromy to be given in the above charts by integrating the vector field ξ which is given by $(p\frac{\partial}{\partial q} - q\frac{\partial}{\partial p})/M_i$ on X_i and by $\frac{1-t}{M_i}(p\frac{\partial}{\partial q} - q\frac{\partial}{\partial p}) + \frac{t}{M_j}(r\frac{\partial}{\partial s} - s\frac{\partial}{\partial r})$ on Y_E.*

Proof It is enough to verify that the given vector field is continuous, and in each piece satisfies $\xi(\log f) = i$, for then it are tangent to the levels of $|f|$ and projects to the standard tangent vector field of S^1.

Continuity is assured by our definition since the parameter t was defined precisely so that at the boundary with a region of type (i) the vectors agree. We next evaluate $\xi(\log f)$ in regions X_i. If V_i is not an arrowhead vertex, this gives

$$\begin{aligned} M_i^{-1}\left(p\frac{\partial}{\partial q} - q\frac{\partial}{\partial p}\right) M_i \log u &= u^{-1}\left(p\frac{\partial}{\partial q} - q\frac{\partial}{\partial p}\right)(p+iq) = \frac{(ip-q)}{(p+iq)} \\ &= i; \end{aligned}$$

if it is, the first summand of the vector field applied to $\log f$ yields $(1-t)i$ and the second gives ti. □

We can now give a model for the Milnor fibre and the action of the monodromy on it.

Theorem 9.3.6 *The Milnor fibre F may be decomposed into pieces as follows.*

For each vertex V_i of Γ_R^+, F_i is a cyclic M_i-fold covering of the punctured exceptional E_i^o or B_j^o.

For each edge $E = V_iV_j$ of Γ_R^+, there is a piece $C_E \times I \subset S^1 \times S^1 \times I$, where the first two factors correspond to the respective projections on E_i and E_j, and C_E is given by the equation $w^{M_i}z^{M_j} = 1$.

The monodromy map h of F may be taken on F_i as a covering transformation of the bundle over E_i^o; on C_E as

$$h(w, z, t) = (e^{2(1-t)\pi i/M_i}w, e^{2t\pi i/M_j}z, t).$$

Proof Since the vector field of Lemma 9.3.5 preserves the pieces described in Proposition 9.3.4, we can treat them in turn.

On X_i, since $f(u, v) = u^{M_i}$, the Milnor fibre $f = \eta$ has just M_i points over each v, given by taking u as an M_i^{th} root of η. Integrating the vector field $(p\partial/\partial q - q\partial/\partial p)/M_i$ leaves v constant and at time T takes u to $ue^{2\pi i T/M_i}$; taking $T = 1$ we see that h has the effect of multiplying u by $e^{2\pi i/M_i}$.

(ii) The relevant part of ∂N^p is given in $\mathbb{C} \times \mathbb{C}$ by $|f| = \eta$ (where $f(u, v) = u_i^{M_i}u_j^{M_j}$) and $|u_i|, |u_j| \le \alpha$. We identify this with $S^1 \times S^1 \times I$, with coordinates (w_i, w_j, t), by setting $w_i = u_i/|u_i|$, $w_j = u_j/|u_j|$ and defining t by (9.1). The Milnor fibre $f = \eta$ is then given by $w_i^{M_i}w_j^{M_j} = 1$. Integrating the vector field of (ii) of 9.3.5 gives the time T map $(u_i, u_j) \to (e^{2\pi i(1-t)T/M_i}u_i, e^{2\pi i t T/M_j}u_j)$. Taking $T = 1$ gives the monodromy map

h as $(u_i, u_j) \to (e^{2\pi i(1-t)/M_i} u_i, e^{2\pi it/M_j} u_j)$ or, in the new coordinates, $(w_i, w_j, t) \to (e^{2\pi i(1-t)/M_i} w_i, e^{2\pi it/M_j} w_j, t)$. □

Since the pieces corresponding to the arrowhead vertices and arrow edges give a collar neighbourhood of the boundary, if we omit all these (so just use the F_i and C_E corresponding to vertices and edges of Γ_R), we still have a model for the Milnor fibre up to homeomorphism and the monodromy up to isotopy. Note that for each arrow attached to V_i we must still retain the corresponding puncture of E_i.

We can extract an interesting result from this description of the monodromy.

Theorem 9.3.7 *The monodromy $h : F \to F$ may be chosen to have no fixed points.*

Proof Consider the above model, indexed by Γ_R, for the monodromy. In each piece F_i it agrees with the covering transformation of a cyclic covering of order M_i. None of these degrees is 1, as follows from the algorithm of Lemma 8.4 to calculate them, so h has no fixed point on F_i; similarly, it has none on $C_E \times I$. □

Corresponding to an edge $E = V_iV_j$ we have a region $Y_E \subset \partial N^p$ which we can write as $T_E \times I$ with T_E a torus; it meets F in $C_E \times I$, where C_E is a union of parallel circles in T_E. The number of components of C_E is the highest common factor (M_i, M_j), which we will denote by M_E.

We shall return to the monodromy in Chapter 10; for the remainder of this chapter we will ignore h and concentrate on the decomposition of the link complement. For this the transition regions (ii) are not necessary: nor indeed are the regions (iii), which are homeomorphic to the product of a torus with an interval.

The local model shows that corresponding to each intersection point $E_i \cap E_j$ the two pieces of ∂N are separated by a torus. If we cut along all these tori, each remaining piece is a (trivial) circle bundle over a punctured E_i. In particular, each of these pieces is Seifert fibred, so we have a torus decomposition.

9.4 The Eggers tree and the resolution tree

According to Theorem 9.1.2 there is an essentially unique minimal decomposition of the complement M of an algebraic link by tori. In Section 9.2 we constructed a decomposition which, after some simplification, had pieces corresponding to the characteristic Puiseux exponents of the separate branches and to their intersection exponents, and hence to the interior vertices of the Eggers tree. In Section 9.3 we constructed a decomposition whose dual graph was $\Gamma_R(C)$. Although this is not minimal, we observed that we obtain a more efficient decomposition by removing the vertices of valence 2 (the opposite procedure to subdivision).

It follows that $\Gamma_E(C)$ and $\Gamma_R(C)$ are closely related. In this section we describe this relation in detail. To formulate the details conveniently, we make some definitions.

Define a resolution of C to be *efficient* if it is good; i.e. two components of $\pi^{-1}(C)$ may only intersect in one point, transversely; and each centre in the sequence of blowing ups lies on the corresponding strict transform of C. The minimal good resolution is efficient, but we need not restrict to quite this case. If we have an efficient resolution, then the effect on $\Gamma_R^+(C)$ of any further blowing up at a point $E_i \cap B_j$ on the strict transform of C is be to subdivide the edge V_iW_j by inserting a new vertex. Thus the homeomorphism type is unchanged.

We have already defined the dual graph $\Gamma_R(C)$ and the augmented dual graph $\Gamma_R^+(C)$. We now define the *doubly augmented dual graph* $\Gamma_R^{++}(C)$ by adding a further edge $W_{\rm in}V_0$ (marked with an inward arrow) to the initial vertex. The *core* of $\Gamma_R^{++}(C)$ is the union of the geodesics from $W_{\rm in}$ to the W_j; the cores of $\Gamma_R^+(C)$ and $\Gamma_R(C)$ are their respective intersections with this.

We recall from Section 4.2 that the interior vertices of the Eggers tree $\Gamma_E(C)$ are the points of discontinuity of ν_C together with the rupture points.

For each vertex V_i of $\Gamma_R(C)$ we choose a curvette ϵ_i. By Lemma 4.2.1, there is a unique point $X_i = X_{\epsilon_i}$ of the Eggers tree $\Gamma_E(C)$ such that for each branch B_j of C, $\mathcal{O}(\epsilon_i, B_j) = v(\inf(X_i, B_j^\infty))$; or equivalently, $(\epsilon_i.B_j)/m(\epsilon_i)m(B_j) = h(\inf(X_i, B_j^\infty))$. We define a map $\phi_C : \Gamma_R^{++}(C) \to \Gamma_E(C)$ as follows. On vertices, $\phi_C(V_i) = X_i$, $\phi_C(W_{\rm in}) = A_0$ and $\phi_C(W_j) = B_j^\infty$. For each edge E, $\phi_C(E)$ is a point if both ends of E have the same image under ϕ_C; otherwise ϕ_C maps E homeomorphically to a geodesic joining the images of the end points.

Lemma 9.4.1 *The map ϕ_C induces a homeomorphism from the core of $\Gamma_R^{++}(C)$ to $\Gamma_E(C)$.*

Proof By Proposition 8.3.2, the function $M_i(C)/m_i$ is strictly increasing along any path in $\Gamma_R^+(C)$ from V_0 to a vertex W_j, and for each edge not on such a path, takes the same value at both ends of the edge. More precisely, by the addendum 8.3.3, if B_j is a branch of C, $M_i(B_j)/m_i$ is strictly increasing along the geodesic V_0W_j, and takes the same value at both ends of an edge not on this.

First suppose C has only one branch. Then $m(C)h(X_i) = \epsilon_i.C/m(\epsilon_i) = M_i(C)/m_i$ is strictly increasing as V_i moves along the core of $\Gamma_R^+(C)$, and takes the same value at both ends of any edge not on it. Hence the core of $\Gamma_R^{++}(C)$ is mapped homeomorphically to $\Gamma_E(C)$, and each other edge is mapped to a point. Note that the double augmentation is required here: W_{in} is mapped to the point at level $v = 0$ on the Eggers tree; the initial vertex V_0 to the point at level 1.

Next suppose C has just two branches B_1, B_2. Then the core of $\Gamma_R^{++}(C)$ is the union of the geodesics $W_{\text{in}}W_1$ and $W_{\text{in}}W_2$; since we have a tree, these intersect in a chain from W_{in} to some vertex $V_{1,2}$. Now as V_i moves from $V_{1,2}$ to W_2,

$$M_i(B_1)/m_i = \epsilon_i.B_1/m(\epsilon_i) = m(B_1)h_{B_1}(\mathcal{O}(\epsilon_i, B_1))$$

is constant, hence $\mathcal{O}(\epsilon_i, B_1)$ is constant, hence is equal to $\mathcal{O}(B_2, B_1)$. Similarly $\mathcal{O}(\epsilon_{1,2}, B_2) = \mathcal{O}(B_1, B_2)$. Thus $\phi(V_{1,2})$ is the rupture point $I_{1,2}$. The assertion follows in this case by applying the result of the preceding paragraph to B_1 and B_2 separately. The same argument now works in general. □

To go further we must analyse the structure of an efficient resolution tree.

Theorem 9.4.2 *The tree $\Gamma_R^{++}(C)$ is obtained from its core by attaching chains of vertices and edges at distinct points, each of which is an infinitely near point giving a rupture vertex of a branch of C.*

Proof We proceed by induction on the number of branches of C, so suppose C formed from C' by adjoining a branch B: a basis for the induction is provided by Lemma 3.6.1. Take a minimal good resolution of C' (for which the result holds, by inductive hypothesis); then the strict transform of B meets the total transform of C at a point P.

If P lies on two components of this total transform, blowing up P has the effect of subdividing $\Gamma_R^{++}(C')$, which does not affect the qualitative description above. We continue blowing up until the contact point P only lies on one component E_f.

If at this stage C is resolved, the core of $\Gamma_R^{++}(C)$ is obtained from that of $\Gamma_R^{++}(C')$ by adding the edge V_fW_B, and the conclusion follows.

Otherwise we must continue resolving B, and all the blow ups already performed at points of the strict transform of B are required for this. The new exceptional curves generated in the course of this further resolution form part of the minimal good resolution of B, and the new part of the graph is attached to the old at V_f; in particular, the geodesic from W_{in} to W_B must pass through V_f. Thus the whole graph consists of the former one, this further part of the core, and (perhaps) chains attached to this new part of the core. The result follows. □

Thus the complement of the core is a union of chains, attached to the core at distinct rupture points. Each of these chains is called a *dead branch* of the tree.

Proposition 9.4.3 *Let ϵ be a curvette at a rupture vertex of the doubly augmented dual tree $\Gamma_R^{++}(C)$ of an efficient resolution of the curve C. Then the point of contact $X(\epsilon)$ on $\Gamma_E(C)$ is an interior vertex. Every interior vertex arises in this way.*

Proof For a single branch B, according to Lemma 8.5.1, the curve there denoted C_q^0 is a curvette at the qth rupture point $V_i = AV_q$. Since $\mathcal{O}(C_q^0, C) = \alpha_q$, $\phi(V_i) = A_q$ is the qth interior vertex of $\Gamma_E(B)$.

It follows from Theorem 9.4.2 that each rupture vertex of $\Gamma_R^{++}(C)$ is either a rupture vertex $V_{j,k}$ of the core, where paths $W_{\text{in}}W_j$ and $W_{\text{in}}W_k$ separate, or (as an infinitely near point) is a rupture vertex corresponding to a single branch (or, indeed, both). We saw in the proof of Lemma 9.4.1 that $\phi(V_{j,k}) = I_{j,k}$; the others are dealt with by the case of a single branch. □

By Theorem 9.4.2, $\Gamma_R^{++}(C)$ is obtained from its core by attaching chains of vertices and edges at distinct points, each of which is an infinitely near point giving a rupture vertex of a branch of C. By Lemma 9.4.1, ϕ induces a homeomorphism from the core of $\Gamma_R^{++}(C)$ to $\Gamma_E(C)$; and it follows from Proposition 8.3.2 that each of the additional chains is mapped to a point. These points are the images of rupture points corresponding to branches of C, so by Proposition 9.4.3 are interior vertices of $\Gamma_E(C)$ where ν is discontinuous. It also follows from the proof of Theorem 9.4.2 that we obtain in this way all such points which are not also rupture points of $\Gamma_E(C)$ and some of those that are: a precise statement will be obtained in Corollary 9.8.3.

We have explored the relation between $\Gamma_R^{++}(C)$ and $\Gamma_E(C)$: to return to $\Gamma_R^{+}(C)$, define the reduced Eggers tree $\Gamma_E^{r}(C)$ by removing the half-open interval $[0, 1)$ from the Eggers tree. Define also the *hairy Eggers tree* $\Gamma_E^{+}(C)$ by attaching an additional edge: a 'whisker': at those vertices

P of type A_q for which $\phi^{-1}(P)$ is a chain, rather than a single point; and the reduced hairy Eggers tree $\Gamma_E^{r+}(C)$ by removing $[0, 1)$ from it. We can summarise as follows.

Theorem 9.4.4 *There is a homeomorphism between $\Gamma_R^{++}(C)$ for an efficient resolution and $\Gamma_E^+(C)$, under which vertices of valence ≥ 3 correspond as above, and the edge $W_{\text{in}}V_0$ corresponds to the interval $[0, 1]$. Hence $\Gamma_R^+(C)$ is mapped to $\Gamma_E^{r+}(C)$.*

In particular, there are natural bijections between:

the set of vertices of valence at least 3 in $\Gamma_R^{++}(C)$ and the set of interior vertices in $\Gamma_E(C)$, and

the complementary pieces in the minimal torus decomposition of the link complement and the vertices of valence ≥ 3 in either $\Gamma_R^+(C)$ or $\Gamma_E^{r+}(C)$.

We established in Proposition 9.4.3 that the rupture points of $\Gamma_R^{++}(C)$ correspond to the set of interior vertices in $\Gamma_E(C)$, so the set of quotients M_i/m_i at the rupture points coincides with the set of values of $\overline{H}_C$ at the interior vertices. By Theorem 4.5.2, the set of interior vertices coincides with the set of contact points with a transverse polar curve P of C. Hence the set of quotients M_i/m_i at the rupture points also coincides with the set of quotients $P_\alpha.C/m(P_\alpha)$ as P_α runs through the set of components of P. This set is an important invariant of the curve C, which we call the set of polar quotients.

In the case when C is a single branch, the polar quotients were calculated in Theorem 8.5.2. If C is an irreducible germ, with Puiseux characteristic $(m; \beta_1, \ldots, \beta_g)$, the values of the invariants $M_k(C)$ and m_k at the ith rupture point of $\Gamma_R(C)$ are, respectively, $M(C) = \frac{e_{i-1}}{e_i}\overline{\beta_i}$ and $m = \frac{m}{e_i}$, so that $\frac{M}{m} = \frac{e_{i-1}}{m}\overline{\beta_i}$. In general, for the ith interior vertex (V_k, say) on B_j, we have $\epsilon_k.B_r/m(\epsilon_j) = \overline{H}_{B_r}(\min(\frac{\beta_i}{m}, \mathcal{O}(B_j, B_r)))$, and must sum over r to obtain the polar quotient; similarly, for $I_{j,k}$, we have $\sum_r \overline{H}_{B_r}(\min(\mathcal{O}(B_j, B_k), \mathcal{O}(B_j, B_r)))$.

We will be particularly interested in Section 11.5 in the largest of the quotients $\frac{M_i(C)}{m_i}$ at a rupture point of $\Gamma_R(C)$, which we will denote by $Q(C)$. If C is irreducible, we have $Q(C) = \frac{e_{g-1}}{m}\overline{\beta_g}$. In general $Q(C)$ is equal to the supremum over j of the value of $\overline{H}_C$ at the greatest interior vertex on the branch $V_0B_j^\infty$ of the Eggers tree.

9.5 Finiteness of the monodromy

We can apply the above results to give a complete analysis of the cases when the monodromy h of the Milnor fibration has finite order (in the set-theoretic sense). First, we give several equivalent conditions on h.

Recall that we have a compact oriented surface F (the Milnor fibre), which apart from the cases where C is smooth or has an A_1 singularity has negative Euler characteristic, and a self-diffeomorphism h of F which can be taken as the identity on the boundary. However in this section we do not restrict h to be the identity on the boundary, though it remains forcedly isotopic to the identity there. Milnor's Fibration Theorem 6.2.2 tells us that the mapping torus of h is homeomorphic to the closed complement M of the singularity link ∂F in S^3.

Lemma 9.5.1 *Let h be a self-homeomorphism of a compact 2-manifold F with $\chi(F) < 0$. For k a natural number, consider the conditions*

(1) *h is isotopic to a homeomorphism h' with ${h'}^k$ the identity,*
(2) *h^k is isotopic to the identity,*
(3) *h^k is homotopic to the identity,*
(4) *h^k_* on $\pi_1(F, x)$ is an inner automorphism,*
(5) *the mapping torus T_h of h is a Seifert fibre space.*

Then conditions (1)–(4) are equivalent. Moreover these hold for some k if and only if condition (5) holds.

Proof Clearly (1) $\Rightarrow$ (2) $\Rightarrow$ (3) $\Rightarrow$ (4).

Since F is aspherical, (4) $\Rightarrow$ (3). It was proved by Epstein [68] Theorem 6.4 that (3) $\Rightarrow$ (2). Finally, that (2) $\Rightarrow$ (1) is a theorem of Nielsen: a full proof is given on [70], p. 219.

It follows from Remark 9.1.4 that if h^k is the identity, T_h is a Seifert fibre space. Conversely, if T_h is a Seifert fibre space, its JSJ decomposition is trivial, hence by Proposition 9.1.5, so is the decomposition of (F, h), and h is isotopic to a diffeomorphism of finite order. □

Theorem 9.5.2 *Let C be a plane curve singularity with Milnor fibre F and monodromy h. The following are equivalent:*

(1) *h is isotopic to a homeomorphism of finite order,*
(2) *the closed link complement M of ∂F in S^3 is a Seifert fibre space,*
(3) *there is only one vertex of valence ≥ 3 in $\Gamma^+_R(C)$,*
(4) *there is only one vertex of valence ≥ 3 in $\Gamma^{r+}_E(C)$,*
(5) *C is equisingular to a weighted homogeneous singularity.*

Proof By Lemma 9.5.1, (1) and (2) are equivalent. Since all the pieces in the minimal torus decomposition of the link complement are Seifert fibre spaces, it follows from Theorem 9.4.4 that (2), (3) and (4) are equivalent.

To show that (5) $\Rightarrow$ (1), suppose C weighted homogeneous – say given by $f = 0$ with $f(t^p x, t^q y) = t^N f(x, y)$. Then we can give a monodromy map explicitly by $e^{i\theta}.(x, y) = (e^{iq\theta} x, e^{ip\theta} y)$; this has finite order.

Recall that the rupture points in $\Gamma_E^+(C)$ are just the interior vertices of the Eggers tree. Deletion of the interval $[0, 1)$ will only decrease the number of rupture points if the point with $v = 1$ is itself a branch point with valence 3, corresponding to V_0 having valence 2 in the $\Gamma_R(C)$.

Now suppose (4) holds, and that there is only one interior vertex V in $\Gamma_E(C)$: set $\alpha = v(V)$. Since $\Gamma_E(C)$, together with the valuation map v on it, determines C up to equisingularity, it suffices to construct a weighted homogeneous curve with the same Eggers tree in order to prove (5). If $\alpha \notin \mathbb{Z}$, say $\alpha = \frac{p}{q}$, and there are r non-smooth branches, then $x^{pr} = y^{qr}$ is such a curve. If in addition there is one smooth branch, with contact α with the rest, (there cannot be two, as their mutual exponent of contact must be an integer), we may take $y(x^{pr} - y^{qr}) = 0$. If $\alpha = r \in \mathbb{Z}$ we have a lot of smooth branches and can take $\prod(y - a_i x^r) = 0$.

If the point $v = 1$ is also an Eggers vertex, there are two groups of branches having exponent of contact 1 with each other (i.e. branches in the same group have the same tangent; those in the other group a different tangent). Each group, unless it consists only of a single curve, has at least one interior vertex in its Eggers tree. Thus one group consists of a single smooth branch, transverse to the other group, which itself must be as in the preceding paragraph. It thus suffices to multiply the equation given above by a factor x. □

9.6 Seifert fibre spaces

We now study the geometry of Seifert fibre spaces: these results will be needed in the next section. Recall that by definition a Seifert fibre space is a 3-manifold X with a fixed-point free action of S^1: write $p : X \to S$ for the quotient map to a surface. Each fibre of p is homeomorphic to S^1. For most points $x \in X$ the isotropy group $G_x := \{t \in G \,|\, t.x = x\}$ is trivial.

At a point where G_x is not trivial, it must be finite of order a, say, since the action is assumed fixed point free, and cyclic, since it is a subgroup of G: choose a generator g. Now we can find a surface S in X, meeting the orbit $G.x$ transversely in x, and invariant under the group G_x: such a surface (which need only exist near x) is called a *slice* for the action. We may suppose S a disc admitting a local chart at x with respect to which G_x acts by rotations, with g giving a rotation through an angle $2\pi s/a$ where s is prime to a. The union $G.S$ of the orbits of points in S, or the image of the map $G \times S \to M$ defined by the action, is homeomorphic to the quotient of the product $G \times S$ by the action of G_x by $t.(g,y) = (gt^{-1}, t.y)$. See Figure 9.5: the arrows at the top and bottom of the figure indicate that the figure should be repeated till the ends join up.

The quotient surface S is often best viewed as an 'orbifold', where a point P_i corresponding to a non-trivial isotropy group G_x is associated with the order a_i of that group: in many ways P_i may be regarded as $1/a_i$ of a point. In particular, the *orbifold Euler characteristic* χ_{orb} is defined in terms of the usual Euler characteristic by

$$\chi_{\text{orb}}(S) = \chi(S) - \sum_i \left(1 - \frac{1}{a_i}\right). \tag{9.2}$$

In general we may refer to such points P_i (with $a_i > 1$) as *cone points*, and to the fibres $p^{-1}(P_i)$ over them as *exceptional fibres*.

In the case where there are no exceptional fibres, X is the total space of a circle bundle ξ and is determined by S and the integer $c(\xi)$. If S has a boundary, there is no obstruction to finding a section, but if we prescribe a section for the restriction to the boundary, then again there is an integer obstruction to extending this to a section defined over all of S.

Now suppose there is no boundary, but there are exceptional points $P_i \in S$. Remove the interiors of disjoint disc neighbourhoods D_i of the points P_i. Then there exist sections over the complement: choose one, σ say. Thus $\sigma(\partial D_i) \subset p^{-1}(D_i)$ is homologous to some multiple b_i of the central curve $p^{-1}(P_i)$ of the solid torus. If g is the genus of the surface S, the 'unnormalised Seifert invariant' is the symbol $(g; (a_1, b_1), \ldots, (a_s, b_s))$. These numbers satisfy $g \geq 0$, $a_i > 1$, $\gcd(a_i, b_i) = 1$. We next describe the extent of their invariance.

Lemma 9.6.1 *For any $(g; (a_1, b_1), \ldots, (a_s, b_s))$ with $g \geq 0$ and $\gcd(a_i, b_i) = 1$, $a_i > 1$ for each i, there exists a Seifert fibre space with these as Seifert invariants.*

Two Seifert manifolds, which have Seifert invariants $(g; (a_1, b_1), \ldots, (a_s, b_s))$ and $(g'; (a'_1, b'_1), \ldots, (a'_{s'}, b'_{s'}))$ respectively, admit a fibre- and orientation-preserving homeomorphism if and only if, after re-indexing

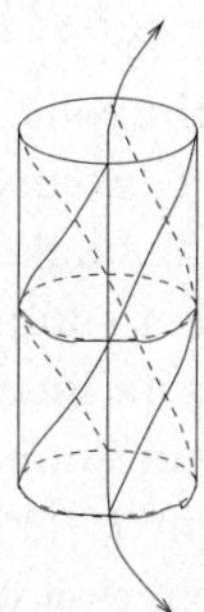

Fig. 9.5. A Seifert fibre space

the pairs (a_i, b_i) *if necessary, we have* $g = g'$, $s = s'$, $a_i = a'_i$ *for each* i, $b_i \equiv b'_i \pmod{a_i}$ *for each* i, *and* $\sum_{i=1}^{s} \frac{b_i}{a_i} = \sum_{i=1}^{s} \frac{b'_i}{a_i}$.

The proof is not difficult and is given in [140]. In most cases it is enough to require that the homeomorphism is orientation-preserving.

The *orbifold normal Euler number* is now defined as

$$e(M) := -\sum_{i=1}^{s} \frac{b_i}{a_i} \tag{9.3}$$

in the above notation: this extends the invariant $c(\xi)$ which is defined in the case where there are no exceptional fibres. The notation may be slightly modified by allowing some a_i to be equal to 1. In particular, if this is so for just one value – say $a_0 = 1$ – then the invariants can be uniquely normalised by $0 < b_i < a_i$ for $i > 0$ with $b_0 \in \mathbb{Z}$: this is essentially the original notation of [162].

We referred in the preceding section to geometric structures. A closed manifold M which is a Seifert fibre space has one of 6 possible types of such structure, and the cases are distinguished by whether $\chi_{\text{orb}}(M)$ is negative, zero or positive, and by whether $e(M)$ is zero or not.

The homology of a Seifert fibre space is calculated in terms of its Seifert invariants as follows.

Proposition 9.6.2 *Suppose* M *a Seifert fibre space with projection* $p : M \to \Sigma$. *Then* $\operatorname{rank} H_1(M) - \operatorname{rank} H_1(\Sigma)$ *is 1 if* $e(M) = 0$ *and 0 if* $e(M) \neq 0$. *If* $H_1(M; \mathbb{Z})$ *is a finite group, it has order* $|e(M) \prod_{i=1}^{s} a_i|$.

Proof Write, as usual, $p : M \to \Sigma$ for the projection, and obtain Σ^o from Σ by removing the interiors of small discs surrounding the exceptional points P_i; set $M^o := p^{-1}(\Sigma^o)$. Since there is a section σ over Σ^o,

M^o is homeomorphic to the product $S^1 \times \Sigma^o$, so $H_1(M^o) \cong \mathbb{Z} \oplus H_1(S^o)$. Now M is the union of M^o and the solid tori $Y_i = p^{-1}(D_i)$. To re-attach a solid torus along its boundary T_i we may attach a 2-disc, then a 3-disc: the effect on H_1 is to factor out a single element.

In the above direct sum, the first summand is represented by the class $[f]$ of an orbit of the S^1-action; its image in $H_1(Y_i)$ was defined to be a_i times a generator. The second summand contains the class $[c_i]$ of the circle $\sigma(\partial D_i)$ whose image in $H_1(Y_i)$ was defined to be b_i times a generator. Thus $b_i[f] - a_i[c_i]$ defines a class in $H_1(\partial Y_i)$ which maps to 0 in $H_1(Y_i)$; since a_i and b_i are coprime it generates this kernel, and so represents the class which is killed by attaching the solid torus Y_i.

For the surface, $H_1(\Sigma^o)$ is obtained from $H_1(\Sigma) \bigoplus_{i=1}^s \mathbb{Z}[c_i]$ by imposing the relation $\sum_i [c_i] = 0$. Hence for the 3-manifold, $H_1(M)$ is isomorphic to the sum of $H_1(\Sigma)$ and the abelian group with generators $[f], [c_i]$ and relations

$$-b_i[f] + a_i[c_i] = 0 \ \ (1 \leq i \leq s), \quad \sum_{i=1}^{s} [c_i] = 0.$$

Now since the a_i are non-zero, the matrix of this presentation

$$A = \begin{pmatrix} 0 & 1 & 1 & 1 & \cdots \\ -b_1 & a_1 & 0 & 0 & \cdots \\ -b_2 & 0 & a_2 & 0 & \cdots \\ -b_3 & 0 & 0 & a_3 & \cdots \\ \vdots & \vdots & \vdots & \vdots & \ddots \end{pmatrix},$$

of size $s+1$, has rank at least s; and its determinant (obtainable by evaluating by the first row) is $\det A = \sum_{i=1}^s \frac{b_i}{a_i} \prod_{i=1}^s a_i = -e(M) \prod_{i=1}^s a_i$. The result follows. □

We will be particularly interested in Seifert fibre spaces M such that $H_1(M)$ is trivial. In this case, S must be homeomorphic to S^2. Since, if $i \neq j$, any common factor of a_i and a_j will divide $\sum \frac{b_i}{a_i} \prod a_i$, a_i and a_j must be coprime. Conversely, we have

Corollary 9.6.3 *Given positive integers a_i with any two coprime, there exists a Seifert fibre space M with $H_1(M) = 0$, unique up to fibre-preserving homeomorphism, with these as its a_i parameters.*

Proof Since any two of the a_i are mutually coprime, the equation

$$1 = -e(M) \prod a_i = b_1 a_2 \ldots a_k + \cdots + a_1 a_2 \ldots a_{k-1} b_k$$

shows that b_i is uniquely determined modulo a_i as the inverse to $\prod_{j \neq i} a_j$. If we choose one such b_i for each i, the above equation will hold as a congruence modulo $\prod_i a_i$, so altering (e.g.) b_1 by a multiple of a_1 we can turn the congruence into an equality. Now by Lemma 9.6.1 there exist corresponding Seifert fibre spaces, and the same lemma also shows that the result is essentially unique.

To deal with the case $\det A = -1$, note that if we change the signs of all the b_i we obtain another Seifert fibre space where the sign of $e(M)$ is changed: this is obtained from the other by changing orientation. □

9.7 The Eisenbud–Neumann diagram

In [65] the authors give a careful analysis of links in the 3-sphere, using the torus decomposition discussed in Theorem 9.1.2, and introduce a notation which completely determines the structure of the decomposition. They proceed to show how to calculate a number of important invariants using their notation. The case of interest here is discussed explicitly in the Appendix to Chapter I in that book. We now recall the details using the notation we have developed in this book.

The essential innovation of [65] is an adaptation of the general theory of torus decompositions to links in the 3-sphere using the notion of splicing. It is convenient to relax the context to allow the ambient manifold to be any homology sphere – i.e. a 3-manifold Σ without boundary whose first homology group $H_1(\Sigma; \mathbb{Z})$ is trivial. An embedded torus T in a homology sphere Σ will always split Σ into 2 components, E_1 and E_2, say, such that the induced map $H_1(T) \to H_1(E_1) \oplus H_1(E_2)$ is an isomorphism. There are thus curves L_i on T, unique up to isotopy and orientation, such that the image of L_1 generates $H_1(E_1)$ and maps to zero in $H_1(E_2)$, while the situation is reversed for L_2.

Now suppose given triples $(\Sigma, \mathbf{L}, C)$ and $(\Sigma', \mathbf{L}', C')$ where Σ is an oriented homology sphere, $\mathbf{L}$ an oriented link in Σ, and C a component of $\mathbf{L}$; and similarly for Σ', etc. Choose a tubular neighbourhood N of C in Σ disjoint from the other components of $\mathbf{L}$, and write T for its boundary $T = \partial N$, so that cutting Σ along T produces two pieces N and E. We have two curves in T: the standard latitude L which maps to 0 in $H_1(E)$ (or equivalently, has zero linking number with C) and to the generator $[C]$ in $H_1(N)$; and the standard meridian M which maps to 0 in $H_1(N)$ and has linking number $+1$ with C. These may be taken to be embedded and to cross each other just once, transversely, as the two factors of the product $L \times M \cong T$. Make corresponding constructions in Σ'.

Then the *splice* of $\mathbf{L}$ and $\mathbf{L}'$ along C and C' is defined to be obtained from $E \cup E'$ by identifying the boundaries so that L is identified to M' and M to L'. Since this means that the orientations of T and T' are reversed, the orientations of E and E' induced from those of Σ and Σ' match up to orient the resulting manifold Σ'', which contains the union $\mathbf{L}''$ of the remaining components of $\mathbf{L}$ and $\mathbf{L}'$, and is necessarily a homology sphere. Even if Σ and Σ' are homeomorphic to S^3, it does not follow that Σ'' is; but this will be the case in our examples below.

The splice construction deals with a closed manifold containing a link rather than with the closed complement of a link. However, the notions of torus splitting and splice decomposition are essentially the same. In fact, now suppose given a link $(\Sigma, \mathbf{L})$ and a torus T embedded in Σ and disjoint from the components of the link. Cut Σ along T to give components E_1 and E_2. We may regard T as a product $L_1 \times L_2$, where the curves $L_1 \times x$ bound in E_2 and the curves $x \times L_2$ bound in E_1. Take discs D_i with $\partial D_i = L_i$, attach $D_1 \times L_2$ along T to E_1 to obtain a homology sphere Σ_1, and provide this with the link $\mathbf{L_1}$ which is the union of those branches of $\mathbf{L}$ contained in E_1 with $0 \times L_2$. Similarly we construct a link $\mathbf{L_2}$ in the homology sphere $\Sigma_2 := E_2 \cup L_1 \times D_2$. Then the result of splicing these gives the original link; and the closed complements are just obtained by splitting the closed complement of the original link along T.

We have seen in Section 9.2 that starting with an algebraic link we can define a sequence of splittings till each resulting link has closed complement an Seifert fibre space. The action of S^1 on this closed complement restricts to a faithful action on each boundary torus. Such an action is determined up to isotopy by the homology class of an orbit, and hence extends to an action on the solid torus. In general this could have a circle of fixed points in the centre of the solid torus.

Lemma 9.7.1 *For each piece of the decomposition constructed in Section 9.2, if we extend the action as above, there are no fixed points.*

Proof We had tori T^{q-1} parametrised by $x = e^{im\theta}$, $y = \sum_{r=1}^{\beta_q - 1} \epsilon_r e^{ir\theta} + \epsilon_q e^{i\phi}$. Consider a piece with this torus as its outer boundary. The solid torus to be attached can then be identified with its exterior in S^3. The circle action on the Seifert fibre space induces the action on T in which θ increases but ϕ is fixed. Thus an orbit does not bound in the exterior region (it is the curves x constant that do). □

We can infer more: the circle in the centre of the solid torus can be identified with the circle $x = 0$ in the 3-sphere; the induced action on this circle is essentially determined by its leading term $e^{ir\theta}$ with $r = \beta_q - 1$. We observe, in particular, that the coefficient of θ is positive.

Our action gives the homology sphere Σ the structure of Seifert fibre space. Thus by Corollary 9.6.3 this structure is determined by the set $\{a_1, \dots, a_k\}$. The link is itself invariant under the action, so is a union of the orbits: this may include some ordinary orbits as well as some of the exceptional ones.

This 'Seifert link' is represented by a diagram in the form of a star with a central node from which there are rays labelled by a_i. If the ith exceptional fibre is a component of the link, the corresponding ray terminates in an arrow; otherwise it terminates in a black dot. We saw in the classification that we could introduce additional parameters a_i equal to 1, and we make use of this for the case that some components of the link correspond to trivial isotropy groups. However, if we have a ray labelled 1 terminating in a black dot, this means that the decomposition chosen is not canonical, since the black dot represents $S^1 \times D^2$ and the '1' shows that the Seifert fibre matches the S^1 factor so that the Seifert fibration extends over the solid torus. In the general notation of [65], the central node is additionally labelled with a sign $\epsilon = \pm 1$ equal to $\sum \frac{b_i}{a_i} \prod a_i$. In our situation, as the homology spheres are all spheres, the Seifert geometry is spherical and the signs are all $+1$, so we omit them.

If two links are represented by diagrams, with in each case each component of the link represented by a ray terminating in an arrowhead, then the link obtained by splicing is represented by amalgamating the two diagrams along the rays representing the components along which splicing takes place. We will call the result the E–N (short for Eisenbud–Neumann) diagram of the link. We will draw E–N diagrams as in Figure 9.6 (we omit ϵ since in all our diagrams the sign will be $+1$).

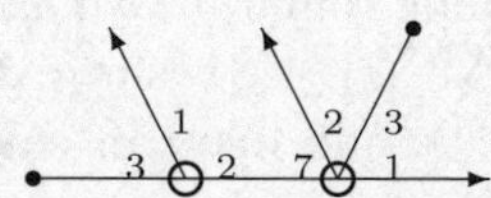

Fig. 9.6. An E–N diagram

Each arrowhead in the diagram represents a component of the final link; each black dot represents a solid torus, the neighbourhood of an exceptional fibre in some Seifert fibration; and each other vertex represents a circle bundle over a compact surface. The edges in the diagram represent tori along which these constituents are attached. For the E–N diagrams representing algebraic links, one of the black dots represents the beginning of the inductive construction. The *core* of the diagram is the union of the geodesics joining this vertex to those marked by arrows.

The treatment in [65] deals with arbitrary diagrams of this kind representing links, and gives a precise algorithm for reducing a diagram corresponding to a decomposition by a non-minimal family of tori to the standard minimal one.

The set of tori corresponding to the edges of the diagram is not quite the same as the minimal set of tori in the sense of Theorem 9.1.2: an edge terminating in a black dot represents a torus bounding a solid torus, and deleting it allows the fibration of the neighbouring region to be extended to a Seifert fibration with just one central fibre inside the solid torus, which is an exceptional fibre unless the edge is marked 1. Deleting all such tori of this type gives a minimal set in the previous sense.

One advantage of the E–N notation is a direct way to calculate the multi-variable Alexander polynomial.

Theorem 9.7.2 *Suppose the link* $\mathbf{L} = (\Sigma, S_1, \dots, S_{p+q})$ *is the result of splicing* $\mathbf{L}' = (\Sigma, S_0', S_1, \dots, S_p)$ *and* $\mathbf{L}'' = (\Sigma, S_0'', S_{p+1}, \dots, S_{p+q})$. *Let* $a_j := Lk(S_0'', S_j)$ *for* $p < j \le p+q$, $b_i := Lk(S_0', S_i)$ *for* $1 \le i \le p$. *Then the Alexander polynomials are related by*

$$\Delta_*^{\mathbf{L}}(t_1, \dots, t_{p+q}) =$$
$$\Delta_*^{\mathbf{L}'}\left(t_{p+1}^{a_{p+1}} \cdots t_{p+q}^{a_{p+q}}, t_1, \dots, t_p\right) \Delta_*^{\mathbf{L}''}\left(t_1^{b_1} \cdots t_p^{b_p}, t_{p+1}, \dots, t_{p+q}\right).$$

For a proof, see [65, Theorem 5.3].

This result contains Theorem 5.5.2 as a special case. For take $q = 1$, $p = 0$; $L' = (\Sigma, S_0')$ as the knot K and $L'' = (\Sigma, S_0'', S_1)$ as the trivial knot and a torus knot cabling round it. We have $a_1 = Lk(S_0'', S_1) = m$. Thus $\Delta^{K'}(t) = \Delta^K(t^m)\Delta^{L''}(t)$. Now since $\Delta^{K'}(t) = (t-1)\Delta_*^{K'}(t)$, $\Delta^K(t^m) = (t^m - 1)\Delta_*^K(t^m)$, and one can calculate $\Delta_*^{L''}(t) = (t^{mp} - 1)/(t^p - 1)$, we have

$$\Delta^{K'}(t) = \frac{(t-1)}{t^m - 1}\Delta^K(t^m)\Delta_*^{L''}(t) = \Delta_{m,p}(t)\Delta^K(t^m)$$

where, in the notation of Section 5.5, $\Delta_{m,p}(t) = (t^{mp} - 1)(t-1)/(t^m - 1)(t^p - 1)$.

9.8 Calculation of E–N diagrams

It follows from the uniqueness clause in Theorem 9.1.2 that any two constructions of a minimal torus decomposition must give essentially the same result. We gave one explicit torus decomposition in Section 9.2, and another explicit decomposition in Section 9.3. In Section 9.7 we did not give a construction, but rather a notation for describing the invariants of a decomposition. In this section we consider how to calculate these invariants using the explicit models of Section 9.2 and Section 9.3. The graph dual to the decomposition of Section 9.2 is closely related to the Eggers tree $\Gamma_E(C)$; the dual graph of that of Section 9.3 is $\Gamma_R^+(C)$; while the underlying graph of E-N diagram is itself dual to the minimal torus decomposition of the closed complement of the link of C. Thus these trees are closely related: we have already noted that inserting a superfluous torus has the effect of subdividing the dual graph or (if the torus is boundary-parallel) of attaching a whisker: and described the relation of $\Gamma_E(C)$ to $\Gamma_R(C)$ as trees.

We can obtain the E–N diagram directly from the geometric approach of Section 9.2 as follows. First consider the case of a single branch.

Theorem 9.8.1 *Let B be a plane curve singularity with just one branch with Puiseux characteristic $\{m; \beta_1, \dots, \beta_g\}$. Then the E-N diagram of the singularity link is*

Proof We refer to the construction in Section 9.2. We had first a solid torus, then, for $1 \leq q \leq g$, a sequence of Seifert fibre spaces with just one exceptional fibre with isotropy group of order e_{q-1}/e_q. Finally we have just a tubular neighbourhood of the knot. This determines the above shape of the diagram; it remains to calculate the other parameters.

In the case $g = 1$ we have a torus knot. If its invariants are (p, q), it is invariant under the action of S^1 on the unit sphere $S^3 \subset \mathbb{C}^2$ given by $t.(x, y) = (t^p x, t^q y)$, which just has two exceptional fibres, whose isotropy groups have orders p and q. The corresponding E–N diagram is thus indeed

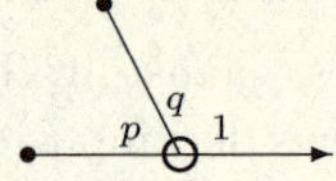

Note that the 'plus' sign comes from a calculation of the Seifert invariants leading to negativity of the 'orbifold normal Euler class' $e(m)$.

We saw in Chapter 5 that the singularity knot is built up inductively by the cabling construction. This is equivalent to taking the splice with a torus knot, and the cabling invariants give the invariants of the torus knot. These invariants were calculated in Section 5.4 (the essential step was obtained in Theorem 5.4.2), and shown to be $(e_{q-1}/e_q, \overline{\beta}_q/e_q)$. The E–N diagram is thus built up from these pieces; the result follows. □

We may describe this in words as follows. The E-N diagram is homeomorphic to the tree obtained from $\Gamma_E(B)$ by attaching a 'side edge' at each interior vertex A_q. In fact, it will be convenient to describe the edges at a vertex of any E-N diagram as incoming (pointing towards V_0); outgoing (along the core); or side (otherwise). In the above example, at the vertex A_q, the incoming edge is marked $\overline{\beta}_q/e_q$, the side edge is marked e_{q-1}/e_q, and the outgoing edge is marked 1.

Example 9.8.1 For a curve with Puiseux characteristic $(4; 6, 7)$, we have already calculated all the relevant parameters, so the diagram is

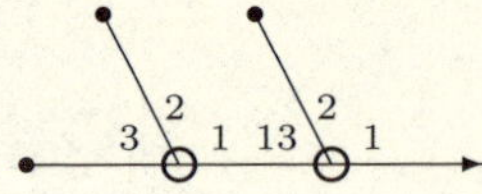

The E-N diagram in the general case comes from those for the separate branches by amalgamating the common parts. In the construction in Section 9.2 we have a non-trivial step each time we reach a characteristic exponent of a branch or an exponent of contact of two branches. These steps correspond to the vertices of the Eggers tree, or rather, of $\Gamma_E^{r+}(C)$.

Theorem 9.8.2 *An E-N diagram of a plane curve singularity is built from the Eggers tree as follows.*

(i) *The core of the E-N diagram is the graph $\Gamma_E^r(C)$; the vertices lying on the core are the same as those of $\Gamma_E^r(C)$.*

(ii) *Let V be a vertex of $\Gamma^r_E(C)$ which is a point of discontinuity of ν; say it is the vertex A_q for B_j. The incoming edge at V is marked $\frac{\overline{\beta}_q}{e_q}$ (where the values are those for B_j). The outgoing edge of $\Gamma_E(C)$ corresponding to B_k is marked 1 if V is a point of discontinuity for ν_{B_k} and $\frac{e_{q-1}}{e_q}$ if not. In addition, if each outgoing edge is marked 1, there is a side edge, marked with $\frac{e_{q-1}}{e_q}$, to a black dot.*

(iii) *Let V be a vertex of $\Gamma^r_E(C)$ at which ν is continuous. Then each outgoing edge at V is marked 1; there is no side edge; and the incoming edge is marked $X = \frac{m^2}{e_q^2}h(V)$, where the parameters are those of any branch through V and $\beta_q < mv < \beta_{q+1}$.*

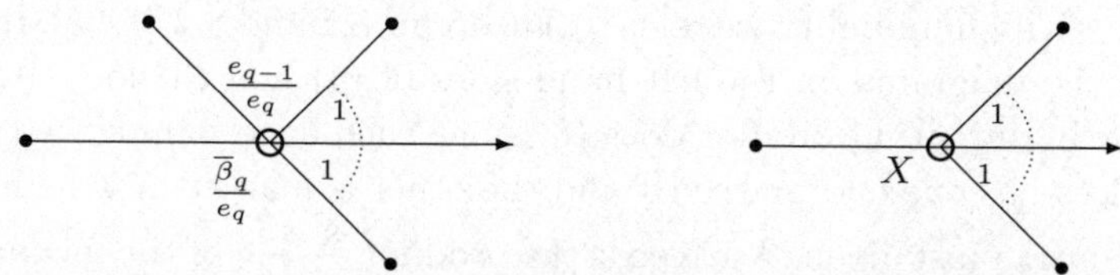

(iv) *The diagram as described is minimal unless the point $V_1 \in \Gamma_E(C)$ with $v(V_1) = 1$ is a vertex. In this case we obtain a minimal diagram by omitting the incoming edge at V_1 and, if there are just 2 outgoing edges V_1V_2 and V_1V_3, amalgamating them to a single edge V_2V_3.*

Proof We follow the description of the decomposition in Section 9.2. For a vertex V which is characteristic in the sense of (ii) of the theorem (Case A of that section), the geometry is the same as described in Theorem 9.8.1. There is a Seifert fibre space with two exceptional fibres, marked $\frac{\overline{\beta}_q}{e_q}$ and $\frac{e_{q-1}}{e_q}$, and the former corresponds to the incoming edge. We see from Figure 9.2 in Section 9.2 that the outgoing edge corresponding to a branch B_k gives an exceptional fibre in the Seifert fibre space if and only if the corresponding coefficient in the Puiseux expansion for B_k vanishes, i.e. ν_{B_k} is continuous at V. If each ν_{B_k} is discontinuous at V, the exceptional fibre requires an edge going to a black dot in the E–N diagram.

It remains to consider Case B, when ν is continuous at V. We have branches with good parametrisations $x = t^m$, $y = \sum_m^k a_r t^r$, where only the values of a_k are different for the different branches. Within the torus U_{k-1} we have unknotted components, but the linking number X of any two of them is still the same as for the original branches. We can think of this as a torus link of type $(1, X)$. There is only one exceptional fibre; it is marked X; and it corresponds to the incoming edge of the diagram; the outgoing edges are marked 1.

Since the components L of this link correspond to curves with multiplicities m/e_q, the linking number X satisfies $X = \frac{m^2}{e_q^2} h(V)$. In terms of Puiseux exponents, $X = mH(v)/e_q^2 = \frac{e_{q-1}\overline{\beta}_q}{e_q^2} + \frac{v-\beta_q}{e_q}$. Since $v = m\kappa$ is *not* a characteristic exponent for B, it is divisible by e_q; thus this formula does give an integer.

We observe that if we substitute $v = \beta_q$ in the formula for X, we obtain $\frac{e_{q-1}\overline{\beta}_q}{e_q^2}$, which is the product of the integers marking the incoming and side edges.

To verify minimality we refer to Theorems 8.1 and 8.2 of [65]. It is clear that the diagrams on the left hand sides of (4) and (5) loc. cit. do not occur in our situation; nor does (6) since each edge joining two rupture vertices has one end marked 1 and the other is marked X where, by the calculation just given, X exceeds the product $\frac{e_{q-1}\overline{\beta}_q}{e_q^2}$ of the incoming and side markings. Case (3) only occurs if $X = 1$, and this in turn occurs only if the vertex in question is at level 1. Applying the rule in Theorems 8.1 of [65] now gives the result. □

This confirms that the minimal E–N diagram is obtained from the reduced Eggers tree $\Gamma_E^r(C)$ by attaching some edges going to black vertices, and assigning integer markings to the edges. It also gives the following

Corollary 9.8.3 *The points V of attachment of the whiskers in $\Gamma_E^{r+}(C)$ are the points for which ν is discontinuous on each branch of the core through V.*

We can reverse the algorithm to deduce the Eggers tree from the E–N diagram.

Theorem 9.8.4 *Suppose given the E–N diagram for a plane curve singularity. Then the Eggers diagram can be found as follows.*

(i) *There is just one rupture point such that two edges join it to vertices of valence 1. Take the edge with the larger marking: its other end is the initial vertex.*

(ii) *As a graph, Γ_E is the union of the geodesics from the initial vertex to the arrowhead vertices (the core).*

(iii) *The value of* ν *for any edge* $V_{i-1}V_i$ *is the product of the markings at vertices interior to the path* V_0V_i *on edges not lying on the path.*

At any vertex V *write* ν_- *for the value of* ν *at the incoming edge, and* ν_+ *for the value of* ν *at an outgoing edge, where we choose* $\nu_+ \neq \nu_-$ *if possible.*

(iv) *The value of* h *at a vertex* V_q *is the incoming marking divided by the product* $\nu_-\nu_+$.

This follows by inspection from the preceding result. The simplicity of this algorithm shows that the E–N diagram is equivalent for practical purposes with the Eggers tree as we have defined it.

Example 9.8.2 Let C be the union of the branches B_1 given by $y = x^{\frac{3}{2}} + x^{\frac{13}{6}}$ and B_2 given by $y = x^{\frac{3}{2}} + x^{\frac{9}{4}}$. Then the resolution graph, with the vertex V_i marked with $\binom{a_i}{m_i}\binom{M_i(B_1)}{M_i(B_2)}$, is

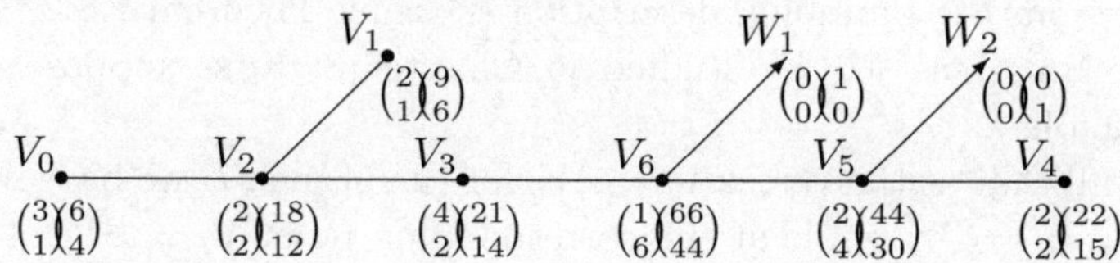

The Eggers tree is

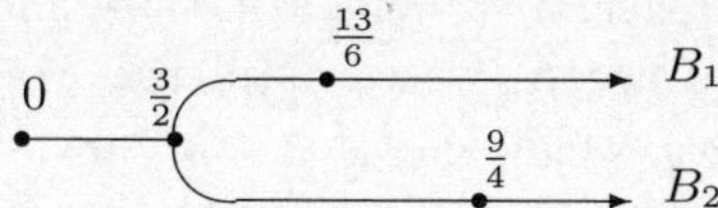

The E–N diagrams of B_1, B_2 and C are:

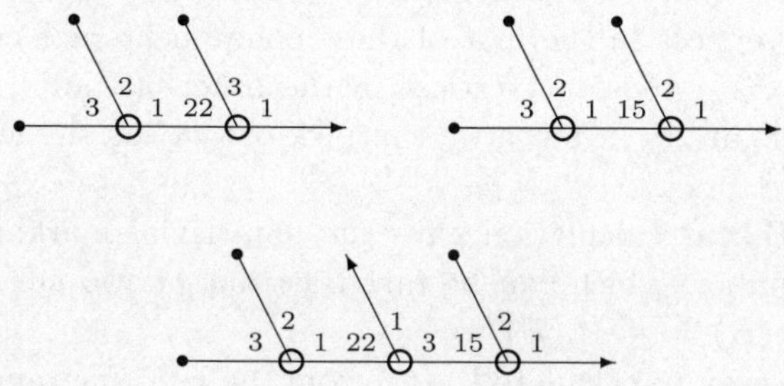

Example 9.8.3 $C = B_1 \cup B_2$, where $B_1: \ y = x^{3/2}$; $B_2: \ y = x^{3/2} + x^2 + x^{9/4}$. Here the resolution graph (marked as in the previous example), Eggers tree and E-N diagram are:

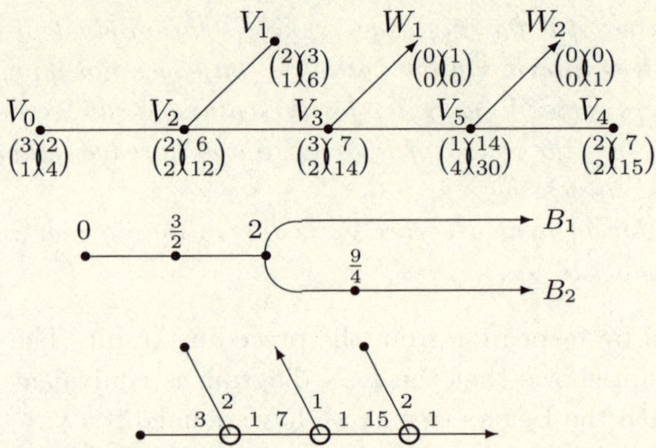

We next compare the plumbing model of Proposition 9.3.4 with the E–N diagram. The quickest way to deduce the markings on the E–N graph from the plumbing description is, using Theorem 9.8.2, in terms of the functions $M_i(B)$ studied in Chapter 8. First suppose $C = B$ irreducible.

Recall that at the vertex $V_i = AV_q$ for the branch B we have $M_i(B) = \frac{e_{q-1}}{e_q}\overline{\beta_q}$, $m_i = \frac{m}{e_q}$, while at the corresponding point $V_j = BV_q$ of valence 1, $M_j(B) = \overline{\beta_q}$ and $m_j = \frac{m}{e_{q-1}}$. Thus the markings on the E–N tree are given by $\frac{\overline{\beta_q}}{e_q} = m_iM_j(B)/m = m_jM_i(B)/m$ and $\frac{e_{q-1}}{e_q} = \frac{m_i}{m_j}$.

In general for Case A, we can still use $M_i(B)$ and m_i, but need a substitute for BV_q, or at least for m_j. Now the good resolution of C can be obtained from that for B by further blowing up. Consider the graph obtained from $\Gamma_R(C)$ by removing the core of $\Gamma_R(B)$: this has one component containing the vertex BV_q, and perhaps other components meeting the core at AV_q. It follows inductively from (8.8) that the values of m_k at all vertices in the first of these components are not less than the value at BV_q; those at vertices in the others (if any) are not less than the value at AV_q. Hence we can pick out m_j as the least of these values.

For Case B, our calculation gave the non-trivial marking as an intersection number, which can be taken as that of two curvettes, or as $\epsilon_i.B.m(\epsilon_i)/m(B) = m_iM_i(B)/m$.

It is not so easy to calculate $\Gamma_R(C)$ from the E-N diagram. To do this we need to reinsert the tori which do not appear in the minimal torus decomposition. Care is needed here, as the natural product decompositions at the two ends of a copy of $S^1 \times S^1 \times I$ will differ. A diffeomorphism of a torus is determined by a matrix of exponents, which also gives the induced map in homology. If the coordinates on the torus are $x = e^{2\pi it}$ and $y = e^{2\pi iu}$, the diffeomorphism $x' = x^a y^b$, $y' = x^c y^d$ is given by the matrix transformation $\begin{pmatrix} t' \\ u' \end{pmatrix} = \begin{pmatrix} a & b \\ c & d \end{pmatrix} \begin{pmatrix} t \\ u \end{pmatrix}$. Abusing notation, we will write $\begin{pmatrix} x' \\ y' \end{pmatrix} = \begin{pmatrix} a & b \\ c & d \end{pmatrix} \begin{pmatrix} x \\ y \end{pmatrix}$.

Suppose V_i a vertex of valence 2 in the dual graph; denote its neighbours by V_{i-1} and V_{i+1}. We need notations for all the coordinates. Decompose E_i into two hemispheres and take coordinates x_i (on E_i) and y_i (on the normal disc) on the hemisphere containing $E_i \cap E_{i+1}$ and similarly (x'_i, y'_i) on that containing $E_i \cap E_{i-1}$. These induce coordinates of the same names on the tori $S^1 \times S^1$, by Lemma 9.3.1 we have

$$\begin{pmatrix} x'_i \\ y'_i \end{pmatrix} = \begin{pmatrix} -1 & 0 \\ a_i & -1 \end{pmatrix} \begin{pmatrix} x_i \\ y_i \end{pmatrix},$$

so since the plumbing construction gives an interchange of the factors,

$$\begin{pmatrix} x_{i-1} \\ y_{i-1} \end{pmatrix} = \begin{pmatrix} 0 & 1 \\ 1 & 0 \end{pmatrix} \begin{pmatrix} x'_i \\ y'_i \end{pmatrix} = \begin{pmatrix} a_i & 1 \\ -1 & 0 \end{pmatrix} \begin{pmatrix} x_i \\ y_i \end{pmatrix} = B^t(a_i) \begin{pmatrix} x_i \\ y_i \end{pmatrix},$$

say, where we identify the tori on which the coordinates are defined by means of the product structure on the copy of $S^1 \times S^1 \times I$ joining them. The same matrix gives the change of basis in homology. Here $B^t(a)$ is the transpose of the matrix $B(a)$ of Section 8.5. Thus, transposing, $(x_{i-1}, y_{i-1}) = (x_i, y_i)B(a_i)$.

Suppose we have a chain $E_0 - E_1 - \cdots - E_s$ (this labelling for the exceptional curves need not correspond to those used earlier) in $\Gamma_R(C)$, where the E_i with $1 \leq i < s$ have valence 2, and the negatives of the self-intersection numbers are $(a_0, a_1, \ldots, a_s)$. Then in the boundary of the plumbed manifold we can delete the tori corresponding to the intersections $E_{i-1} \cap E_i$ for $1 \leq i \leq s$, and it follows from the above that

$$(x_0 . y_0) = (x_s, y_s)B(a_1)B(a_2) \cdots B(a_s).$$

We can see how this works in the case of a single branch, of genus 1. Use the notation of Section 8.5. Recall that we used the continued fraction decomposition of a/b to define three sequences of numbers and obtained the identities: $A(q_1)A(q_2) \cdots A(q_f)U_s$ ($s = 0$ or 1) is equal, according as $f + s$ is even or odd, to $U_0 B(2)^{q_1 - 1} B(q_2 + 2) B(2)^{q_3 - 1} \cdots$ or $U_1 B(q_1 + 2) B(2)^{q_2 - 1} B(q_3 + 2) \cdots$. Since the negative self-intersection at the end is $q_f + 1$ instead of $q_f + 2$ we correct using the identity $B(q + 1)V = B(q)$.

Now $P := A(q_1)A(q_2)\cdots A(q_f)$ has entries a and b in the first column, so we may write $P = \begin{pmatrix} a & c \\ b & d \end{pmatrix}$. Thus the product of the matrices B from the vertex V_0 up to (but not including) the rupture vertex, namely $B(q_1+2)B(2)^{q_2-1}B(q_3+2)\cdots$, is equal, if f is even, to $U_1^{-1}PU_1$; if f is odd, allowing for the correction factor, we obtain $U_1^{-1}PU_0V$. Similarly, the product from the other vertex of valence 1 up to the rupture vertex, namely $B(2)^{q_1-1}B(q_2+2)B(2)^{q_3-1}\cdots$, is equal to $U_0^{-1}PU_0V$ if f is even, and to $U_0^{-1}PU_1$ if f is odd.

First suppose f even. Evaluating these products gives

$$\begin{pmatrix} a+b & -a-b+c+d \\ b & -b+d \end{pmatrix} \quad \text{and} \quad \begin{pmatrix} a & -c \\ a-b & -c+d \end{pmatrix}.$$

Multiplying on the left by $(1,0)$ gives the coefficients of that homology class in the torus on the boundary of the central piece of the decomposition which becomes trivial when we attach the chain of pieces going out to V_0 or to the other end. Thus this defines the Seifert invariants. In the notation of Corollary 9.6.3 we have $a_0 = a+b$, $b_0 = -a-b+c+d$, $a_1 = a$ and $b_1 = -c$. We also have $a_2 = 1$; setting $b_2 = 1$ we verify that $\prod a_i \sum \frac{b_i}{a_i} = 1$ since $ad - bc = 1$.

Similarly if f is odd, we obtain

$$\begin{pmatrix} a+b & -c-d \\ b & -d \end{pmatrix} \quad \text{and} \quad \begin{pmatrix} a & -a+c \\ a-b & -a+b+c-d \end{pmatrix}.$$

Multiplying on the left by $(1,0)$ gives the Seifert invariants $a_0 = a+b$, $b_0 = -c-d$, $a_1 = a$ and $b_1 = -a+c$. We also have $a_2 = 1$; setting $b_2 = 1$ we verify that $\prod a_i \sum \frac{b_i}{a_i} = 1$ since $ad - bc = -1$.

In general we need to use the continued fraction algorithm on each edge of the E-N graph. We will not develop this here: a good description of the procedure is given in [65] Section 22.

9.9 The polar discriminant

In this section we present a very brief outline of an alternative approach to the geometry of the singularity which can also be used to obtain the decomposition.

Suppose C given by an equation $f(x,y) = 0$. Choose a linear function – which we shall take as x – such that $x = 0$ is not tangent to any branch of C at $(0,0)$, so that the curve C_y given by $\partial f/\partial y = 0$ is a transverse polar of C.

Consider the map $F = (x, f) : \mathbb{C}^2 \to \mathbb{C}^2$. We write (X, Y) for the coordinates in the target $\mathbb{C}^2$. The map F has critical points where

$$0 = \frac{\partial(x, f)}{\partial(x, y)} = \begin{vmatrix} 1 & f_x \\ 0 & f_y \end{vmatrix} = f_y,$$

giving the polar curve C_y. Its image $\Delta(F) = F(\Sigma_F)$ is the *polar discriminant.* Different branches P_i of C_y project to (in general) different branches Δ_i of $\Delta(F)$.

Lemma 9.9.1

(i) $x = 0$ *is not tangent to* P_i *and* $X = 0$ *is not tangent to* Δ_i.

(ii) *The multiplicity* $m(P_i) = m(\Delta_i)$.

(iii) *The intersection numbers* $P_i.C = \Delta_i.\{Y = 0\}$.

Proof Since the line $x = 0$ is not tangent to C, the sum of the terms of degree m in f is not divisible by x. Thus the sum of terms of degree $m - 1$ in $\partial f/\partial y$ is also not divisible by x, so no P_i has $x = 0$ as tangent line. Now suppose P_i parametrised by $(\phi(t), \psi(t))$, so that $\operatorname{ord}\phi \leq \operatorname{ord}\psi$. Then Δ_i is parametrised by $(\phi(t), f(\phi(t), \psi(t)))$, and since $\operatorname{ord} f(\phi(t), \psi(t)) \geq \min(\operatorname{ord}\phi, \operatorname{ord}\psi) \geq \operatorname{ord}\phi$, it follows that Δ_i is not tangent to $X = 0$.

Since $x = 0$ is transverse to P_i, and $X = 0$ to Δ_i,

$$\begin{aligned} m(\Delta_i) &= \Delta_i.\{X = 0\} = F(P_i).\{X = 0\} = P_i.F^{-1}\{X = 0\} \\ &= P_i.\{x = 0\} = m(P_i), \end{aligned}$$

by Lemma 1.2.1.

Similarly, we have

$$\Delta_i.\{Y = 0\} = F(P_i).\{Y = 0\} = P_i.F^{-1}\{Y = 0\} = P_i.\{f = 0\} = C.P_i.$$

□

The lemma shows that the multiplicity m_i of P_i is also the multiplicity of Δ_i, and the intersection number $M_i = P_i.C$ is equal to the intersection number $\Delta_i.\{Y = 0\}$. Thus the Puiseux expansion for Δ_i has the form

$$X = t^{m_i}, \quad Y = at^{M_i} + \text{ higher terms.}$$

We have $m_i < M_i$, so the polar quotient $\rho_i := M_i/m_i > 1$ and the tangent to Δ_i is $Y = 0$. We obtain a carousel – the *Lê carousel* – similar to that in Section 5.3.

It follows from Proposition 9.4.3 that the nodes of the Eggers graph correspond naturally to those of the E–N diagram, and hence to the pieces into which the link complement is decomposed. One of the objectives of the approach using the polar discriminant is to gain a direct geometrical understanding of this correspondence. Indeed, a quite different approach to the JSJ decomposition of the link complement, developed in [112] using earlier work of Lê, depends on the use of the polar discriminant.

The curve Δ_i is contained in a region $R_i : A_i|X|^{\rho_i} \leq |Y| \leq B_i|X|^{\rho_i}$. As in the case of the other carousel, we think of Y as fixed and draw figures in the X-plane. We see that if $|X|$ is taken small enough, the

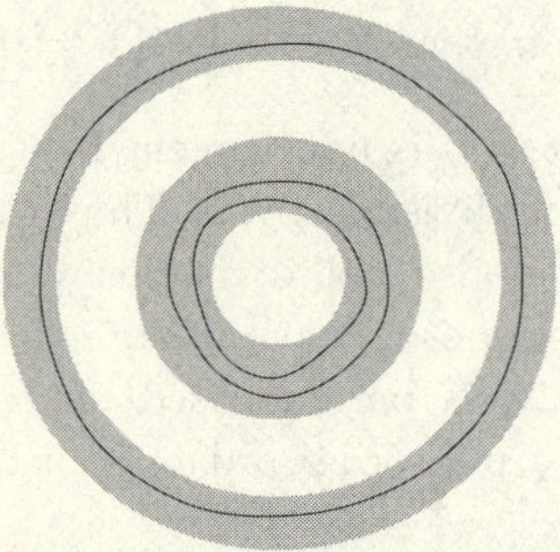

Fig. 9.7. The Lê carousel, with regions R_i shaded

regions R_i for different values of ρ_i become well separated from each other, and their union contains the polar discriminant. The idea is to filter $B_\epsilon \cap f^{-1}(\partial D_\eta)$, and hence the Milnor fibre F, by the preimages of discs $|Y| \leq |X|^{\tau_j}$, where the τ_j are chosen to interpolate between the (distinct) quotients ρ_i (we can take $\tau_{\min} = 0$, $\tau_{\max} = \infty$). Observe that there is one piece for each polar quotient.

We now state the main theorem of [112]: first we need some notation. There are some numbers ϵ, η, θ chosen small enough in a sense similar to that of Theorem 6.2.1. Set

$$M_B := \{(x,y) : |x|^2 + |y|^2 \leq \epsilon,\ |f(x,y)| = \eta,\ |x| \leq \theta\}$$

and $Z_j = \{|X|^{\tau_{j-1}} \leq |Y| \leq |X|^{\tau_j}\}$ Then every component of $M_B \cap F^{-1}Z_j$ is a Seifert fibre space.

The key step in the proof is as follows. For each component Δ_i pick a Puiseux expansion $X = aY^{1/\rho_i} + \cdots$, a number s_i between $1/\rho_i$ and the next higher exponent appearing, and a value Y_0 with $|Y_0| = \eta$, and write

$$W_i := \{(X,Y) : |X - aY_0^{1/\rho_i}| \le \eta^s\}.$$

Then the result states that $F^{-1}(W_i) \cap M_B \cap f^{-1}(Y_0)$ is a disjoint union of discs.

To deduce the main theorem, define Z_j' by deleting from Z_j the interiors of those W_i lying in it. There is an obvious fibration of this by circles, and this can be lifted to $M_B \cap F^{-1}Z_j'$ which is an unbranched cover of it, since all the branching takes place over the $\Delta_i \subset W_i$. On the other hand, since $F^{-1}(W_i) \cap M_B \cap f^{-1}(Y_0)$ is a disjoint union of discs, $F^{-1}(W_i) \cap M_B$ is a disjoint union of solid tori, and the circle fibration on the boundary of each of these solid tori can be extended to the interior as a Seifert fibration.

Since the components of $M_B \cap F^{-1}Z_j$ are Seifert fibre spaces, we have indeed a decomposition of the type of Theorem 9.1.3. We have not shown that the decomposition is minimal, and indeed in general it is not: some tori may need to be deleted to give a minimal decomposition. This approach does not appear to lead so easily to direct calculations, and we will not pursue it.

9.10 Notes

Section 9.1 The main references are given in the text. The proofs of [92] and [93] are long. An alternative account has been given by Neumann and Swarup [141].

The Thurston programme is incomplete: as well as the Poincaré conjecture, it is not established in general that an atoroidal irreducible manifold admits a geometric structure. I had intended to give a reference discussing what is known on this topic, but in view of the work of Perelman, at present in the form of preprints on the archive, while there is hope for a complete account, this is not yet available.

A manifold which can be built up from Seifert fibre spaces by attaching along tori is called a *graph manifold.* These were first introduced and studied by Waldhausen in papers [186], [187] giving the first results on these topics.

Section 9.2 Although this approach is well known, I do not know a reference which treats the construction in detail.

Section 9.3 A combinatorial model for the Milnor fibration was first given by A'Campo in [3], which is the main reference for this approach. Theorem 9.3.6 is essentially due to A'Campo [5]. Theorem 9.3.7 is due to Lê [109].

All these papers treat the general case of an isolated singularity of a holomorphic function on $\mathbb{C}^n$, and do not consider the special features of the case $n = 2$. It was first observed by Lê, Michel and Weber [112] that in the case $n = 2$ this model gives the JSJ decomposition of the link complement: a remark which underlies our whole approach.

Section 9.4 The relation between the Eggers tree and the resolution tree was hinted at in the calculations of polar quotients by Lê, Michel and Weber in [111] and their result determining which components of the resolution graph meet components of a transverse polar. This was further clarified by García Barroso in [75]; in addition, her characterisation of topological type by matrices of partial polar quotients is essentially equivalent to the rewriting in terms of the numerical properties of an Eggers tree. A more explicit result was given in the thesis of Popescu-Pampu [151], but our homeomorphism is new.

The essential parts of these results were generalised by Popescu-Pampu [151] to the case of 'quasi-ordinary' singularities. This was further developed by García Barroso and González Pérez [77].

Polar quotients were introduced by Teissier [175] for isolated hypersurface singularities in all dimensions. He conjectured, and proved in [174], that they are equisingularity invariants. Their use was emphasised in several papers of Lê; for plane curves they are studied in detail in [111], where it was shown that if C has a unique tangent at O, the quotients $m_i/M_i(C)$ at rupture points equal the polar quotients; if C has two tangents, polar quotients include these and also $1/m$. Lê's terminology differs from ours in that his quotients are obtained by subtracting 1 from those defined in the text.

The terminology of [111] is 'contact coefficients in the sense of Hironaka' for the quotients $C.B/m(B)$ which we calculated in Section 4.2, 'polar quotients' for the special cases when B is a component of a transverse polar of C, and 'coefficients of insertion' for the special case when B is a curvette.

Section 9.5 The question of finite order was first considered by A'Campo. Theorem 9.5.2 seems to be new. According to A'Campo [3], on the assumption that the curve has only one branch, a necessary condition for h to have finite order is that it has only one Puiseux pair: a condition equivalent to being equisingular to a weighted homogeneous singularity.

Section 9.6 Seifert fibre spaces were introduced in Seifert [162]. For their classification we follow Neumann and Raymond [140]; Lemma 9.6.1 is also due to them.

For further discussion and references on Seifert manifolds see [126], chapter 9.

Section 9.7 Many further details (including a treatment of arbitrary graph links) are given in [65]. In particular, they include a generalisation to 'multilinks', where each component is assigned a (positive integer) multiplicity. This may be regarded as the topological counterpart of allowing f to have repeated factors, or the curve C to be non-reduced. Many of the results in this book could also be treated in this generality.

The fact that resolution can be achieved in fewer steps if toric blow-ups are used seems be related to the fact that the resolution tree is simplified by amalgamating edges to give the E-N diagram: presumably we can achieve the same diagram directly by toric blowings up. This is well known for the case of irreducible curves: see e.g. Oka [146]. The case of curves with several components was explored in [110], but the results are not decisive.

Section 9.8 The result in Case A of Theorem 9.8.2 is essentially as described in [65], pp. 51–54. However, Eisenbud and Neumann do not discuss Case B explicitly, though X can be determined using the equation of [65]10.1.

Section 9.9 Some references are given in the text. The polar discriminant is often called Cerf diagram after [37], where it was first introduced.

9.11 Exercises

Exercise 9.11.1 Determine the Eggers tree and the resolution tree for the following singularities and in each case describe the map $\phi : \Gamma_R^+(C) \to \Gamma_E(C)$.

(i) $y^8 = x^{11}$,
(ii) $y = x^{5/2} + x^{17/6} + x^{13/4}$,
(iii) $(y^2 - x^5)(x^2 - y^5) = 0$.

Exercise 9.11.2 Calculate the E–N diagrams for the following curve singularities:

(i) $(x, y) = (t^4, t^{10} + t^{11})$,
(ii) $(x, y) = (t^6, t^{15} + t^{20})$,
(iii) $(y^2 - x^3)(y^3 - x^4) = 0$,
(iv) $(y^3 - x^5)(y^4 - x^7) = 0$,
(v) $(y - x^{6/5})(y - x^{14/5}) = 0$,
(vi) $(y - x^{10/7})(y - x^{10/3}) = 0$.

Exercise 9.11.3 Calculate the E–N diagrams for the curve $C = B \cup B'$, where

(i) B is given by $(x, y) = (t^4, t^{10} + t^{11})$, and B' by $(x, y) = (t^4, t^{10} + t^{11} + t^{17})$,

(ii) B is $(x, y) = (t^8, t^{12} + t^{18} + t^{23})$, and B' is $(x, y) = (t^8, t^{12} + t^{18} + t^{20} + t^{21})$.

Exercise 9.11.4 Show that for any curve C and vertex V_i of the Eggers tree, the multiplicity m_i of a curvette with contact point V_i is equal to $\nu_+(V_i)$ (where ν_+ is defined as in Theorem 9.8.4 (iii)).

Exercise 9.11.5 Show that the $M_i(B_j)$ at rupture vertices of Γ_R^+ can be calculated from the E–N diagram as follows. Take a geodesic γ from V_i to W_j in Γ_R^+, and the corresponding path in the E–N diagram, and form the product of all markings at vertices of γ along edges not lying in γ. (Hint: use the preceding exercise, and first consider the case when V_i lies on the geodesic from V_0 to W_j.)

Exercise 9.11.6 The intersection of the lattice

$$\{(r, s) \in \mathbb{Z}^2 \,|\, r \equiv 3s \pmod{11}\}$$

with the positive quadrant $r \geq 0$, $s \geq 0$, excluding the origin, forms an additive semigroup. The Newton polygon formed by the elements of the semigroup has successive vertices $P_0 := (11, 0)$, $P_1 := (3, 1)$, $P_2 := (1, 4)$ and $P_3 := (0, 11)$. The area of the triangle OP_kP_{k+1} is $\frac{1}{2}11$. The P_i generate the semigroup, and satisfy $4P_1 = P_0 + P_2$, $3P_2 = P_1 + P_3$. The negative continued fraction $4 - \frac{1}{3} = \frac{11}{3}$ gives the parameters defining the original lattice.

Formulate and prove a corresponding statement for the lattice $r \equiv bs \pmod{a}$, where a and b are coprime positive integers.

Exercise 9.11.7 Give an example of a curve such that two distinct rupture points of $\Gamma_R(C)$ yield the same polar quotient.

0

Exercise 9.11.8 Show that if f is NPND, every polar quotient is equal either to $m(f)$ or to the length cut out on a coordinate axis by (the extension of) some side of the Newton polygon.

10

The monodromy and the Seifert form

Central to the study of the topology attached to the Milnor fibration is understanding the monodromy. As well as the monodromy map on homology we consider the Seifert form. These two, together with the intersection form, form a single algebraic structure which gives a rather fine invariant of the topology, and enables a number of numerical invariants to be picked out. We will use the decomposition of the Milnor fibre obtained in the preceding chapter: this permits simplified proofs of a number of basic results.

A Seifert form can be defined for any knot or link provided with a spanning Seifert surface, but in the case of fibred knots such a surface is canonically provided, so that the Seifert form is intrinsic in this case. Both Seifert forms and monodromy can be defined and studied in a higher dimensional situation, but there the canonical decomposition is lacking, so proofs are more sophisticated.

The chapter opens by defining the Seifert form and eliciting its algebraic properties. Next we derive the special features of Seifert forms for the case of fibred knots using the JSJ decomposition established in Chapter 9. Using the model constructed in Section 9.3 we obtain an analysis of algebraic properties of the monodromy.

We then investigate Seifert forms in the abstract in sufficient detail to obtain in principle enough invariants to classify them with rational coefficients. This section involves algebraic technicalities, and the reader may choose to omit it since the Seifert forms arising for curve singularities have special properties.

We would like to be able to calculate these invariants in terms of the Eggers diagram, or equivalently the E-N diagram, or the resolution graph. Such formulae are not currently known for all the invariants of the Seifert form of a plane curve singularity. The final sections are devoted to a study of signature invariants of Seifert forms over $\mathbb{R}$, and their explicit calculation.

10.1 Definition of Seifert forms

In this section we define the Seifert form, and obtain the basic properties of the Seifert form attached to a curve singularity.

We take as our starting point the Milnor fibration of Chapter 6. We recall the notation. We have a reduced function-germ $f : (\mathbb{C}^2, O) \to (\mathbb{C}, 0)$, and for ϵ small enough (see proofs of Theorems 6.2.1, 6.2.2) we write B_ϵ for the disc $\{(x, y) \mid |x|^2 + |y|^2 \leq \epsilon^2\}$, with boundary sphere S_ϵ, and K for the link $f^{-1}(0) \cap S_\epsilon$. For suitably small η we write D_η for the disc $\{z \mid |z| \leq \eta\}$, and have a tubular neighbourhood $N(K) = f^{-1}(D_\eta \cap S_\epsilon)$ of K, with closed complement M. Then $f/|f|$ induces a fibration $\pi : S_\epsilon - K \to S^1$ (its restriction to M, also a fibration, was denoted f_2 in Theorem 6.2.2), and we write $\tilde{F}_\theta$ for the closure of the preimage of a point θ of S^1: this is obtained from its intersection F_θ with M by attaching a collar on the boundary, so the two are diffeomorphic, and we will not always distinguish between them. Usually we pick a particular point θ and just write F for the Milnor fibre F_θ.

The fibration provides a homeomorphism $h : F \to F$ whose restriction to the boundary ∂F is the identity: this is the monodromy, and is determined up to isotopy. We can identify M with the quotient of $F \times I$ obtained by identifying $(x, 1)$ with $(h(x), 0)$ for each $x \in F$. If we do the same for $\tilde{F} \times I$, and also identify (y, t) with $(y, 0)$ for all $y \in \partial\tilde{F}$ and all $t \in I$, we obtain a copy of S^3.

We shall be primarily interested in the map h_* induced by h on the homology group $H := H_1(F; \mathbb{Z})$; for the remainder of this section we denote this map also by h. The group H is free abelian, and we denote its rank by μ (the Milnor number).

Intersection numbers on the surface F give a skew-symmetric bilinear form $H \times H \to \mathbb{Z}$, which we denote by $(x, y) \mapsto \langle x, y\rangle$. The radical of this form (i.e. $\{x \in H \mid (\forall y \in H)\langle x, y\rangle = 0\}$) consists of the image of $H_1(\partial F) \to H_1(F)$. Since h is a homeomorphism, h_* preserves intersection numbers. Since H is the first homology group of a compact oriented surface, the isomorphism class of the pair $(H, \langle,\rangle)$ is determined by the homeomorphism class of the surface, and hence by the number r of boundary components together with the genus, or equivalently the Euler characteristic $1 - \mu$.

Choose a number t with $0 < t < 1$ (we need not take t small, though to define the Seifert form for a knot in general one does take nearby

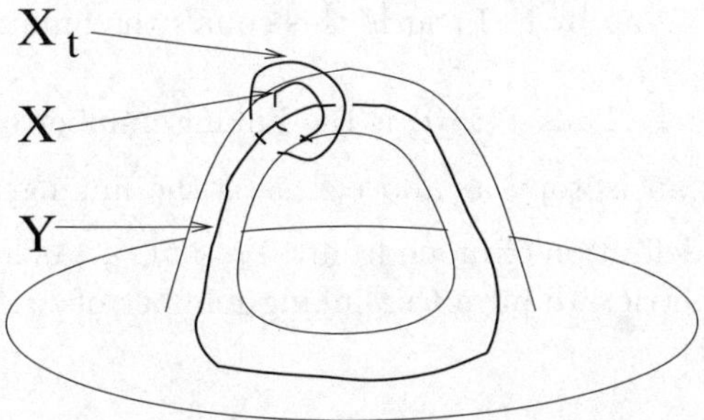

Fig. 10.1. Deformation for Seifert pairing

copies) and write F_t for the image of $F \times \{t\}$ in S^3 under the above identification: this is a 'parallel' copy of F, so any cycle $X \subset F$ can be moved to a parallel cycle $X_t \subset F_t$. Choose cycles X, Y representing $x, y \in H_1(F)$, and write $S(x, y)$ for the linking number of X and Y_t in S^3. This does not depend on the choices of the cycles X and Y. For if X' is another choice, and the 2-chain W satisfies $\partial W = X' - X$, then since W does not intersect Y_t it follows that the linking number of X' with Y_t is the same as that of X. A similar argument holds with the roles of X and Y reversed. We also see that the result does not depend on the choice of t. Likewise, it is clear that $S(x, y)$ is bilinear. This S is called the Seifert form.

Lemma 10.1.1

(i) *For all* $x, y \in H$, $\langle x, y\rangle = S(x, y) - S(y, x)$.

(ii) *For all* $x, y \in H$, $S(x, y) = S(h(y), x)$.

Proof Take $t < \frac{1}{2}$ and construct another parallel F_{-t} of F on the other side. Using a product structure $F \times [-t, t]$, we can move cycles in F in either direction. Now for 1-cycles X, Y in F we have the 2-chain $Z = Y \times [-t, t]$ in S^3, so the linking number of X and ∂Z is the intersection number of X and Z in S^3. As Z meets F transversely, this is equal to the intersection number in F of X and $Z \cap F = Y$, i.e. to $\langle x, y\rangle$.

On the other hand, $\partial Z = Y_t - Y_{-t}$. The linking number of X and Y_t is, by definition, $S(x, y)$. The linking number of X and Y_{-t} is equal to that of X_t and Y, since we may perform an isotopy of the whole of S^3 which 'moves everything up by t'. In turn, this equals the linking number of Y and X_t, i.e. $S(y, x)$.

Here take $t = \frac{1}{2}$. Thus $S(x, y)$ is the linking number of X and $Y_{\frac{1}{2}}$ in S^3. Performing an isotopy as above, this is the linking number of $X_{\frac{1}{2}}$ and Y_1. By the definition of monodromy, $Y_1 = h(Y)$. As linking numbers in S^3 are symmetric, we have the linking number of $h(Y)$ and $X_{\frac{1}{2}}$, i.e. $S(h(y), x)$. □

It follows from the geometric interpretation that h preserves the intersection form on H. It follows from the lemma that h also preserves S. For

$$S(h(x), h(y)) = S(h(y), x) = S(x, y).$$

The relations may be expressed in matrix terms. Choose a basis $\{e_i\}$ for H, and set $a_{i,j} = \langle e_i, e_j \rangle$, $s_{i,j} = S(e_i, e_j)$ and $h_*(e_i) = \sum_j m_{i,j} e_j$, defining matrices A, S, X. Then (i), (ii) and invariance of intersections by h_* translate to the respective matrix equations

$$A = S - S^t, \quad S^t = XS, \quad A = XAX^t. \tag{10.1}$$

Roughly speaking, these equations show that S determines the pair A and X and conversely. More precisely,

Lemma 10.1.2 *Consider triples related by* (10.1). *Then*

(i) *The matrix S determines A and, if S is non-degenerate, X. Any matrix may be chosen for S such that S^t is a left multiple of S.*
(ii) *If 1 is not an eigenvalue of X, the matrices A and X determine S. A skew-symmetric matrix A and invertible X with $(I - X)$ non-degenerate and $XAX^t = A$ may occur if A is a right multiple of $I - X$.*

Proof The first two assertions follow from the identities $A = S - S^t$ and $S^t = XS$. Given S, if we define A by the first of these, and X satisfies the second, then $XAX^t = (XS)X^t - X(S^t X^t) = S^t X^t - XS = S - S^t = A$, so all the relations hold.

The first assertion follows from the relation $A = (I - X)S$. Now suppose A and X are given with A skew-symmetric, $A = XAX^t$, and $A = (I - X)S$ for some S. Then $A^t = S^t(I - X^t)$ so $(I - X)(S^t - XS)(I - X^t) = (I - X)A^t - XA(I - X^t) = -A + XA - XA + XAX^t = 0$ and since $(I - X)$ (and hence also $(I - X^t)$) is non-degenerate, $S^t - XS = 0$. It follows that $A = S - S^t$. □

We will see in Proposition 10.2.2 that the eigenvalue 1 plays a special role also in the geometry.

Although the coefficient group $\mathbb{Z}$ is understood above, the argument is valid with any group of coefficients. In particular, if the coefficient group is a field, the condition 'A is a right multiple of $I - X$' in (ii) is superfluous.

The precise classification of possible Seifert forms up to isomorphism involves delicate arithmetic considerations, in which it is easy to lose sight of the overall structure. In section 10.4 we will derive the classification of Seifert forms over a field.

10.2 Use of the Thurston decomposition

We recall from Theorem 9.1.3 that a diffeomorphism h of a closed surface F induces a decomposition of F by an h-invariant family C of disjoint simple closed curves, unique up to isotopy. The characteristic property of C is that the restriction of h to each complementary component is (up to isotopy) either of finite order or pseudo-Anosov. In turn, by Proposition 9.1.5, this decomposition induces the decomposition given by Theorem 9.1.2 of the total space M of the bundle over S^1 with fibre F and monodromy h. Moreover, the pseudo-Anosov pieces of F correspond to atoroidal pieces of M and the pieces with h of finite order to the Seifert fibre space pieces of M.

In Section 9.2 we exhibited a torus decomposition of the closed complement M of an algebraic link, and showed in particular that all the pieces into which it is decomposed are Seifert fibre spaces. It follows that in the decomposition of the Milnor fibre F, the restriction of h to each piece is homotopic to a map of finite order. We now exploit this fact to obtain algebraic information about the monodromy: we will establish the so-called 'monodromy theorem' in our situation, and also define a canonical filtration of the homology group $H_1(F)$.

Write $C^+ = C \cup \partial F$. Write N_C for a product neighbourhood of C^+ in F and F_C for its closed complement: up to diffeomorphism, this may be identified with the disjoint union of the pieces into which F is cut by C. Define subgroups of $H = H_1(F)$ as follows.

$$W_0 := \text{Im } (H_1(C^+) \to H_1(F)),$$
$$W_1 := \text{Im } (H_1(F_C) \to H_1(F)),$$
$$W_2 := H_1(F),$$

where the maps are those induced by the inclusions. Then $W_0 \subseteq W_1 \subseteq W_2$ is a filtration of H by subgroups invariant under h_* (submodules). It is known as the *weight filtration*. Notice that if the family C is augmented to a non-minimal family, the extra components are parallel to some existing components, and the additional components of F_C are annuli, so the filtration given by this definition is unaffected.

Theorem 10.2.1 *(Monodromy Theorem) All eigenvalues of h_* on H are roots of unity. All blocks in the Jordan decomposition of h_* on $H_1(F;\mathbb{C})$ have size at most 2.*

Proof We first show that there is an exact sequence

$$0 \to H_1(C) \to H_1(F_C) \to H_1(F) \to H_0(C) \to H_0(F_C) \to H_0(F) \to 0, \tag{10.2}$$

where all the maps are compatible with the action of h. In fact we will identify this with the homology exact sequence of the pair (F, F_C): we thus need to identify $H_i(F, F_C)$ with $H_{i-1}(C)$. By excision, $H_i(F, F_C) \cong H_i(N_C, \partial N_C)$, and since N_C is a product, $H_i(N_C, \partial N_C) \cong H_i(C \times I, C \times \partial I) \cong H_{i-1}(C)$.

Since the restriction of h to F_C is homotopic to a map of finite order, the action of h_* on $H_1(F_C)$ is semi-simple, with all eigenvalues of finite order. The same follows for the action on W_1. By the exact sequence, the quotient $H_1(F)/W_1$ is isomorphic to a submodule of $H_0(C)$. But as h is isotopic to a map of finite order on C, the same argument applies here.

It follows that all eigenvalues are roots of unity and all Jordan blocks have size at most 2, as required. □

We will see later that a rather complete description of the monodromy can be obtained using these ideas. Before plunging into calculations, however, we slightly sharpen the above conclusions.

Proposition 10.2.2

(i) *The radical R of the intersection form on H is the image of $H_1(\partial F)$.*

(ii) *The fixed set of h_* on H is equal to R.*

(iii) *The submodules W_0 and W_1 are mutual orthogonal complements with respect to the intersection form.*

(iv) *The Jordan blocks of h belonging to the eigenvalue 1 have size 1.*

Proof Lefschetz duality for surfaces tells us that the adjoint to the intersection form is the map $H_1(F) \to H_1(F, \partial F)$ induced by inclusion.

So the radical of the form is the kernel of this map, which is the image of $H_1(\partial F)$.

By construction, h acts as the identity on ∂F. Conversely, if $y \in H$ satisfies $h(y) = y$, then for all $x \in H$ we have

$$\langle x, y\rangle = S(x, y) - S(y, x) = S(h(y), x) - S(y, x) = S(h(y) - y, x) = 0,$$

so y is in the radical and the conclusion follows from (i).

The orthogonal complement of W_0 consists of homology classes having zero intersection number with each component of C (also – but this is automatic – with components of ∂F). Let X be a 1-cycle, transverse to C, with this property. Then if P, Q are two intersection points of X with some component of C which cancel numerically, we may join PQ by an arc α on C and choose a tubular neighbourhood N_α of α whose ends are arcs of X. Adding or subtracting ∂N_α to X gives a homologous 1-cycle having 2 fewer intersection points with C. We may proceed until we find a homologous cycle disjoint from C, and an isotopy will then deform it into F_C. Hence the orthogonal complement of W_0 is contained in W_1.

Now W_0 and W_1 are orthogonal, since C^+ and F_C are disjoint. Hence W_1 is equal to the orthogonal complement of W_0. The converse follows since W_0 contains the radical (it was for this reason that we defined W_0 using C^+ rather than C).

Here it is convenient to work over a field, say in $H_{\mathbb{Q}} = H_1(F; \mathbb{Q})$. Since the restriction of h_* to W_1 is semi-simple, the eigenspace belonging to $+1$ is equal to the fixed subspace of h_*, and by (ii) this is contained in W_0. The intersection form induces a dual self-pairing of W_1/W_0 and a dual pairing between W_2/W_1 and W_0/R. Since 1 does not appear as eigenvalue on W_0/R (by (i) and (ii)), it does not appear on W_2/W_1 either. Hence the $+1$-eigenspace is contained in W_1 and the assertion follows. □

Thus R is the $+1$ eigenspace of h on H, so on R, h acts as the identity, S is symmetric, and the intersection form vanishes.

Since R is the image of $H_1(\partial F)$, it is spanned by the classes $[L_i]$ of the components L_i of the link ∂F, which correspond to the branches B_i of Γ; and since F is an orientable surface, the only relation between these is that $\sum_i [L_i] = 0$. For $i \neq j$, $S([L_i], [L_j])$ is the linking number of L_i and L_j, i.e. the intersection number of B_i and B_j. The values $S([L_i], [L_i])$ can now be inferred from the relation $\sum_i [L_i] = 0$.

Lemma 10.2.3 *The form S is negative definite on R.*

Proof Since $\sum_i [L_i] = 0$ we have, for any (integer or real) coefficients a_i,

$$\begin{aligned} S\left(\sum_i a_i[L_i], \sum_j a_j[L_j]\right) &= \sum_i a_i S\left([L_i], \sum_{j\neq i}(a_j - a_i)[L_j]\right) \\ &= \sum_{i\neq j} a_i(a_j - a_i) S([L_i],[L_j]), \end{aligned}$$

which is equal to $-\frac{1}{2}\sum_{i\neq j}(a_i - a_j)^2 B_i.B_j$. Since the intersection numbers $B_i.B_j$ are positive, this expression is non-positive, and vanishes only if all $a_i - a_j$ are zero, so we have a multiple of the zero class $\sum_i [L_i]$. □

Write temporarily $P := \{x \in H \mid S(x,y) = 0 \text{ for all } y \in R\}$.

Proposition 10.2.4

(i) *If $x \in R$ and $y \in P$, then $S(x,y) = S(y,x) = \langle x,y\rangle = \langle y,x\rangle = 0$.*
(ii) *Over $\mathbb{Q}$ we have $H = R \oplus P$ and $P = \mathrm{Im}\,(h-1) : H \to H$.*
(iii) *The form $\langle , \rangle$ is nonsingular on P.*

Proof We have $S(x,y) = 0$ by definition of P, and $S(y,x) = S(h(x),y) = S(x,y) = 0$. The other identities follow using Lemma 10.1.1.

Since the restriction of S to R is nonsingular over $\mathbb{Q}$, by Lemma 10.2.3, we have $R \cap P = \{0\}$. Using the same result again, we see that for any $x \in H$, the linear map $s_x : R \to K$ given by $s_x(r) = S(x,r)$ determines a unique $y \in R$ such that $s_x = s_y$ and hence $x - y \in P$. The final assertion follows using (iv) of Proposition 10.2.2.

follows from (ii) since R is the radical of $\langle , \rangle$. □

Thus $H = R \oplus P$, R is the $+1$ eigenspace of h on H, and $h-1$ is invertible on P. We thus introduce the notation $H^{\neq 1}(F)$, or $H^{\neq 1}$ for short, for P.

10.3 Calculation of the monodromy

In this section we study H together with the monodromy operator $h : H \to H$. Although many of the arguments will apply to this integer monodromy, for the more precise results we will work over a field K, which we will usually think of as the rational field $\mathbb{Q}$. We obtain a complete description of the isomorphism class of (H,h) over $\mathbb{Q}$ in terms of the invariants studied in Chapter 8.

The classification of pairs (H,h) over a field is equivalent to the classical theory of the 'rational canonical form', so we briefly recall how this works (cf. e.g. MacLane and Birkhoff [122]). Let t be an indeterminate over K. Then the polynomial ring $K[t]$ is a principal ideal ring, and we may regard H as a module over it, by letting t act as h_*. The theory of modules over principal ideal rings tells us that every module is a direct sum of cyclic modules $K[t]/\langle\phi(t)\rangle$ for polynomials ϕ. Since multiplying ϕ by a non-zero constant does not change this module, we may assume ϕ to be monic – i.e. to have 1 as coefficient of the highest power of t occurring. If ϕ and ψ are coprime, we have

$$K[t]/\langle\phi(t)\psi(t)\rangle \cong K[t]/\langle\phi(t)\rangle \oplus K[t]/\langle\phi(t)\rangle,$$

so we may assume that the polynomials ϕ which occur are powers of polynomials which are prime in $K[t]$. The list of these polynomials – which are called *invariant factors* – is then uniquely determined by the module. The direct summands themselves are not unique, but if we collect together all those for which ϕ is a power of a particular prime $p(t)$, we obtain a well defined subspace $H_{(p)}$ of H.

Another approach is to begin with the characteristic polynomial $\Delta(t)$, defined as the determinant of the operator $tI - h$ on the free module $V \otimes_K K[t]$. This is the product of the polynomials ϕ corresponding to the terms of the decomposition. If we factorise it as $\Delta(t) = \prod p(t)^{r_p}$, then to find the invariant factors we require, for each p, a partition of r_p (into positive integers).

Since the characteristic polynomial is the easiest invariant to calculate, we make a few comments about it here. If we have a graded finite dimensional vector space $V = \oplus V_i$, and a grade-preserving endomorphism h of it, we form the alternating product $\Delta_V(t) := \prod_i (\det{(tI - h)}|V_i)^{(-1)^i}$, which we may call the graded characteristic polynomial. This has the usual property of Euler characteristics, that if we have three graded vector spaces V', V and V'', and a long exact sequence $\dots V'_i \to V_i \to V''_i \to V'_{i-1} \to V_{i-1} \to V''_{i-1} \dots$ on which h acts, then we have $\Delta_V(t) = \Delta_{V'}(t)\Delta_{V''}(t)$. For example, if h is a self-map of the pair (X, Y) of spaces, we have $\Delta_{H_*(X)}(t) = \Delta_{H_*(Y)}(t)\Delta_{H_*(X,Y)}(t)$.

The simplest case is when $K = \mathbb{C}$: then prime polynomials all have degree 1, so we have $p(t) = t - \lambda$ for some $\lambda \in \mathbb{C}$. The subspace $H_{(p)}$ is then the (generalised) eigenspace H_λ, and to give the structure of the $\mathbb{C}[t]$-module H we only need to determine, for each λ, how often each polynomial $(t - \lambda)^r$ occurs in the list of invariant factors.

The invariant factors for $\mathbb{C}$ determine those for $\mathbb{Q}$, or for intermediate fields such as $\mathbb{R}$. For an irreducible monic polynomial $p(t)$ over $\mathbb{Q}$ factors over $\mathbb{C}$ into linear factors $\prod_i (t - \lambda_i)$, and an invariant factor

$p(t)^r$ arising over $\mathbb{Q}$ thus gives rise to the collection of invariant factors $(t - \lambda_i)^r$ arising over $\mathbb{C}$. Since $p(t)$ is irreducible over $\mathbb{Q}$, the λ_i form a complete set of algebraic conjugates over $\mathbb{Q}$. The possibility of grouping the invariant factors over $\mathbb{C}$ into complete sets of algebraic conjugates in this way is a necessary and sufficient condition for an endomorphism h over $\mathbb{C}$ to arise by tensoring from an endomorphism defined over $\mathbb{Q}$.

In particular, for the cases in which we are interested, it follows from Theorem 10.2.1 that for $K = \mathbb{C}$ the λ_i are roots of unity, or equivalently that for $K = \mathbb{Q}$, the polynomials p are among the cyclotomic polynomials ϕ_r. Moreover, the only values of q that occur are 1 and 2. To complete the determination of (H_C, h_*) up to isomorphism we need to determine the list of eigenvalues with multiplicities, or equivalently, the characteristic polynomial of h_*, and also to list those corresponding to blocks of size 2.

The calculations depend on the model for the resolution of the curve singularity given in Section 9.3, and in particular on Theorem 9.3.6. Recall that $\pi : T \to S$ is a good resolution, and the total transform $\pi^{-1}(C)$ is the union of the strict transforms B_j of the branches of C and the exceptional curves E_i. We can take T as a neighbourhood N of $\pi^{-1}(C)$ formed as a union of tubular neighbourhoods of the E_i and the B_j with certain 'plumbing' identifications. Moreover, we wrote E_i^o, B_j^o for the result of removing the interiors of plumbing discs from E_i, B_j so that ∂N is the union of circle bundles over the E_i^o (or B_j^o), which intersect along certain tori; to obtain a convenient model we thicken up these tori.

Theorem 9.3.6 states that the Milnor fibre F may be decomposed into pieces indexed by $\mathcal{V}(\Gamma_R^+)$: for each vertex V_i a piece F_i and for each edge E a connecting piece $C_E \times I$. As the pieces lying over B_j^o are annuli, omitting these will not affect the homeomorphism type of the fibre. Thus from now on we may suppose pieces to be indexed by $\mathcal{V}(\Gamma_R)$. The pieces are as follows.

For a vertex V_i a cyclic M_i-fold covering F_i of E_i^o, and for an edge $E = V_iV_j$ a piece $C_E \times I \subset S^1 \times S^1 \times I$, given by the equation $w_i^{M_i} w_j^{M_j} = 1$. The monodromy h may be taken on F_i, as a covering transformation over E_i^o; and on $C_E \times I$ as given by $h(w, z, t) = (e^{2\pi i(1-t)/M_i} w, e^{2\pi i t/M_j} z, t)$. This decomposition has the disadvantage that the pieces are not necessarily connected, but the advantage that each of them is invariant by the monodromy.

In Section 10.2 we had a decomposition of F by curves with union C, and wrote $C^+ = C \cup \partial F$, N_C for a product neighbourhood of C^+ in F, and F_C for its closed complement. Thus we can take N_C as the union of the $C_E \times I$ and the F_i corresponding to arrowhead vertices; F_C as the union of the other F_i.

We have the exact sequence (10.2), where all the maps are compatible with the action of h. We approach the calculation of $H_1(F)$ by describing the other terms in this sequence: we begin with $H_k(C)$. Each group $H_k(C)$ is a sum $\oplus_{ij} H_k(C_E)$ over edges E. The number of components of C_E is the highest common factor $M_E := (M_i, M_j)$. Since h permutes these components transitively, we have an isomorphism of $\mathbb{Z}[t]$-modules

$$H_0(C_E) \cong H_1(C_E) \cong \mathbb{Z}[t]/\langle t^{M_E} - 1\rangle. \tag{10.3}$$

Summing over edges E, we have

Lemma 10.3.1 *There is an isomorphism*

$$H_0(C) \cong H_1(C) \cong \bigoplus_{E \in \mathcal{E}(\Gamma_R)} \mathbb{Z}[t]/\langle t^{M_E} - 1\rangle. \tag{10.4}$$

as $\mathbb{Z}[t]$-modules

It follows, in particular, that $\Delta_{H_*(C)}(t) = 1$, and hence from the multiplicative property of Δ that $\Delta_{H_*(F)}(t) = \Delta_{H_*(F_C, \partial F_C)}(t)$.

Next consider F_i: we know that it is a cyclic M_i-fold cover of E_i^o: write t for the preferred generator of the covering group. The boundary components of E_i^o correspond to the edges $E = V_i V_j$ of Γ_R^+ incident to V_i. We denote by v_i the valence of V_i in Γ_R^+, which is equal to the number of these components.

Over the component corresponding to E we can take coordinates z tangent to E_i and w normal to E_i such that the Milnor fibre meets the torus $|z| = |w| = 1$ in the curve $w^{M_i} z^{M_j} = 1$. Set $w = e^{2\pi i a}$, $z = e^{2\pi i b}$: then the equation becomes $M_i a + M_j b = 0$. Thus as z goes round the circle ∂E_i^o, b runs from 0 to 1, so a runs from 0 to $-\frac{M_j}{M_i}$. We can say that the monodromy round the component acts as t^{-M_j}. Since the components of ∂E_i^o generate $\pi_1(E_i^o)$, this is enough to determine the structure of F_i, and also of the monodromy h, which acts as t on it.

Since the monodromy group is cyclic, hence abelian, and the sum of the C_E bounds F_i, the product of the corresponding monodromies must be trivial. Hence $\sum M_j$ is divisible by M_i. In fact we saw in (8.12) that $\sum M_j = a_i M_i$ (where $a_i = -[E_i].[E_i]$).

To calculate the homology of F_i we introduce a construction. Observe first that if the valence $v_i = 1$ then E_i^o is a disc, so the M_i-fold cover F_i must be trivial. Thus in this case, $H_1(F_i) = 0$ and $H_0(F_i) \cong \mathbb{Z}[t]/\langle t^{M_i} - 1\rangle$ as $\mathbb{Z}[t]$-module. If $v_i = 2$ then E_i^o is an annulus and deformation retracts on each boundary component. Correspondingly, F_i is homeomorphic to the product $C_E \times I$; we have observed above that such components may be deleted.

Otherwise we may suppose $v_i \geq 3$. Choose one of the edges $E = V_iV_j$ and regard the corresponding component c_E of ∂E_i^o as 'exterior', and E_i^o as a disc with $v_i - 1$ punctures corresponding to the others, giving boundary components c_r; write $\sum'_r$, $\oplus'_r$ and $\bigcup'_r$ to denote sums and unions indexed by the remaining edges through V_i. Choose a tree X in E_i^o meeting the boundary at just one point Q_r in each 'interior' boundary component c_r; write $\tilde{X}$ for its preimage in F_i, $\tilde{Q}_r$ for the preimage of Q_r and C_r for the preimage of c_r.

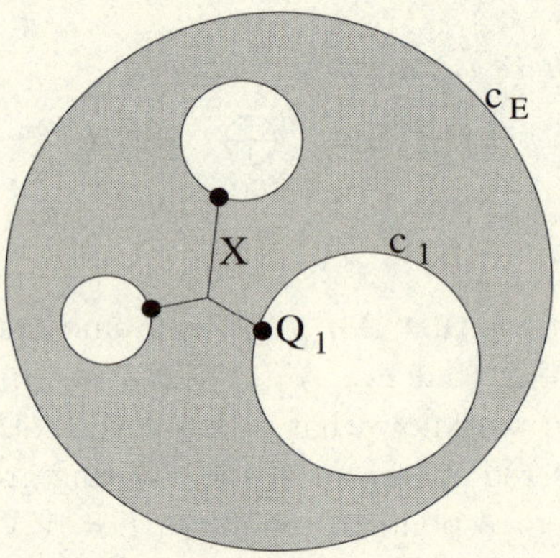

Fig. 10.2. The model for E_i^o

We see that there is a deformation retraction of E_i^o on the union $X \cup'_r c_r$; and this is covered by a t-equivariant deformation retraction of F_i on $\tilde{X} \cup'_r C_r$. Hence we have isomorphisms of $\mathbb{Z}[t]$-modules

$$H_k(F_i, \bigcup_r{}' C_r) \cong H_k\left(\tilde{X} \bigcup_r{}' C_r, \bigcup_r{}' C_r\right) \cong H_k(\tilde{X}, \cup'_r \tilde{Q}_r).$$

Moreover, since each inclusion $\tilde{Q}_r \hookrightarrow \tilde{X}$ is an equivariant homotopy equivalence (each component is contractible), these groups are zero for $k \neq 1$, while $H_1(F_i, \bigcup'_r C_r)$ is isomorphic to a sum of $v_i - 2$ copies of $\mathbb{Z}[t]/\langle t^{M_i} - 1\rangle$. In particular, we have already proved

Theorem 10.3.2

(i) *We have* $\Delta_{H_*(F_i, \partial F_i)}(t) = (t^{M_i} - 1)^{2-v_i}$.

(ii) *Hence* $\Delta_{H_*(F)}(t) = \prod_{V_i \in \mathcal{V}(\Gamma_R)} (t^{M_i} - 1)^{2-v_i}$.

Note that although the sum is over vertices of Γ_R, the valences are those in the augmented tree Γ_R^+.

Proof Let the chosen component of ∂E_i^o correspond to P_{is}. We have seen that $\Delta_{H_*(F_i,\bigcup'_r C_r)}(t) = (t^{M_i} - 1)^{v_i-2}$. But $\Delta_{H_*(C_{is})}(t) = 1$. The result now follows from multiplicativity of Δ and the exact homology sequence of $(F_i, \bigcup'_r C_r)$ relative to C_E.

This now follows from (i) and the remarks preceding the theorem. □

Thus the homology exact sequence of F_i relative to $\bigcup'_r C_r$ yields

$$H_1(F_i) \to \bigoplus^{v_i-2} \mathbb{Z}[t]/\langle t^{M_i} - 1\rangle \to \bigoplus_r^{\prime} \mathbb{Z}[t]/\langle t^{M_E} - 1\rangle \to H_0(F_i) \to 0.$$

We denote by hM_i the highest common factor of M_i and all the M_r such that V_r is a vertex of $\Gamma_R^+(C)$ and E_r intersects E_i. Since $\sum_r M_r = a_i M_i$ is divisible by M_i we can omit one term M_r, e.g. that corresponding to the outside boundary component, without affecting the highest common factor.

Lemma 10.3.3 *F_i has hM_i components, and $H_0(F_i) \cong \mathbb{Z}[t]/\langle t^{hM_i} - 1\rangle$ as $\mathbb{Z}[t]$-modules.*

Proof Since the tree X is contractible, we may choose a lift of X to F_i. This gives preferred lifts of the points Q_r, whose classes we take as preferred generators θ_r of the $H_0(\tilde{Q}_r)$ (each isomorphic to $\mathbb{Z}[t]/\langle t^{M_i}-1\rangle$). Then $H_1(F_i, \bigcup'_r C_r)$ maps isomorphically to the submodule generated by the differences $\theta_r - \theta_s$. On the other hand, the image of θ_r may also be taken as generator of $H_0(C_r) \cong \mathbb{Z}[t]/\langle t^{hcf(M_i,M_r)} - 1\rangle$. Thus $H_0(F_i)$ is the quotient of the sum of these by the relations $\theta_r = \theta_s$, and hence is isomorphic to the quotient of $\mathbb{Z}[t]$ by the ideal generated by the $t^{hcf(M_i,M_r)} - 1$ for all interior boundary components r.

The result now follows from the fact that, if (a, b) is the highest common factor of the positive integers a and b, the polynomial $t^{(a,b)} - 1$ is the highest common factor of $t^a - 1$ and $t^b - 1$, so the ideal $\langle t^a - 1, t^b - 1\rangle = \langle t^{(a,b)} - 1\rangle$. □

The above result gives an exact description of the integer homology. For $H_1(F_i)$ we will be less precise, and from now on work over $\mathbb{Q}$.

Corollary 10.3.4 *The characteristic polynomial of h on $H_1(F_i)$ is*

$$(t^{M_i} - 1)^{v_i - 2}(t^{hM_i} - 1).$$

Proof In the homology exact sequence of F_i relative to ∂F_i, the alternating product of the characteristic polynomials of t on the non-trivial terms must equal 1. Those on H_0 and H_1 of ∂F_i coincide, so cancel in the alternating product. The result follows from the calculations in Theorem 10.3.2(i) $\Delta_{H_*(F_i, \partial F_i)}(t) = (t^{M_i} - 1)^{2 - v_i}$ and in Lemma 10.3.3 that $H_0(F_i) \cong \mathbb{Z}[t]/\langle t^{hM_i} - 1\rangle$. □

A similar argument now gives

Proposition 10.3.5 *The characteristic polynomial of h on $H_1(F)$ is*

$$(t-1) \prod_{V_i \in \mathcal{V}(\Gamma_R)} (t^{M_i} - 1)^{v_i - 2}.$$

Proof We consider the alternating product of the characteristic polynomials of t on the non-trivial terms of (10.2). Again the 'boundary' terms cancel out, so we have the characteristic polynomial $t - 1$ on $H_0(F) \cong \mathbb{Z}$ multiplied by the product over i of the polynomial over $H_1(F_i)$ divided by that on $H_0(F_i)$. □

We have determined the characteristic polynomials of the action of h on all the terms of (10.2): see (10.4) for $H_0(C)$ and $H_1(C)$; Lemma 10.3.3 for $H_0(F_C)$, Corollary 10.3.4 for $H_1(F_C)$ and Proposition 10.3.5 for $H_1(F)$. We recall the weight filtration of Section 10.2. $W_0 := \operatorname{Im} H_1(C^+) \to H_1(F)$, $W_1 := \operatorname{Im} H_1(F_C) \to H_1(F)$, $W_2 := H_1(F)$. It follows that

Corollary 10.3.6 *The characteristic polynomial of h on W_1 is*

$$\prod_{V_i \in \mathcal{V}(\Gamma_R)} (t^{M_i} - 1)^{v_i - 2}(t^{hM_i} - 1) \Big/ \prod_{E \in \mathcal{E}(\Gamma_R)} (t^{M_E} - 1).$$

We saw in Theorem 10.2.1 that the action of h_* on $W_1 \subset H$ is semi-simple, so blocks of size 2 can arise, at most, from the eigenvalues of h_* on H/W_1, or equivalently on its dual W_0/R. Choose a natural number M divisible by all M_i (e.g. their least common multiple). Then by Proposition 10.3.5 all the eigenvalues arising are roots of $t^M - 1$. Thus t has finite order on H if and only if $t^M - 1$ is zero. We determine the invariant factors of t via a calculation of $t^M - 1$.

Lemma 10.3.7 *(Twist formula) For any $y \in H$,*

$$h^M(y) - y = -\sum_r \langle y, C_r\rangle \frac{MM_E}{M_iM_j} C_r.$$

where $E = V_iV_j$ runs through the edges of Γ_R, and C_r through the components of C_E.

Proof We use the model of Theorem 9.3.6, and represent y by a cycle transverse to all the curves which are components of any $C_E \times \frac{1}{2}$. By an isotopy, we may suppose that the cycle intersects N_C in a union of lines $P \times I \subset S^1 \times S^1 \times I$. Each such line (if oriented according to the usual orientation of I) has intersection number $+1$ with C_r, and the signed number of such lines is thus just $\langle y, C_r\rangle$.

On each component of F_C, h^M acts as the identity. Thus $h^M(y) - y$ is represented by the cycle which is the sum, over intersection points P, of the cycle $h^M(P \times I) - P \times I$ in $C_r \times I$. Recall that in the local model of Theorem 9.3.6, C_r is a component of $w_i^{M_i} w_j^{M_j} = 1$ and h is given on $C_r \times I$ by $h(w_i, w_j, s) = (e^{2\pi i(1-s)/M_i} w_i, e^{2\pi is/M_j} w_j, s)$.

We may parametrise C_r as $(\xi e^{2\pi i(M_j/M_E)\theta}, e^{-2\pi i(M_j/M_E)\theta})$ for $\theta \in [0, 1]$. We have $h^M(P, s) = (e^{2\pi i(1-s)M/M_i} w_i, e^{2\pi isM/M_j} w_j, s)$. The projection on $S^1 \times S^1$ thus runs round C_r exactly $-(M/M_i)/(M_j/M_E)$ times. It follows that the homology class of $h^M(P \times I) - P \times I$ is $-(M/M_i)/(M_j/M_E)$ times that of C_r.

The lemma follows by collecting these over all components $P \times I$. □

Theorem 10.3.8

(i) *A class $y \in H$ is in* Ker $(t^M - 1)$ *if and only if y is orthogonal (under the intersection form) to W_0.*

(ii) *We have $W_1 =$* Ker $(t^M - 1)$.

(iii) *Over $\mathbb{Q}$, $(t^M - 1)H = H^{\neq 1} \cap W_0$.*

(iv) *Multiplication by $(t^M - 1)$ induces an isomorphism of W_2/W_1 on W_0/R.*

Proof It follows from Lemma 10.3.7 that if $\langle y, C_r\rangle = 0$ for each r, then $(t^M - 1)y = 0$. The converse is not immediate, since the cycles C_r are not independent. However, if we intersect each side of the twist formula with y, we obtain

$$\langle t^M(y) - y, y\rangle = -\sum_r \langle y, C_r\rangle \frac{MM_E}{M_i M_j}\langle C_r, y\rangle$$

$$= \sum_r \frac{MM_E}{M_i M_j}(\langle C_r, y\rangle)^2,$$

and since all the coefficients are strictly positive, this can vanish only if $\langle C_r, y\rangle = 0$ for each r. Thus if $y \in \operatorname{Ker} t^M - 1$, y is orthogonal to all the components of all C_E, and hence – since the remaining components of C^+ are in the radical – to the subgroup W_0.

now follows since, by Proposition 10.2.2, the orthogonal complement of W_0 is W_1.

It follows from the same result that there is a dual pairing of H/W_1 and W_0/R, and hence that these two have the same dimension. Since $W_1 = \operatorname{Ker}(t^M - 1)$, H/W_1 has the same dimension as $\operatorname{Im}(t^M - 1)$. Now by Lemma 10.3.7, $\operatorname{Im}(t^M - 1) \subseteq W_0$, since it is contained in the subspace spanned by the C_r. Also, $\operatorname{Im}(t^M - 1) \subseteq \operatorname{Im}(t-1)$, which equals $H^{\neq 1}$ by (ii) of Proposition 10.2.4. Thus $\operatorname{Im}(t^M - 1) \subseteq W_0 \cap H^{\neq 1}$, and since these two have the same dimension, they coincide. This argument also proves (iv). □

Corollary 10.3.9 *The action of h on $H_1(F)$ is semi-simple if and only if $W_1 = H_1(F)$. An equivalent condition is that*

$$(t-1)\prod_{E\in\mathcal{E}(\Gamma_R)}(t^{M_E} - 1) = \prod_{V_i\in\mathcal{V}(\Gamma_R)}(t^{hM_i} - 1).$$

Proof The first assertion follows at once from the Theorem. The second then follows by comparing the calculations of characteristic polynomials in Proposition 10.3.5 and Corollary 10.3.6. □

Since Γ_R is a tree we can associate to each vertex V_i other than the initial vertex V_0 the first edge E_i in a geodesic joining V_i to V_0, and this gives each edge just once. The condition can thus be rewritten

$$\prod_{i\neq 0}\{(t^{M_{E_i}} - 1)/(t^{hM_i} - 1)\} = (t^{hM_0} - 1)/(t-1). \tag{10.5}$$

Theorem 10.3.10
(i) *For Γ a single branch, the characteristic polynomial of h on H is*

$$\frac{(t-1)\prod_1^g(t^{\frac{e_{k-1}}{e_k}\overline{\beta}_k}-1)}{(t^m-1)\prod_1^g(t^{\overline{\beta}_k}-1)}.$$

(ii) *The action of h on H is semi-simple in this case.*

Proof By Proposition 10.3.5, the characteristic polynomial of h on $H = H_1(F)$ is $(t-1)\prod_i(t^{M_i}-1)^{v_i-2}$. Thus the vertices of valence 2 in the resolution tree make no contribution. By Theorem 8.5.2, the values of the invariants $M_k(C)$ at the ith point B_i of valence 3 and the ith point A_i of valence 1 of $\Gamma_R(C)$ are, respectively, $\frac{e_{i-1}}{e_i}\overline{\beta_i}$ and $\overline{\beta_i}$, while of course M_0 is the multiplicity m of C. Substituting these values gives the result.

We use the criterion 10.5, and recall the calculation of Proposition 8.5.3. At any vertex of valence 1 or 2, the value of M_E at any adjacent edge is equal to hM_i, thus the factor $(t^{M_{E_i}}-1)/(t^{hM_i}-1)$ is equal to 1. For each edge between B_q and B_{q+1} we have $M_E = e_q$, hence the value of hM_i at B_q is the highest common factor of e_{q-1}, e_q and $\overline{\beta_q}$, namely e_q. Thus the factor $(t^{M_{E_i}}-1)/(t^{hM_i}-1)$ corresponding to B_q reduces to $(t^{e_{q-1}}-1)/(t^{e_q}-1)$. Taking the product over $1 \le q \le g$ gives $(t^{e_0}-1)/(t^{e_g}-1)$, which is indeed equal to $(t^{hM_0}-1)/(t-1)$. □

The characteristic polynomial as studied here is the same as the Alexander polynomial defined in Chapter 5. The formula we have just obtained agrees with that of Corollary 5.5.6. The multi-variable Alexander polynomial for a curve C with branches B_j can be expressed in similar terms.

Proposition 10.3.11 *The multi-variable Alexander polynomial of C is equal to*

$$\prod_{V_i\in\mathcal{V}(\Gamma_R)}\left(\prod_j t_j^{M_i(B_j)}-1\right)^{v_i-2}.$$

This result follows from Theorem 12.1 of [65] in view of the identification of the parameters there denoted ℓ_{ij} in Exercise 9.11.5.

The result about semi-simplicity does not extend to the case of curves with more than one branch. The argument shows that in the criterion

(10.5), the factors on the right hand side corresponding to vertices of valence 1 or 2 reduce to 1, so it suffices to take the factors corresponding to rupture points. There does not seem to be a simple way to state the condition in terms of the invariants of the curve: although we can give formulae for values of $M_i(B_j)$ on a tree, $M_i(C)$ is obtained by adding these together, and we cannot predict how highest common factors behave when we add terms up.

The cancellation argument above can also be presented as follows. The terms $(t^{hM_i} - 1)$ and $(t^{M_E} - 1)$ along the dead branches cancel out. In the rest of the tree, the value of hM_i at a rupture point equals M_E for the next higher edge; for the top rupture point we get 1. Suppose C has two branches, and $\Gamma_R(C)$ consists of a graph Γ together with chains joining its top vertex V_v to the two arrowhead vertices: let the values of M_E along these two chains be P and Q. Then the expression for the characteristic polynomial of H/M_1 reduces to $(t^P - 1)(t^Q - 1)/(t^{hM_v} - 1)(t - 1)$. Since, by Lemma 5.5.3, the polynomials $t^r - 1$ are multiplicatively independent, this expression can only be trivial if the factors in the numerator coincide with those in the denominator, thus (up to order) $P = 1$ and $Q = hM_v$. But then hM_v, which divides P, must also equal 1. Hence also $Q = 1$. Thus the necessary and sufficient condition for semi-simplicity is that each of P and Q must equal 1.

Example 10.3.1 For a singularity of type A_{2k}, Γ_R^+ and the values of the $M_r(C)$ are displayed in Example 8.4.2. We have $v_i \neq 2$ only for $v_0 = 1$, $v_k = 1$ and $v_{k+1} = 3$; the Alexander polynomial is thus $(t - 1)(t^{4k+2} - 1)/(t^2 - 1)(t^{2k+1} - 1)$, agreeing with earlier results.

For a singularity of type A_{2k-1}, Γ_R^+ is again pictured in Example 8.4.2. Here $v_i \neq 2$ only for $v_0 = 1$ and $v_{k-1} = 3$; the multi-variable Alexander polynomial is $(t_1^k t_2^k - 1)/(t_1 t_2 - 1)$ and the characteristic polynomial of h is $(t^{2k} - 1)/(t^2 - 1)$.

Example 10.3.2 Consider the curves $(y - x^2)(y^2 - x^7) = 0$ and $(y - x^3)(y^2 - x^7) = 0$. The resolution graphs, with the vertex V_i marked with $\binom{a_i}{M_i}$, are

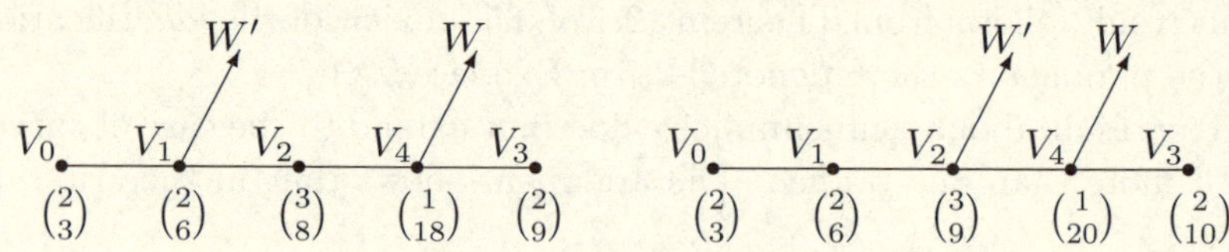

In the first case, the 'V_v' above is V_1; we have $(M_1, M_2) = 2$, and the characteristic polynomial of h on H/M_1 is $t+1$: semi-simplicity fails. In the second case, the separating vertex V_v is V_2, and since $(M_2, M_4) = 1$, the monodromy is semi-simple.

10.4 Algebraic classification of Seifert forms

Define an *isometric triple* over a field K to consist of a K-vector space H, a *non-singular* skew-symmetric pairing $\langle , \rangle : H \times H \to K$, and an automorphism $h : H \to H$ which preserves $\langle , \rangle$. We regard H as a $K[t]$-module, where multiplication by t is given by the action of h. In this section we analyse the algebraic classification of isometric triples.

By Lemma 10.1.2, if 1 is not an eigenvalue of h, we can define a Seifert form S by $S(x,y) := \langle (1-t)^{-1}x, y\rangle$, so we do not need to include the form S in the definition. In the situation of Section 10.2, taking $H = H^{\neq 1}$, $\langle , \rangle$ given by intersection numbers on F, and h the Milnor monodromy, say with $K = \mathbb{Q}$, defines an isometric structure: our interest is prompted by this example.

If p is prime, it generates a prime ideal in $K[t]$, and the quotient $L_p := K[t]/\langle p\rangle$ is a field, whose degree over K is that of the polynomial p. The summand of H corresponding to a prime invariant factor p is isomorphic to the field L_p. We will reduce the classification of isometric triples over K to that of Hermitian forms over the extension fields L_p.

The module H is canonically split as a direct sum of modules $H_{(p)}$. In turn, $H_{(p)}$ is a direct sum $\bigoplus_i H_{p^i}$ of h-invariant subspaces such that H_{p^i} is a free module over $K[t]/\langle p(t)^i\rangle$, but this expression as a direct sum is not unique. Write, for each q, $K(q)$ for the kernel of the endomorphism of H given by $p(t)^q$. Then H_{p^i}/pH_{p^i} is naturally isomorphic to the invariantly defined $K[t]$-module

$$V_{p,i} := K(i)/K(i-1) + pK(i+1), \tag{10.6}$$

which is a vector space over the field L_p: its dimension is the number of invariant factors of H equal to p^i.

Define the 'dual' of the polynomial p by reversing the order of the coefficients and normalising. More precisely, as p is monic, we write $p(t) = t^k + \sum_1^k a_i t^{k-i}$, and set

$$p^\vee(t) := t^k p(t^{-1})/a_k = t^k + \sum_1^{k-1} \frac{a_{k-i}}{a_k} t^{k-i} + \frac{1}{a_k}.$$

Proposition 10.4.1 *The generalised eigenspaces $H_{(p)}$ and $H_{(q)}$ are orthogonal unless $q = p^\vee$.*

Proof Since t acts as h, which is an isometry, we have, for any $u, v \in H$, $\langle u, t^{-1}v\rangle = \langle tu, v\rangle$, and hence for any polynomial $\phi(t)$, $\langle u, \phi(t^{-1})v\rangle = \langle \phi(t)u, v\rangle$. Now if n is large enough, $p(t)^n$ annihilates $H_{(p)}$. Thus if $u \in H_{(p)}$,

$$\langle u, p(t^{-1})^n v\rangle = \langle p(t)^n u, v\rangle = 0.$$

But if $q \neq p^\vee$, $p^\vee$, and hence also $p(t^{-1})$ and $p(t^{-1})^n$, are invertible endomorphisms of the generalised eigenspace $H_{(q)}$. Thus any $w \in H_{(q)}$ may be expressed in the form $p(t^{-1})^n v$, and hence is orthogonal to any $u \in H_{(p)}$. □

This proposition allows us to reduce the classification problem of isometric triples as follows. We may split H as a direct sum of the subspaces $H_{(p)} \oplus H_{(p^\vee)}$ (for $p \neq p^\vee$) and $H_{(p)}$ (for $p = p^\vee$). Then h preserves each summand, and two distinct summands are orthogonal, so it suffices to consider one at a time.

Since the inner product on H is non-degenerate, so is that on $H_{(p)} \oplus H_{(p^\vee)}$ (with $p \neq p^\vee$). Each of $H_{(p)}$ and $H_{(p^\vee)}$ is self-orthogonal, and so these two must be dually paired, and we can identify the second of these with the space dual to the first. Thus up to isomorphism, the isometric structure on $H_{(p)} \oplus H_{(p^\vee)}$ is determined by the behaviour of h on $H_{(p)}$, i.e. by the invariant factors which are powers of p.

It remains to consider the cases $p = p^\vee$. Before proceeding, we consider which of these cases arise.

Lemma 10.4.2 *Suppose that $p(t)$ is an irreducible polynomial of degree k, and that $p = p^\vee$. Then either $k = 1$ and $p(t) = t \pm 1$ or k is even and $a_k = 1$.*

Proof Since $p = p^\vee$, equating constant terms gives $a_k = a_k^{-1}$, so $a_k = \pm 1$.

If $a_k = -1$, then for $1 \leq i \leq \frac{k}{2}$, $a_{k-i} = -a_i$, so $a_{k-i}t^{k-i} + a_i t^i = a_i t^i (t^{k-2i} - 1)$ is divisible by $(t-1)$. Hence so is $p(t)$: since p is irreducible, it must coincide with $t - 1$.

Similarly if $a_k = 1$ and k is odd, each pair of terms $a_{k-i}t^{k-i} + a_i t^i = a_i t^i (t^{k-2i} + 1)$ is divisible by $(t+1)$; hence $p(t) = t + 1$. Note that this argument holds even if K has characteristic 2. □

If $\deg p = 2d$ is even, we simplify notation by setting $s(t) := t^{-d}p(t)$, so that $s(t^{-1}) = s(t)$.

First consider the special case when all the invariant factors coincide with the irreducible polynomial p – equivalently, h has irreducible minimal polynomial p. This is known as the *semi-simple* case, and is in some sense generic. If $p = t \pm 1$, the given form $\langle , \rangle$ over K is symmetric or skew-symmetric, and no further reduction is required.

Otherwise, as we saw above, we can regard H as a vector space over the field L_p. Write τ for the image of t in L_p, so that $p(\tau) = 0$, hence $s(\tau) = 0$ and $s(\tau^{-1}) = s(\tau) = 0$. Thus $p(\tau^{-1}) = 0$, and the map $\tau \to \tau^{-1}$ induces a field automorphism of L_p. We denote this by an overline. Clearly we have $\overline{\overline{e}} = e$ for any $e \in L_p$ and $\overline{e} = e$ if $e \in K$. Since h preserves the form $\langle , \rangle$ we have an identity $\langle u, \phi(t^{-1})v \rangle = \langle \phi(t)u, v \rangle$, which may now be written $\langle u, \overline{e}v \rangle = \langle eu, v \rangle$.

We next seek to translate the inner product on H into a structure defined over L_p. The usual procedure for doing this uses a function $T : L \to K$ (we will drop the subscript p for the general discussion), and the bilinear form b_T over K given by

$$b_T(\ell, m) := T(\ell m) \quad \text{for all } \ell, m \in L. \tag{10.7}$$

We require that b_T is nonsingular, and that, for any $y \in L_p$, we have $T(\overline{y}) = T(y)$.

If L_p is a separable extension of K, we can take T as the usual trace: that it has these properties follows from the standard theory of field extensions. However in the present situation we have a simpler alternative.

Let p have degree $2d$; then we can take as basis for L_p as vector space over K the monomials $\{\tau^{1-d}, \tau^{2-d}, \ldots, \tau^{-1}, 1, \tau, \tau^2, \ldots, \tau^{d-1}, \tau^d\}$ and define T to be the unique K-linear map $L_p \to K$ which vanishes on all these except τ^d, where it takes the value 1. Applying T to the equation $s(\tau) = 0$, we deduce that $T(\tau^{-d}) = -1$. It follows, since this holds on all basis elements, that for any $y \in L_p$, we have $T(\overline{y}) = -T(y)$. Moreover if we denote the basis elements by $e_i = \tau^{i-d}$ for $1 \leq i \leq 2d$, then the matrix $(a_{i,j}) = (T(e_i e_j)) = (T(\tau^{i+j-2d}))$ of the bilinear form b_T has $a_{i,j} = 0$ if $|i + j - 2d| < d$. Thus if we partition the matrix by separating the first $d - 1$ rows or columns from the rest, the upper left and lower right blocks are unitriangular, and the upper left and lower right blocks are identically zero. Hence in the usual expansion of the determinant there is just one non-zero term, which is equal to 1; so indeed b_T is nonsingular.

If $\epsilon = \pm 1$, we will say that a form $\phi : V \times V \to L$ is ϵ-hermitian if it is hermitian for $\epsilon = +1$ and skew-hermitian for $\epsilon = -1$; similarly for ϵ-symmetric.

Lemma 10.4.3 *Let K be a field, L a finite extension field, admitting an involution 'bar' over K, and V a vector space over L. Let $\langle , \rangle$ be a ϵ-symmetric form on V regarded as a vector space over K, such that $\langle u, \bar{\ell} v\rangle = \langle \ell u, v\rangle$ for all $\ell \in L$, $u, v \in V$.*

Let $T : L \to K$ be a K-linear map such that $T(\overline{y}) = \eta T(y)$ for any $y \in L$ (where $\eta = \pm 1$) and the bilinear form b_T is nonsingular over K.

Then there is a unique $\epsilon\eta$-hermitian inner product $(u, v) \mapsto u.v$ on H over L such that

$$\langle u, v\rangle = T(u.v). \tag{10.8}$$

If $\langle , \rangle$ is nonsingular, then this form also is.

Proof For each $u, v \in H$, the map $\ell \mapsto Y(\ell) := \langle \ell u, v\rangle$ from L to K is K-linear. Since b_T is nonsingular, there is a unique element $m \in L$ such that, for each $\ell \in L$, $Y(\ell) = T(m\ell)$. Define $u.v$ to be this element m. Thus by definition we have the identity

$$\langle \ell u, v\rangle = T(\ell(u.v)). \tag{10.9}$$

Taking $\ell = 1$ in (10.9) gives (10.8). We next check linearity of $u.v$ in the first variable. Additivity is immediate, and using (10.9) twice: first with $\ell_1\ell_2$ for ℓ and second with ℓ_1 for ℓ and $\ell_2 u$ for u, we obtain

$$T(\ell_1\ell_2(u.v)) = \langle \ell_1\ell_2 u, v\rangle = T(\ell_1(\ell_2 u.v)).$$

Since this holds for all ℓ_1, it follows from non-degeneracy of b_T that, for all u, v,

$$\ell_2(u.v) = (\ell_2 u.v). \tag{10.10}$$

Next we obtain the hermitian property. Since $T(\overline{y}) = -T(y)$, we may calculate

$$T(\ell\overline{u.v}) = \eta T(\bar{\ell} u.v) = \eta\langle \bar{\ell} u, v\rangle = \eta\langle u, \ell v\rangle = \epsilon\eta\langle \ell v, u\rangle = \epsilon\eta T(\ell(v.u)).$$

Since this holds for all $\ell \in L$, non-degeneracy of b_T implies $\overline{u.v} = \epsilon\eta(v.u)$, as required.

For any inner product satisfying (10.8) and (10.10) we have $T(\ell(u.v)) = T((\ell u).v) = \langle \ell u, v\rangle$, giving the relation (10.9) which characterised the hermitian form.

If $u.v = 0$ for all u then $\langle u, v\rangle = 0$ for all u, so if $\langle , \rangle$ is nonsingular it follows that $v = 0$. □

We return to the general discussion. Then $H_{(p)}$ is a direct sum of $K[t]$-modules $H_{p,j}$, where $H_{p,j}$ is a sum of a_j copies of $K[t]/\langle p(t)^j\rangle$. We wrote $K(i)$ for the kernel of the endomorphism of H given by $p(t)^i$ and defined $V_{p,i}$ as $K(i)/K(i-1) + pK(i+1)$, so that $V_{p,i}$ is isomorphic to $H_{p,i}/pH_{p,i}$, and hence is a vector space of rank a_i over L_p. Write q for the largest value of i such that $a_i \neq 0$.

Lemma 10.4.4 *Suppose* $\deg p$ *even. Then there is a unique* $(-1)^i$-*hermitian inner product* $(u, v) \mapsto u.v$ *on* $V_{p,i}$ *over* L_p *such that* $\langle u, s(t)^{i-1}v\rangle = T(u.v)$. *If* $i = q$, *this form is nonsingular.*

Proof First consider the scalar product $\langle u, s(t)^{i-1}v\rangle$ for $u, v \in K(i)$. It is $(-1)^{i-1}$-symmetric since

$$\begin{aligned} \langle v, s(t)^{i-1}w\rangle &= \langle s(t^{-1})^{i-1}v, w\rangle = \langle (-1)^{i-1}s(t)^{i-1}v, w\rangle \\ &= (-1)^{i-1}\langle w, s(t)^{i-1}v\rangle. \end{aligned}$$

It vanishes if $w \in K(i-1)$, while if $w \in p(t)K(i+1) = s(t)K(i+1)$, write $w = s(t)u$: then it equals $\langle v, s(t)^i u\rangle = \langle s(t)^i v, u\rangle = 0$. It thus induces an inner product on $V_{p,i}$. Now applying Lemma 10.4.3 gives a form as required.

It remains to show that if $i = q$ this form is nonsingular. Suppose the class of w is in the radical. Then $s(t)^{q-1}w$ is orthogonal to $K(q)$, and hence to all of H. Since $\langle , \rangle$ is nonsingular, it follows that $s(t)^{q-1}w = 0$, so the class of w is zero. □

Nonsingularity for other values of i will follow indirectly from the results below.

Lemma 10.4.5 *Suppose* H *a free module over* $K[t]/\langle p(t)^q\rangle$. *Then the isomorphism class of the product defined in Lemma 10.4.4 determines the isomorphism class of the isometric triple.*

Proof Suppose H, H' support isometric triples, and $\overline{\phi} : H/pH \to H'/pH'$ an isomorphism of Hermitian forms. Lift $\overline{\phi}$ to an isomorphism $\phi_1 : H \to H'$ of $K[t]$-modules. Then since $\overline{\phi}$ is an isomorphism of Hermitian forms, the following statement holds for $i = 1$:

$$\langle \phi_i(s^{q-i}u), \phi_i(v)\rangle = \langle s^{q-i}u, v\rangle \quad \text{for all } u, v \in H.$$

We will construct in turn isomorphisms ϕ_i for $i = 1, \dots, q$ such that this holds; then ϕ_q gives the desired isomorphism of triples.

Assume ϕ_i as above. Then $\phi_i(s^{q-i-1}u), \phi_i(v)\rangle - \langle s^{q-i-1}u, v\rangle$ vanishes for $u \in sH$, hence also for $v \in sH$, so depends only on the classes $\overline{u}, \overline{v}$ of u and v in H/pH. By Lemma 10.4.3, we have a hermitian form H_i on H/pH such that this expression is equal to $T(H_i(\overline{u}, \overline{v}))$. Choose a sesquilinear form A_i such that

$$H_i(\overline{u}, \overline{v}) = A_i(\overline{u}, \overline{v}) + \overline{A_i(\overline{v}, \overline{u})}.$$

Since the form $u.v$ is nonsingular, there is an L_p-linear map $\psi_i : H/pH \to H/pH$ such that

$$A_i(u, v) = \psi_i(u).v = \langle s^{q-1}\psi_i(u), v\rangle,$$

and so $\overline{A_i(v, u)} = \langle s^{q-1}u, \psi_i(v)\rangle$. Now set $\phi_{i+1} := \phi_i + s^i\psi_i$. Then for all $u, v \in H$, $\langle \phi_{i+1}(s^{q-i-1}u), \phi_{i+1}(v)\rangle$ is equal to

$$\langle \phi_i(s^{q-i-1}u), \phi_i(v)\rangle + \langle \psi_i(s^{q-1}u), \phi_i(v)\rangle + \langle \phi_i(s^{q-1}u), \psi_i(v)\rangle,$$

since the final term vanishes, and hence to $\langle s^{q-i}u, v\rangle$. □

We are now ready to state the classification theorem for isometric structures.

Theorem 10.4.6 *Let $(H, h, \langle,\rangle)$ be an isometric triple over a field K of characteristic $\neq 2$. Split H as a sum of generalised eigenspaces $H_{(p)}$ corresponding to the irreducible factors p of the characteristic polynomial of h on H; and write each $H_{(p)}$ as a direct sum $\bigoplus_i H_{p,i}$ of h-invariant subspaces such that $H_{p,i}$ is a free module over $K[t]/\langle p(t)^i\rangle$. Then*

(i) *The spaces $H_{(p)}$ and $H_{(q)}$ are orthogonal unless $q = p^\vee$, and $\langle,\rangle$ induces a dual pairing of $H_{(p)}$ and $H_{(p^\vee)}$.*

(ii) *For each self-dual p and each i, the isometric structure induces a nonsingular sesquilinear form $\phi_{p,i}$ on $V_{p,i} \cong H_{p,i}/p(t).H_{p,i}$ over $L_p := K[t]/\langle p(t)\rangle$. The form $\phi_{p,i}$ is hermitian if $\deg p$ is even and $(-1)^{i-1}\epsilon$-symmetric if $p(t) = t + \epsilon$.*

(iii) *There is an induced isometric structure on $H_{p,i}$, which is determined up to isomorphism by $\phi_{p,i}$. Morever, we can choose the $H_{p,i}$ to be orthogonal.*

Thus the isometric structure is determined up to isomorphism by the $K[t]$-module H and the forms $\phi_{p,i}$.

Proof We established the orthogonality in Lemma 10.4.1; that the pairing is dual follows from nonsingularity of $\langle , \rangle$. In particular, for p self-dual, we have an induced isometric structure on $H_{(p)}$.

This was proved in the case deg p even in Lemma 10.4.4. For the case $p(t) = t + \epsilon$, where $\epsilon = \pm 1$, we argue as for Lemma 10.4.4, but using $S(t) := t - t^{-1}$ in place of $s(t)$. This affects the following points. First, like $s(t)$ in the previous case, $S(t)$ is the product of $p(t)$ with an operator which is invertible on H. (Note that this is the only point at which we assume that the characteristic is not 2.) Second, we have $L = K$.

Here we fix p, and induct on the largest integer $i = q$ such that $H_{p,i} \neq 0$. For any splitting as in (i) we saw in Lemma 10.4.4 that $\langle , \rangle$ is nonsingular on $H_{p,q}$ so that we have an induced isometric structure there, and in Lemma 10.4.5 that this structure is determined up to isomorphism by $\phi_{p,q}$.

In view of the nonsingularity, there is an orthogonal direct sum splitting $H_{(p)} = H_{p,q} \oplus H'_{(p)}$. We have an induced isometric structure on $H'_{(p)}$, and can split this as a (non-orthogonal) sum $H'_{(p)} = \bigoplus_{i<q} H_{p,i}$ (thus redefining the $H_{p,i}$ for $i < q$).

It now follows, by induction on q, that we can choose the direct sum splitting $H_{(p)} = \bigoplus_i H_{p,i}$ to be orthogonal, and also that, for each i, $\langle , \rangle$ is nonsingular on $H_{p,i}$ so that we have an induced isometric structure there, and that this structure is determined up to isomorphism by $\phi_{p,i}$. □

10.5 Hermitian forms

We have seen in the preceding section that the classification of triples $(H, \langle . \rangle, h_*)$ over a field K is equivalent to classifying a series of ± 1-Hermitian forms on vector spaces V over finite extensions L of K. To use this result effectively, it is necessary to know something about the classification of non-singular ± 1-hermitian forms.

In this chapter we first develop the theory of Hermitian forms up to Witt's cancellation theorem, then define Witt equivalence, and recall known results about the classification of forms up to Witt equivalence over certain fields K. We then define Witt equivalence for isometric structures, and show that Witt equivalence is much easier to study than isomorphism.

For any subspace $W \subset V$ we have its orthogonal complement

$$W^\perp := \{x \in V \mid (\forall y \in W)\ \langle x, y\rangle = 0\}.$$

It follows from non-singularity that $\dim W^\perp = \dim V - \dim W$. Since $\langle y, x\rangle = \overline{\langle x, y\rangle}$, $W \subseteq W^{\perp\perp}$; equality of dimensions now forces equality. The restriction of the form to W is non-singular if and only if $W \cap W^\perp = \{0\}$; in this case V is decomposed as the orthogonal direct sum of W and $W^\perp$.

At the opposite extreme, we say that W is sublagrangean if the restriction of the form to W vanishes identically, so $W \subseteq W^\perp$, and Lagrangean if $W = W^\perp$, or equivalently, $W \subseteq W^\perp$ and $\dim V = 2 \dim W$. For example, a form with basis $\{e_i, f_i\}$ such that all $\langle e_i, e_j\rangle$, $\langle f_i, f_j\rangle$ and $\langle e_i, f_j\rangle$ are zero except $\langle e_i, f_i\rangle = 1$ is called hyperbolic: the subspace spanned by the e_i is Lagrangean. Any skew-symmetric form is hyperbolic, so we do not need to discuss these further.

If the involution on L is non-trivial, there exist elements $\iota \in L$ such that $\bar{\iota} = -\iota$. Choose such an element; then multiplying all values of a form by ι converts a hermitian form to a skew-hermitian form and conversely. Thus it suffices to consider hermitian forms defined on vector spaces V over L. First we show that there exists $x \in V$ such that $\langle x, x\rangle \neq 0$ (this fails, of course, for skew-symmetric forms). For suppose not: then, for any $x, y \in L$,

$$\langle x, y\rangle + \overline{\langle x, y\rangle} = \langle x, y\rangle + \langle y, x\rangle = \langle x + y, x + y\rangle - \langle x, x\rangle - \langle y, y\rangle = 0.$$

In the case of symmetric bilinear forms it follows, since the characteristic is not 2, that $\langle , \rangle$ is identically zero, contradicting non-singularity. Otherwise,

$$\begin{aligned} 0 &= \langle x, y\rangle + \overline{\langle x, y\rangle} - i(\langle ix, y\rangle + \overline{\langle ix, y\rangle}) \\ &= \langle x, y\rangle + \overline{\langle x, y\rangle} + \langle x, y\rangle - \overline{\langle x, y\rangle} = 2\langle x, y\rangle, \end{aligned}$$

and we conclude as before.

Choose $e_1 \in V$ with $\langle e_1, e_1\rangle \neq 0$: then the subspace Le_1 is non-singular, so V is the orthogonal direct sum $Le_1 \oplus (Le_1)^\perp$. Since $(Le_1)^\perp$ is non-singular, we can now apply the argument to choose an element e_2 with $\langle e_2, e_2\rangle \neq 0$. Continuing inductively, we obtain an orthogonal basis $\{e_i\}$ for V.

The first important general theorem is

Theorem 10.5.1 *Any isometry between non-singular subspaces W_1, W_2 of V extends to an isometry of V on itself.*

Proof First consider the case $\dim W_1 = 1$, so the hypothesis can be reformulated: we have two vectors in V with $\langle e_1, e_1\rangle = \langle e_2, e_2\rangle = a$, say. We consider two cases.

If the subspace $W = Le_1 + Le_2$ is non-singular, define $\phi(e_1) = e_2$, $\phi(e_2) = \frac{\langle e_1, e_2\rangle}{\langle e_2, e_1\rangle} e_1$. It is immediate that this defines an isometry of W on itself: we extend to an isometry of V which is the identity on $W^{\perp}$.

Otherwise we seek an e_3 with $\langle e_3, e_3\rangle = a$ and such that $Le_1 + Le_3$ and $Le_2 + Le_3$ are both non-singular: we can then use what we have just proved. Since W is singular, its radical $W \cap W^{\perp}$ contains a non-zero vector $pe_1 + qe_2$: now $q \neq 0$, so $f_1 := xe_1 + e_2 \in W$, where $x := p/q$.

Since $f_1 \in (Le_1)^{\perp}$, which is non-singular, there exists $f_2' \in (Le_1)^{\perp}$ such that $\langle f_1, f_2'\rangle = 1$. Choose λ such that $\langle f_2', f_2'\rangle = \lambda + \overline{\lambda}$ (e.g. $\lambda = \frac{1}{2}\langle f_2', f_2'\rangle$); then $f_2 = f_2' - \lambda f_1$ satisfies $\langle f_1, f_2\rangle = 0$, $\langle f_2, f_2\rangle = 1$.

Now take $e_3 = f_1 + \frac{1}{2} a f_2$. Since $a = \overline{a}$ we have $\langle e_3, e_3\rangle = a$; moreover $\langle e_1, e_3\rangle = 0$ and $\langle e_2, e_3\rangle = \langle f_1, e_3\rangle = \frac{1}{2}a$. Thus $Le_1 + Le_3$ is non-singular, and so is $Le_2 + Le_3$ provided $a^2 \neq a^2/4$, i.e. except in characteristic 3. We leave it to the reader to adapt the argument to cover this case.

We return to the general case of the theorem: we prove the result by induction on $\dim W_1$. Choose an orthogonal basis $\{e_i\}$ for W_1: let this correspond by the isometry to a basis $\{f_i\}$ for W_2. By what we have already shown, there is an isometry α_1 of V such that $\alpha_1(e_1) = f_1$. Then $V_1 := (Lf_1)^{\perp}$ contains the subspace W_1' with basis $\{\alpha_1(e_i) \mid i > 1\}$, and the map $W_1' \to W_2$ taking this basis to $\{f_i \mid i > 1\}$ is an isometry. By the induction hypothesis, this extends to a self-isometry of V_1, and this in turn extends to a self-isometry α_2 of V with $\alpha_2(f_1) = f_1$. Then $\alpha_2 \circ \alpha_1$ gives an isometry as required. □

Corollary 10.5.2 *Given non-singular hermitian forms on spaces V_1, V_2, V_3 and an isometry from $V_1 \oplus V_3$ to $V_2 \oplus V_3$, there is an isometry from V_1 to V_2.*

We now use the set of isometry classes of forms to define a group. Define $(V, \langle,\rangle)$ and $(V', \langle,\rangle')$ to be *Witt equivalent* if $(V, \langle,\rangle) \oplus (V', -\langle,\rangle')$ possesses a Lagrangean subspace.

Proposition 10.5.3 *Witt equivalence classes of forms, with (orthogonal) direct sum, form a group $W(L, \bar{.})$. Any form V is isometric to an orthogonal sum $H \oplus V'$ such that H is hyperbolic and $\{0\}$ is the only sublagrangean subspace of V'. Two forms $H_1 \oplus V_1'$ and $H_2 \oplus V_2'$ are Witt equivalent if and only if V_1' and V_2' are isometric.*

Proof First we show that if V admits a Lagrangean subspace W, it is isomorphic to the hyperbolic form $H(W)$. Since V is non-singular, the composite $V \xrightarrow{A} V^\vee \to W^\vee$ is surjective, where $A(f)(e) := \langle e, f\rangle$. Since W is Lagrangean, this map has kernel W. Thus if W has a basis $\{e_i\}$, V has a basis $\{e_i, f'_i\}$ such that $\langle e_i, f'_j\rangle$ is equal to 1 if $i = j$, 0 if not. Now define $x_{i,j} \in L$ by $x_{i,i} = \frac{1}{2}\langle f'_i, f'_i\rangle$, $x_{i,j} = \langle f'_i, f'_j\rangle$ for $i < j$ and $=0$ for $i > j$: then, for all i, j, $x_{i,j} + \overline{x_{j,i}} = \langle f'_i, f'_j\rangle$. Hence if we set $f_i := f'_i - \sum x_{i,j}e_j$ we have a basis $\{e_i, f_i\}$ such that $\langle e_i, f_j\rangle$ is equal to 1 if $i = j$, 0 if not and $\langle e_i, e_j\rangle = \langle f_i, f_j\rangle = 0$. Thus V is hyperbolic.

If W is a sublagrangean subspace of V, we argue as in the preceding paragraph to choose elements f'_i and f_i, but no longer have a basis of V: instead, we have a hyperbolic subspace H. As this is non-singular, V is an orthogonal direct sum $H \oplus V'$. If V' has a non-trivial sublagrangean subspace W', then $W \oplus W'$ is also sublagrangean. Thus if we choose W sublagrangean of maximal dimension, V' has no non-trivial sublagrangean subspace.

Since any hyperbolic space has a Lagrangean, so is Witt equivalent to zero, it remains only to show that if V_1 and V_2 are Witt equivalent, and neither has a non-trivial sublagrangean, they are isometric. Thus suppose $(V, \langle,\rangle) \oplus (V', -\langle,\rangle')$ possesses a Lagrangean subspace Λ. Since neither V nor V' has a non-trivial sublagrangean, $\Lambda \cap V = \Lambda \cap V' = \{0\}$. It follows that $\dim \Lambda \leq \dim V', \dim V$: but since $2\dim\Lambda = \dim V + \dim V'$, all three dimensions are equal. Hence the projections π, π' from $\Lambda \subset V \oplus V'$ to the summands are both isomorphisms. We claim that $(\pi)^{-1} \circ \pi' : V \to V'$ is an isometry. For if $(e, e') \in \Lambda$, then $0 = \langle (e, e'), (e, e')\rangle = \langle e, e\rangle - \langle e', e'\rangle'$.

It follows that Witt equivalence is an equivalence relation. To obtain a group, the only non-trivial item to verify is existence of inverses. But the diagonal subspace $\Delta(V) \subset (V, \langle,\rangle) \oplus (V, -\langle,\rangle)$ is a Lagrangean subspace, so $(V, -\langle,\rangle)$ is inverse to $(V, \langle,\rangle)$. □

The arguments above, applied to non-singular skew-symmetric forms, show that they are all hyperbolic.

We now define some invariants of hermitian forms. The first invariant is the rank of the vector space V over L. Next we define the discriminant. If V has basis $\{e_i\}$, and the form is ϕ, then we define its matrix to be $X := (\phi(e_i, e_j))$, and take the determinant. Use an asterisk to denote conjugate transpose. Then $X^* = X$, so $\overline{\det(X)} = \det(X)$. If $\{f_i\}$ is another base, with $f_i = \sum a_{ij}e_j$, and A denotes the matrix

(a_{ij}), then the matrix of ϕ with respect to the new basis is A^*XA, and $\det(A^*XA) = \overline{\det(A)}\det(X)\det(A)$.

Write L_0 for the fixed field of the involution 'bar' on L, and write $L^\times$ for the multiplicative group of non-zero elements of L (and similarly for L_0). Then $x \mapsto x\bar{x}$ defines a homomorphism $N : L^\times \to L_0^\times$. We have seen that $\det(X) \in L_0^\times$, and that a change of basis has the effect of multiplying this by an element of the image of N. Thus its class in the cokernel $L_0^\times/N(L^\times)$ is an invariant of the isomorphism class of the form. This invariant is called the *discriminant* of the form. Since if $x \in L_0$, $N(x) = x^2$, the quotient group $L_0^\times/N(L^\times)$ has exponent 2. Since the discriminant of a hyperbolic plane is -1, it is necessary to modify this definition a little to obtain an invariant defined on the Witt group.

For Hermitian forms in the usual sense, where $L = \mathbb{C}$, we have the signature. We can choose the basis so that $\phi(e_i, e_j)$ vanishes for $i \neq j$: then if $\phi(e_i, e_i)$ is positive for n^+ values of i and negative for n^- values, the signature is $\sigma := n^+ - n^-$. We recall briefly the proof of its invariance: there is a subspace of dimension n^+ on which the form is positive definite, and one of dimension n^- on which the form is negative definite, so if a different basis produced (say) $n'^+ > n^+$ positive terms, the subspaces of dimensions n'^+ and n^- would have to intersect non-trivially, giving a contradiction.

For symmetric forms there is an additional, more technical invariant, defined using the Clifford algebra.

The full classification of hermitian forms depends on the field L.

If L is finite, a hermitian form is determined up to isomorphism by its rank; a symmetric form by its rank and discriminant.

Next suppose L is a local field, i.e. a field complete with respect to a discrete valuation: e.g. a finite extension of the field of p-adic numbers, or of the field of formal power series in one variable over a finite field (a detailed study of these can be found in [166]). Then a hermitian form over L is determined up to isomorphism by its rank and discriminant (the latter taking values in a group of order 2). For a symmetric form we need an additional 2-valued invariant, known as the Hasse-Witt invariant.

If L is $\mathbb{R}$ or $\mathbb{C}$, then a symmetric form over $\mathbb{R}$ or a hermitian form over $\mathbb{C}$ is determined by its rank and signature; a symmetric form over $\mathbb{C}$ by its rank.

These lead up to the case when L is a global field, i.e. a finite extension of $\mathbb{Q}$ or of the field $k(t)$ of rational functions over a finite field.

A hermitian form is determined up to isomorphism by its rank, its discriminant and, for each embedding of L_0 in $\mathbb{R}$ such that $L \otimes_{L_0} \mathbb{R} \cong \mathbb{C}$, a signature. For a quadratic form we require, as well as rank, discriminant, and signature at each real embedding, a Hasse-Witt invariant for each discrete valuation.

We can now apply these results to isometric triples. First, to clear up a possible misconception, consider isometric triples over $K = \mathbb{C}$. Then every irreducible polynomial has degree 1, the only self-dual polynomials are thus $t \pm 1$; for each, we have $L = L_0 = \mathbb{C}$; and the symmetric and skew-symmetric forms contribute no invariants.

A complete set of invariants for isometric triples over $K = \mathbb{Q}$ consists of: a list of prime polynomials $p \in \mathbb{Q}[t]$ (each p determines a field $L_p := \mathbb{Q}[t]/\langle p \rangle$); for each p, a partition π_p (so that the generalised eigenspace $H_{(p)}$ is the sum of $\pi_p(i)$ copies of $\mathbb{Q}[t]/\langle p^i \rangle$ for all i; for each p with $p = p^{\perp}$ and of even degree and each i with $\pi_p(i) \neq 0$, a discriminant and a set of signatures; and for $p = t \pm 1$ and each odd i with $\pi_p(i) \neq 0$, a discriminant, a signature, and a set of Hasse–Arf invariants.

Now consider the problem of calculating these invariants for the isometric triple given by the Seifert form on the homology of the Milnor fibre in terms of the invariants of the singularity. No direct answer is known to the full problem. We observe, however, that certain simplifications take place. By the monodromy Theorem 10.2.1, all eigenvalues of h are roots of unity, so the only prime polynomials p that occur are cyclotomic (but $p = p^{\perp}$ for each cyclotomic polynomial). All blocks in the Jordan decomposition have size at most 2, so the partitions π_p have $\pi_p(i) = 0$ for $i > 2$ (and in 'most' cases also for $i = 2$). We excluded the eigenvalue 1 from the discussion, in order to construct an isometric triple, but in this case by Proposition 10.2.2 the generalised eigenspace on $H_1(F)$ is equal to the eigenspace R; intersection numbers vanish on it; and the Seifert form S is determined by the intersection numbers of the branches and, by Lemma 10.2.3, is a negative definite quadratic form.

The notion of Witt equivalence extends to give a stronger equivalence relation than isomorphism for isometric triples, which is characterised by weaker, but more flexible invariants. A subspace Λ of H is a *Lagrangean subspace* for the isometric triple $(H, \langle , \rangle, h)$ if $h(\Lambda) = \Lambda$ and $\Lambda = \Lambda^{\perp}$. We define a triple to be equivalent to 0 if it possesses a Lagrangean subspace. A subspace Λ such that $\Lambda \subseteq \Lambda^{\perp}$ and $\dim_K H = 2 \dim_K \Lambda$ is Lagrangean.

If Λ' is a subspace of H with $h(\Lambda') = \Lambda'$ and $\Lambda' \subset \Lambda'^{\perp}$, we can define an inner product on $\Lambda'^{\perp}/\Lambda'$ by $([x], [y]) := \langle x, y \rangle$. Then the image of $\Lambda'^{\perp}$

in $(H, \langle,\rangle, h) \oplus (\Lambda"^{\perp}/\Lambda', -(,), h_*)$ by the diagonal map is a Lagrangean subspace.

We say that triples $(H, \langle,\rangle, h)$ and $(H', \langle,\rangle', h')$ are *Witt equivalent* if $(H, \langle,\rangle, h) \oplus (H', -\langle,\rangle', h')$ possesses a Lagrangean subspace. It is not difficult to show that this is an equivalence relation.

Lemma 10.5.4 *For the submodules $H_{p,i}$ of Lemma 10.4.4, we have Witt equivalences $H_{p,2k} \sim 0$, $H_{p,2k+1} \sim V_{p,2k+1}$.*

Proof If $H = H_{p,2k}$, the subspace $\Lambda := K(k)$ is Lagrangean, since it is self-orthogonal and of half the dimension.

If $H = H_{p,2k+1}$, we take $\Lambda := K(k+1)$: then $\Lambda^{\perp} = K(k)$. We have an isomorphism of

$$V_{p,2k+1} = H_{p,2k+1}/pH_{p,2k+1}$$

on $\Lambda^{\perp}/\Lambda = p^k H_{p,2k+1}/p^{k+1}H$ given by multiplication by $s(t)^k$, and $\langle u, s(t)^{2k}v\rangle = \langle s(t)^k u, s(t)^k v\rangle$ since $s(t)$ is self-adjoint; thus the form we have obtained on $V_{p,2k+1}$ is the same as that referred to in Lemma 10.4.4. □

Thus any isometric triple over a field K is Witt equivalent to a semisimple one. Witt equivalence for semisimple forms reduces to the case of hermitian forms over a field L, which we have already discussed.

We now give an additivity theorem, which will allow us to utilise the decomposition already used for the monodromy. We begin to formulate the hypotheses.

Suppose we have a surface F decomposed as $F_1 \cup F_2$, with $\partial_m F := F_1 \cap F_2 = \partial F_1 \cap \partial F_2$ disjoint from ∂F. Write $\partial_i F$ for the union of those components of ∂F_i which do not belong to $\partial_m F$ $(i = 1, 2)$.

Let $h : F \to F$ be a diffeomorphism which respects this decomposition. Write $H^{\neq 1}(-)$ for the sum of the generalised eigenspaces of h on $H_1(-)$ corresponding to all eigenvalues $\neq 1$ (the coefficients may be taken as $\mathbb{C}$ or as $\mathbb{Q}$). We consider the induced isometric structure on $H^{\neq 1}(F)$; similarly for F_1 and F_2. Note that by Lemma 10.1.2, we can define a Seifert form S uniquely on $H^{\neq 1}(F)$ to satisfy the properties of Lemma 10.1.1: indeed, given x there is a unique $z \in H^{\neq 1}$ with $z - h(z) = x$ and we set $S(x, y) := \langle z, y\rangle$.

Since the radical of the intersection form on $H_1(F)$ is equal to the image of $H_1(\partial F)$, the same follows for $H^{\neq 1}$. Choose an h-invariant additive complement $J^{\neq 1}(F)$ in $H^{\neq 1}(F)$ to the image of $H^{\neq 1}(\partial F)$: then $J^{\neq 1}(F)$ maps isomorphically onto the image of $H^{\neq 1}(F)$ in $H^{\neq 1}(F, \partial F)$, and the restriction to $J^{\neq 1}(F)$ of the intersection form is non-degenerate. We thus have an isometric structure on $J^{\neq 1}F$; similarly for F_1 and F_2.

Lemma 10.5.5 *In the above situation, there is a Witt equivalence of* $[J^{\neq 1}(F)]$ *to* $[J^{\neq 1}(F_1)] \oplus [J^{\neq 1}(F_2)]$.

Proof The image of $H^{\neq 1}(F_i) \to H^{\neq 1}(F)$ is the sum of the images of $J^{\neq 1}(F_i)$ and $H^{\neq 1}(\partial F_i) = H^{\neq 1}(\partial_i F) \oplus H^{\neq 1}(\partial_m F)$; since $\partial_i F \subseteq \partial F$, $H^{\neq 1}(\partial_i F)$ maps to zero, so we have the sum of the images of $J^{\neq 1}(F_i)$ and $H^{\neq 1}(\partial_m F)$. The same follows for the images of $H^{\neq 1}(F_i)$ in $H^{\neq 1}(F, \partial F)$.

Since the intersection form is non-degenerate on $J^{\neq 1}(F_i)$, the map from $J^{\neq 1}(F_i)$ to $H^{\neq 1}(F, \partial F)$ is injective. Moreover, since $J^{\neq 1}(F_1)$ and $J^{\neq 1}(F_2)$ are mutually orthogonal (a cycle in the interior of F_1 is disjoint from one in F_2) their images form a direct sum in $H^{\neq 1}(F, \partial F)$, so we can choose $J^{\neq 1}(F)$ to contain this sum. Write X for the intersection of the image of $J^{\neq 1}(F)$ in $H^{\neq 1}(F, \partial F)$ with the orthogonal complements of $J^{\neq 1}(F_1)$ and $J^{\neq 1}(F_2)$. Then X is h-invariant, and we can identify

$$J^{\neq 1}(F) \cong J^{\neq 1}(F_1) \oplus J^{\neq 1}(F_2) \oplus X,$$

an orthogonal direct sum. The desired additivity property is equivalent to the assertion that the induced isometric structure on X is Witt equivalent to zero.

Now the image of $H^{\neq 1}(\partial_m F)$ in $H^{\neq 1}(F_i)$ is contained in the radical. Hence its image Λ in $H^{\neq 1}(F, \partial F)$ is orthogonal to $J^{\neq 1}(F_i)$, and hence is contained in X. The image of $H^{\neq 1}(F_1) \oplus H^{\neq 1}(F_2)$ in $H^{\neq 1}(F, \partial F)$ is equal to $J^{\neq 1}(F_1) \oplus J^{\neq 1}(F_2) \oplus \Lambda$. Its orthogonal complement consists of the cycles mapping to 0 in each $H^{\neq 1}(F_i, \partial F_i) \cong H^{\neq 1}(F, F_{3-i} \cup \partial_i F)$. These lie in the image of $H^{\neq 1}(F_2 \cup \partial_1 F) = H^{\neq 1}(F_2) \oplus H^{\neq 1}(\partial_1 F)$, and hence in the image of $H^{\neq 1}(F_1) \oplus H^{\neq 1}(F_2)$. But the form on $J^{\neq 1}(F_1) \oplus J^{\neq 1}(F_2)$ is non-degenerate, so the orthogonal complement is contained in Λ.

Thus Λ coincides with its own orthogonal complement in X, so is Lagrangean. This completes the proof. □

Corollary 10.5.6 *The isometric structure defined by the Seifert form on the Milnor fibre, away from the eigenvalue* 1, *is Witt equivalent to the sum of those on the pieces* F_i *of the decomposition of Theorem 9.3.6.*

This follows by repeated application of the lemma.

10.6 Signatures

It follows from the results of the preceding section that the Witt equivalence class of the isometric structure defined by a Seifert form over $\mathbb{R}$ is determined by a collection of signatures of skew-hermitian forms over $\mathbb{R}$. In this section we give several alternative versions of this set of signatures. We then proceed to calculate them explicitly, using a couple of basic results whose proofs are discussed in the following section.

We can define signatures of a Seifert form directly as follows. Consider the (real) matrix S of the Seifert form as defining a sequilinear form on the complex vector space $H_{\mathbb{C}} := H \otimes_{\mathbb{Z}} \mathbb{C}$. Here, by *sesquilinear*, we mean that the pairing $(x, y) \mapsto S(x, y)$ is $\mathbb{C}$-linear in x and antilinear in y; thus $S(\lambda x, \mu y) = \lambda\overline{\mu}S(x, y)$. The identities become

$$\langle x, y\rangle = S(x, y) - \overline{S(y, x)}, \qquad S(x, y) = \overline{S(h(y), x)};$$

in matrix terms they are unaltered, but now we replace the transpose S^t by the conjugate transpose (adjoint) S^* (as S has real entries, this is indeed the same).

For any ξ with $|\xi| = 1$, $\xi S - \overline{\xi}S^*$ defines a skew-Hermitian form on $H_{\mathbb{C}}$. Any skew-Hermitian form may be diagonalised, and will have pure imaginary elements on the diagonal. If r^+ of these have positive imaginary part and r^- have negative imaginary part, the signature σ_ξ is defined to be $i(r^+ - r^-) \in i\mathbb{Z}$. That it is well defined follows since skew-Hermitian forms are obtained from Hermitian forms by multiplying by i, and the result holds in the Hermitian case. Observe that $\sigma(-\xi) = -\sigma(\xi)$.

As ξ moves round the unit circle, the signature σ_ξ will change only when ξ passes through a value for which the form $\xi S - \overline{\xi}S^*$ is singular. From (10.1) we have $\xi S - \overline{\xi}S^* = (\xi I - \overline{\xi}X)S$; now since S is nonsingular and $\overline{\xi} = \xi^{-1}$, the condition is equivalent to $X - \xi^2 I$ being singular, i.e. to ξ^2 being an eigenvalue of the monodromy matrix X.

Lemma 10.6.1 *The signatures σ_ξ with ξ^2 not an eigenvalue of X are invariant under Witt equivalence.*

Proof It is sufficient to show that if $(H, \langle,\rangle, h)$ is an isometric structure over $\mathbb{Q}$ (such that 1 is not an eigenvalue of h) possessing a Lagrangean subspace Λ, then the signatures σ_ξ vanish.

Since 1 is not an eigenvalue, for any $x \in H$ there exists a unique $x' \in H$ such that $x = x' - h(x')$: the Seifert form is then given by

$S(x, y) = \langle x', y \rangle$. As Λ is h-invariant, if $x \in \Lambda$ then also $x' \in \Lambda$; so if $x, y \in \Lambda$, $S(x, y) = \langle x', y \rangle = 0$.

We now consider the skew-Hermitian forms on $H_{\mathbb{C}}$ defined, for any ξ with $|\xi| = 1$, by $\xi S - \overline{\xi} S^*$; provided ξ^2 is not an eigenvalue of h, this form is non-singular. Since Λ is Lagrangean, for any $x, y \in \Lambda_{\mathbb{C}} := \Lambda \otimes_{\mathbb{Z}} \mathbb{C}$, the form vanishes on (x, y). Since also $\dim H = 2 \dim \Lambda$, $\Lambda_{\mathbb{C}}$ is a Lagrangean subspace for the skew-Hermitian form. Hence its signature vanishes. □

If ξ^2 is an eigenvalue of X, the assertion fails. See Exercise 10.9.11.

To compare this definition with the general theory of Section 10.4, take the base field K there to be the real field $\mathbb{R}$. Irreducible polynomials have degree 1 or 2; the self-dual ones have the form $p(t) = t^2 - 2ct + 1$ with $|c| < 1$. Write $c = \cos 2\phi$: we may suppose $0 < 2\phi < \pi$. Thus over $\mathbb{C}$ we have $t^2 - 2ct + 1 = (t - e^{2i\phi})(t - e^{2i\phi})$: this case corresponds to the complex eigenvalues $e^{\pm 2i\phi}$. Choosing the eigenvalue $e^{2i\phi}$ gives an identification of the field L_p with $\mathbb{C}$. We have $s(t) = t - 2c + t^{-1}$.

The $\mathbb{R}$-linear map $T : \mathbb{C} \to \mathbb{R}$ has $T(1) = 0$, $T(e^{2i\phi}) = 1$, so for any complex number $a + ib$ we have $T(a + ib) = \frac{b}{\sin 2\phi}$. It will affect no signatures if we multiply T by the positive constant $\sin 2\phi$, thus replacing T by the imaginary part $\Im$. It seems more natural to use the real part: multiplying the form $u.v$ by $-i$ in Lemma 10.4.4 gives a unique $(-1)^{i-1}\epsilon$-Hermitian inner product $(,)_h$ on $V_{p,i}$ with $\langle u, s(t)^{i-1} v \rangle = \Re(u, v)_h$.

Now restrict to the semisimple case: write V for $V_{p,1}$, so that $pV = 0$. Then $V \otimes_{\mathbb{R}} \mathbb{C}$ splits into eigenspaces $V_{\pm}$ corresponding to the eigenvalues $e^{\pm 2i\phi}$ for t; there is an $\mathbb{R}$-isomorphism $V \to V_+$ taking the action of t to multiplication by $e^{2i\phi}$. We extended $\langle , \rangle$ by complexification to a skew-Hermitian form on $V \otimes_{\mathbb{R}} \mathbb{C}$. The restriction of this form to V_+ coincides with $(,)_h$, since each of these is a skew-Hermitian form such that taking the real part coincides with $\langle , \rangle$.

By Lemma 10.1.2, we have $S(u, v) = \langle (1 - t)^{-1} u, v \rangle$ on V, and thus $S(u, v) = ((1 - e^{2i\phi})^{-1} u, v)_h$ on V_+. Hence, if $\xi = e^{i\theta}$,

$$(\xi S - \overline{\xi} S^*)(u, v) = \xi((1 - e^{2i\phi})^{-1} u, v)_h - \overline{\xi((1 - e^{2i\phi})^{-1} v, u)_h}$$

$$= \frac{\xi}{1 - e^{2i\phi}}(u, v)_h - \frac{\xi^{-1}}{1 - e^{-2i\phi}}\overline{(v, u)_h} = C(u, v)_h,$$

where

$$C = \frac{\xi}{1 - e^{2i\phi}} + \frac{\xi^{-1}}{1 - e^{-2i\phi}} = \frac{-e^{i\theta} e^{i\phi} + e^{-i\theta} e^{i\phi}}{e^{i\phi} - e^{-\phi}} = \frac{\sin(\phi - \theta)}{\sin \phi}.$$

So $\xi S - \overline{\xi} S^*$ is obtained from $(,)_h$ by multiplying by $\frac{\sin(\theta - \phi)}{\sin \phi}$. Since

$\sin\phi > 0$, the signature of $\xi S - \overline{\xi} S^*$ on V_λ is obtained from the signature, τ_λ say, of $(\,,)_h$ by multiplying by $+1$ if $0 < \theta - \phi < \pi$ and by -1 if $-\pi < \theta - \phi < 0$.

The eigenvalue $\lambda = e^{2i\phi}$ of the monodromy determines the points $\lambda^{\frac{1}{2}} = \pm e^{i\phi}$ on the unit circle, which partition it in to two semicircles. We attach the numbers $\pm\tau_\lambda$ to these semicircles as above (and 0 to the boundary points), and sum over all eigenvalues λ to obtain the function σ. Thus as ξ moves round the circle, σ_ξ jumps (by $\pm 2\tau_\lambda$) every time ξ^2 is an eigenvalue.

The above analysis covers the summands $H_{p,1}$. Provided ξ^2 is not an eigenvalue of X, each summand $H_{p,2k}$ contributes 0 to σ_ξ; each summand $H_{p,2k+1}$ contributes a signature. This follows from Lemma 10.5.4.

We now proceed to the results allowing us to calculate these signatures for the monodromy action on the Milnor fibre. In this section we present the calculations as deductions from certain key results: in the next we will sketch proofs of these.

For the rest of the chapter, we make the convention that 'manifold' means compact oriented manifold, 'G-manifold' is a manifold with a free orientation-preserving action of the group G; a G-n- manifold is a G-manifold of dimension n.

We begin by recalling that by Corollary 10.5.6, the Witt class of the isometric structure on H^* is equal to the sum of those corresponding to the pieces F_i of the Thurston decomposition. A key property of this decomposition is that $h|F_i$ is isotopic to a diffeomorphism of finite order. We may thus suppose that h generates a (finite, cyclic) group G of diffeomorphisms of F_i, i.e. that F_i is a compact G-manifold. We now again reformulate the definition of signatures.

Given an action of a finite group G on a $\mathbb{R}$-vector space V preserving a quadratic form q, we can equivalently

(a) find (by averaging over G) an equivariant orthogonal splitting $V = V_+ \oplus V_0 \oplus V_-$ where q is positive definite on V_+, zero on V_0 and negative definite on V_-, and take the class $[V_+] - [V_-]$ in the representation ring of G. Its character is the function $\sigma(V, q)$ on G whose value at $g \in G$ is the difference of traces $\sigma^g(V, q) := \mathrm{Tr}\,(V_+, g) - \mathrm{Tr}\,(V_-, g)$.

(b) (assuming G abelian) complexify V, and extend q to $V_{\mathbb{C}}$ as a Hermitian form. Split $V_{\mathbb{C}} = \bigoplus V_\chi$ into isotypic pieces corresponding to the characters χ of G, so that for $g \in G, \chi \in \widehat{G}, x \in V_\chi$ we have $gx = \chi(g)x$. Observe that if $g \in G$, $x \in V_\chi$ and $x' \in V_{\chi'}$ we have

$$\langle x, x'\rangle = \langle gx, gx'\rangle = \langle \chi(g)x, \chi'(g)x'\rangle = \overline{\chi(g)}\chi'(g)\langle x, x'\rangle,$$

and hence vanishes unless, for all $g \in G$, $1 = \overline{\chi(g)}\chi'(g) = \chi(g)^{-1}\chi'(g)$, i.e. $\chi = \chi'$. Thus the splitting is an orthogonal splitting, the summands are non-degenerate, and we may define $\sigma_\chi(V, q)$ to be the signature of the induced Hermitian form on V_χ.

The two versions of $\sigma(V, q)$ are related by

$$[V_+] - [V_-] = \sum_\chi \sigma_\chi(V, q)[\mathbb{C}_\chi],$$

where $\mathbb{C}_\chi$ is the irreducible module with character χ, and hence

$$\sigma^g(V, q) = \sum_\chi \sigma_\chi(V, q)\chi(g). \tag{10.11}$$

It follows from character orthogonality relations that this is equivalent to

$$\overline{\psi}(g)\sigma^g(V, q) = \sum_\chi \sum_g \overline{\psi}(g)\chi(g)\sigma_\chi(V, q) = |G|\sigma_\psi(V, q). \tag{10.12}$$

If the quadratic form q is replaced by a skew-symmetric bilinear form ϕ on V we can proceed as in (b): first complexify and extend ϕ to a skew-Hermitian form $\phi_\mathbb{C}$ on $V_\mathbb{C}$; multiply by i to obtain a Hermitian form $\langle , \rangle = i\phi_\mathbb{C}$. We now obtain signatures $\sigma_\chi(V, \phi)$ as above and define $\sigma^g(V, \phi) := -i \sum_\chi \chi(g)\sigma_\chi(V, \phi)$.

We are interested in the case where G acts freely on a compact oriented surface M and hence on the homology group $H_1(M; \mathbb{R})$, respecting the intersection self-pairing $\cap$ to $\mathbb{R}$. Write $\sigma(M)$ for $\sigma(H_1(M), \frown)$.

The first result asserts that these invariants are determined by the G-action on the boundary.

Proposition 10.6.2 *There is an invariant $\rho(C) : G \to \mathbb{Q}$ defined for G-1-manifolds C, such that if M is a G-2-manifold, $\sigma^g(M) = \rho^g(\partial M)$. The invariant ρ is additive for disjoint unions.*

We can use (10.11) and (10.12) (with ρ replacing σ) to pass freely between alternative forms ρ^g and ρ_χ. The invariant ρ is calculated as follows.

Proposition 10.6.3 *Let G be abelian, C' an oriented G-1-manifold with a component C such that $C' = G.C$. Let T generate the stabiliser of C and act on it by rotation through $2\pi/r$; let χ be a character of G, and let $\chi(T) = e^{2\pi i s/r}$ with $0 \leq s < r$. Then*

$$\rho_\chi(C') = \begin{cases} 1 - 2\frac{s}{r} & \text{if } s \neq 0, \\ 0 & \text{if } s = 0 \end{cases}$$

If $g \in G$ does not stabilise C, then $\rho^g(C') = 0$;
if g rotates C by an angle 2θ, we have $\rho^g(C') = \cot\theta$.

If we do not normalise s by $0 \le s < r$, the term $\frac{s}{r}$ is replaced by $\{\frac{s}{r}\}$, where the braces denote the fractional part: $\{\frac{s}{r}\} := \frac{s}{r} - \lfloor\frac{s}{r}\rfloor$.

We now combine this result with Proposition 10.6.2. Let G be a cyclic group of order m with preferred generator T, and F be a G-2-manifold. Choose one component C_r from each orbit of the action of G on $\pi_0(\partial F)$; let the stabiliser of C_r have order n_r, and let T^{m/n_r} act on C_r by rotation through $2\pi b_r/n_r$. Choose an inverse a_r to b_r mod n_r such that a_r is prime to m. Write χ_s for the character of G with $\chi_s(T) = e^{2\pi i s/m}$.

To apply Proposition 10.6.3 to the component C_r we use the generator T^{a_r} of G, so that its $(m/n_r)^{\text{th}}$ power $U_r := T^{a_r m/n_r}$ acts on C_r by rotation through $2\pi/n_r$, thus takes the place of the T in Proposition 10.6.3. We have $\chi_s(U_r) = e^{2\pi i a_r s/n_r}$. The contribution of $G.C_r$ to ρ vanishes if $a_r s$ is divisible by n_r, or equivalently, if s is divisible by n_r. Otherwise the contribution is $1 - 2\{\frac{sa_r}{n_r}\}$. Hence

Lemma 10.6.4 *With the above notation we have*

$$\sigma_{\chi_s}(F) = \sum_{n_r \nmid s} \left(1 - 2\left\{\frac{sa_r}{n_r}\right\}\right),$$

$$\sigma^{T^k}(F) = \sum_{m|kn_r} \cot(kb_r/m).$$

We apply this to the Seifert form on the Milnor fibre. First condider the pieces F_i of the decomposition of Theorem 9.3.6.

Theorem 10.6.5 *The G-signature defined by the action of the monodromy on F_i, corresponding to the character $\chi(h) = e^{2\pi i s/M_i}$, is*

$$\sum\left\{\left(1 - 2\left\{\frac{sM_j}{M_i}\right\}\right) \,\middle|\, V_iV_j \in \mathcal{E}(\Gamma_R^+(C)),\ M_i \nmid sM_j\right\}.$$

Proof Recall that the valence of the vertex V_i of $\Gamma_R^+(C)$ is denoted v_i; thus there are v_i edges $E = V_iV_j$. The numerical parameter associated to V_i is M_i and the highest common factor of M_i and M_j is M_E. Write $M_i = a_iM_E$, $M_j = a_jM_E$ and choose b_i, b_j with $a_ib_i + a_jb_j = 1$.

The piece F_i of the decomposition corresponding to V_i is an M_i fold covering of a v_i times punctured 2-sphere E_i^0; the monodromy h acts as a covering translation. At the part of the boundary corresponding to E we have a torus $S^1 \times S^1$ with coordinates (w, z), say, where the projection to the boundary component of E_i^0 is given by $(w,z)\mapsto z$; the Milnor fibre meets this where $w^{M_i}z^{M_j}=1$; and the monodromy acts by $T(w,z)=(e^{2\pi i/M_i}w,z)$.

Thus we have a union C_E of M_E boundary components, permuted transitively by the monodromy. Choose the component through $(1,1)$ and parametrise it by (t^{a_j},t^{-a_i}). The M_E^{th} power of the monodromy acts on this component by $T^{M_E}(t)=e^{2\pi i b_j/a_i}t$; for this carries (t^{a_j},t^{-a_i}) to $(e^{2\pi i a_j b_j/a_i}t^{a_j},e^{-2\pi i b_j}t^{-a_i})=(e^{2\pi i/a_i}t^{a_j},t^{-a_i})$.

We can thus apply Lemma 10.6.4, with the following substitutions:

F	r	C_r	m	n_r	a_r	b_r	s
F_i	j	C_E	M_i	a_i	a_j	b_j	s

This gives $\sigma_{\chi_s}(F_i)=\sum_{a_i\not\,|s}\left(1-2\{\frac{sa_j}{a_i}\}\right)$, which equals

$$\sum_{M_i\not\,|sM_E}\left(1-2\left\{\frac{sM_j}{M_i}\right\}\right),$$

since the condition $M_i\not\,|sM_E$ can be restated as $M_i\not\,|sM_j$. □

We make a consistency check: this should be an integer. Calculating modulo $\mathbb{Z}$, the above is congruent to $\sum_E -2\frac{sM_j}{M_i}=(-2s/M_i)\sum_E M_j$. But by (8.12) we have $\sum_E M_j=a_iM_i$, so our sum is congruent to $-2sa_i\in\mathbb{Z}$.

Now, by Lemma 10.5.5, we just have to sum over all pieces F_i to obtain the G-signature of F.

Proposition 10.6.6 *Let* $\chi(h)=e^{2\pi i\alpha}$. *Then*

$$\sigma_\chi(F_i)=\sum\{1-2\{\alpha M_j\}\mid V_iV_j\in\mathcal{E}(\Gamma_R^+(C)),\ M_i\alpha\in\mathbb{Z},\ M_j\alpha\notin\mathbb{Z}\}.$$

The same formula, extended over all edges of $\Gamma_R^+(C)$, *gives* $\sigma_\chi(F)$.

10.7 Proof of 10.6.2 and 10.6.3

The full proofs of the results we need depend on facts from bordism theory which are beyond the scope of this book. We offer a discussion to explain just which facts are used and how: the general principles show that there must exist a formula. The main part of this section consists of an explicit calculation which establishes the precise formula. For M a G-$2n$-manifold, write $\sigma(M)$ for $\sigma(H_n(M),\frown)$.

Proposition 10.7.1 *If M is a closed G-$2n$-manifold then $\sigma^g(M) = 0$ for all $g \in G$ if n is odd; for $g \neq 1$ if n is even.*

The proof is outside the scope of this book. We outline it here so that the reader can have some idea what is involved and follow up the arguments if he or she wishes.

One first shows that if W is a G-bordism between M and M', then $\sigma^g(M) = \sigma^g(M')$. Thus we have an invariant of the bordism group, and this group may be written as the oriented bordism group $\Omega_{2n}(K(G,1))$ of the classifying space $K(G,1)$. If n is odd, this is a finite group, and it suffices to note that any homomorphism from a finite group to $\mathbb{Z}$ is trivial.

If n is even, $\Omega_{2n}(K(G,1))$ is the sum of Ω_{2n} and a finite group. It is now sufficient to consider the case of Ω_{2n}, and hence the case when M is the disjoint union of $|G|$ copies of a manifold L, which are permuted regularly by the action of G. In this case, the equivariant splitting $H_n(M) = V_+ \oplus V_-$ has each of V_+ and V_- a free module over $\mathbb{C}G$, so the trace of $g \neq 1$ vanishes on each, hence on their difference.

This proof does not simplify essentially for the case when M is a surface – though the calculations do simplify:

$$\Omega_2(K(G,1)) \cong H_2(K(G,1);\mathbb{Z}) \cong G^{ab}.$$

Since $\sigma^g(M)$ vanishes when $\partial M = \emptyset$ has empty boundary, we may hope to express it in terms of ∂M. In fact, Proposition 10.6.2 generalises as follows.

Proposition 10.7.2 *There is an invariant $\rho^g(C) : G \to \mathbb{Q}$ (ρ^1 is undefined if n is even), defined for G-$(2n-1)$-manifolds C, such that if M is a compact G-$2n$-manifold, then $\sigma^g(M) = \rho^g(\partial M)$. The invariant ρ^g is additive for disjoint unions.*

Proof This depends on two key results: the additivity property Corollary 10.5.6 which implies additivity of the G-signature, and the fact that for any closed G-$(2n-1)$-manifold C, there exist an integer N and a G-$2n$-manifold M such that $\partial M \cong NC$ (this denotes a union of N copies of C).

The second of these will be proved below by explicit construction when $n = 1$. In general, the argument is similar to that of Proposition 10.7.1: the bordism group $\Omega_{2n-1}(K(G,1))$ of G-$(2n-1)$-manifolds is finite: choose an integer N that annihilates it.

Now define $\rho(C) := N^{-1}\sigma(M)$. This is well defined since if N', M' is an alternative choice we can form a closed manifold M'' by attaching $N'M$ to NM' (with orientation reversed) along NC; then $\sigma(M'') = N'\sigma(M) - N\sigma(M')$ by Corollary 10.5.6, but vanishes by Proposition 10.7.1. The invariant is additive since if $\partial M \cong NC$ and $\partial M' \cong NC'$ we can choose the disjoint union $M \cup M'$ with $\partial(M \cup M') \cong N(C \cup C')$.

Now for any $G-2n-$manifold M, since $N(\partial M) = \partial NM$ we have, by definition, $\rho(\partial M) := N^{-1}\sigma(NM) = \sigma(M)$. □

Observe that in this technique of calculation we introduce denominators. Thus the method which works for signatures cannot be used for discriminants, which take values in groups of exponent 2.

We now begin our calculation of $\rho(C)$ in the case when C is 1-dimensional. It turns out that a detailed calculation for an explicit surface F will allow us to obtain the general formula.

Suppose G cyclic of order r, generated by T. We have an action on the smooth surface $\hat{F}$ defined in $P^2(\mathbb{C})$ by the equation $x^r + y^r + z^r = 0$. The group G acts by $T(x, y, z) = (e^{2\pi i/r}x, y, z)$. There are r fixed points where the line $x = 0$ meets the curve. Removing neighbourhoods of these will give our surface F. Observe that the actions of G on these neighbourhoods are all equivalent. The surface $\hat{F}$ has genus $\frac{1}{2}(r-1)(r-2)$, so $H_1(\hat{F})$ has rank $r^2 - 3r + 2$ and $H_1(F)$ has rank $r^2 - 2r + 1$. The projection from $(1, 0, 0)$ onto the line $x = 0$ takes the G-orbits to points on the line; the quotient of F by G is thus identified with the $P^1(\mathbb{C})$ given by $x = 0$ with neighbourhoods of the r points where $y^r + z^r = 0$ deleted.

We also note the action of another cyclic group of order r by $U(x, y, z) = (x, e^{2\pi i/r}y, z)$: we have $TU = UT$. This passes to an action on the quotient $P^1(\mathbb{C})$ and permutes the r fixed points of T.

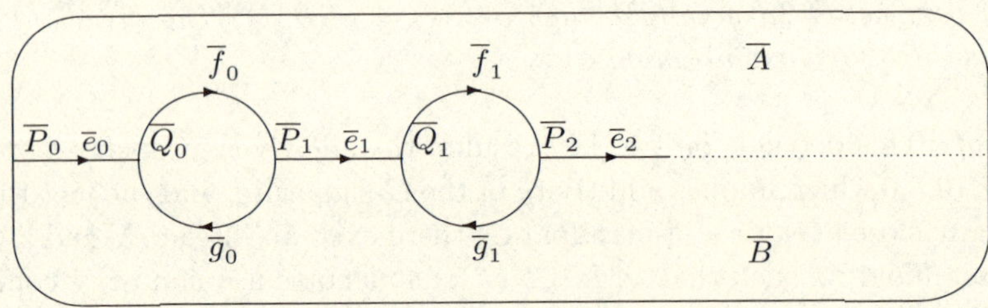

Fig. 10.3. The model for F

Rather than work directly with this, we construct a combinatorial model for F. Our model for the quotient consists of two $2r$-gons $\overline{A}$ and $\overline{B}$

attached along alternate edges. We thus have $2r$ vertices $\overline{P}_i$, $\overline{Q}_i$ (where we regard the suffices i as integers modulo r) and $3r$ edges $\overline{e}_i, \overline{f}_i, \overline{g}_i$: orient these so that

$$\partial\overline{e}_i = \overline{Q}_i - \overline{P}_i, \quad \partial\overline{f}_i = \overline{P}_{i+1} - \overline{Q}_i, \quad \partial\overline{g}_i = \overline{Q}_i - \overline{P}_{i+1},$$

$$\partial\overline{A} = \sum(\overline{e}_i + \overline{f}_i), \quad \partial\overline{B} = \sum(\overline{g}_i - \overline{e}_i).$$

The orientations of $\overline{A}$ and $\overline{B}$ are given by the complex structure, and $\partial(\overline{A}+\overline{B})$ is the sum of the boundary cycles $\overline{f}_i+\overline{g}_i$. We regard the action of U as increasing the subscripts by 1.

We lift this to a model for F consisting of $2r$ n-gons T^jA, T^jB, $3r^2$ edges T^je_i, T^jf_i, T^jg_i and $2r^2$ vertices T^jP_i, T^jQ_i with boundary maps given by

$$\partial e_i = Q_i - P_i, \quad \partial f_i = P_{i+1} - Q_i, \quad \partial g_i = TQ_i - P_{i+1}$$

$$\partial A = \sum(e_i + f_i), \quad \partial B = \sum T^{-i}(g_i - e_{i+1}).$$

The action of U lifts as $UA = A$, $UB = T^{-1}B$, $Ue_i = e_{i+1}$, $Uf_i = f_{i+1}$, $Ug_i = g_{i+1}$, $UP_i = P_{i+1}$ and $UQ_i = Q_{i+1}$.

Figure 10.3 can be used to illustrate A by removing the overlines. We picture B as follows:

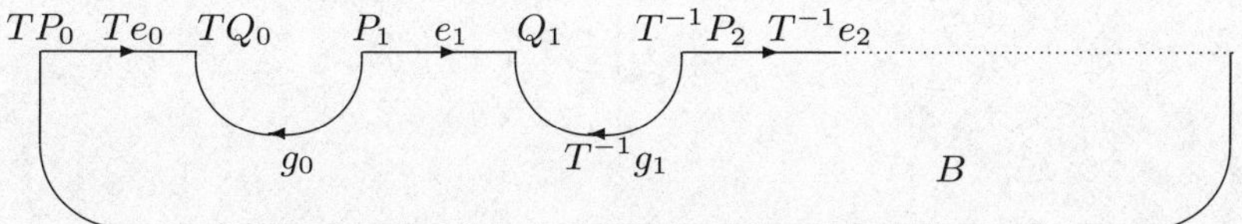

The 1-cycles include $T^j\partial A$, $T^j\partial B$ and the boundary cycles $N(f_i+g_i)$, where $N = \sum T^j$. We also have the cycle $\alpha_1 = e_1+f_1+g_1-Te_1-Tf_0-g_0$ and the cycles $T^j\alpha_i = T^jU^{i-1}\alpha_1$, giving r^2+3r in all. Not all these are independent: we have the relations

$$\begin{aligned}
N\alpha_i &= N(f_i+g_i) - N(f_{i-1}-g_{i-1}),\\
\sum \alpha_i &= \sum(e_i+f_i) - T\sum(e_i - f_i) = (1-T)\partial A,\\
\sum T^{-i}\alpha_i &= \sum T^{-i}e_i + \sum T^{-i}g_i - \sum T^{1-i}e_i - \sum T^{-i}g_{i-1}\\
&= (1-T^{-1})\partial B,
\end{aligned}$$

and the syzygy $N(\sum \alpha_i) = 0$. There are thus r^2+1 independent cycles. Of these, the $T^j\partial A$ and $T^j\partial B$ are nullhomologous, so the homology group has rank $r^2 - 2r + 1$. The r boundary cycles (whose sum is nullhomologous) span the radical; factoring these out leaves $r^2 - 3r + 2$ which was indeed the rank of $H_1(\hat{F})$.

To illustrate the cycle α_1 we need two pictures: see Figure 10.4. The cycle starts in B and follows the dotted line in the first picture through A to TB; we then pick it up again in the second picture and follow it from TB through TA and back into B. Note that the heavy vertical line in each picture represents a cut.

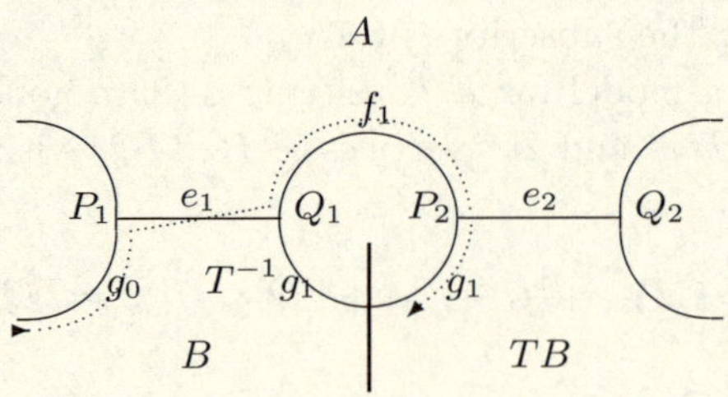

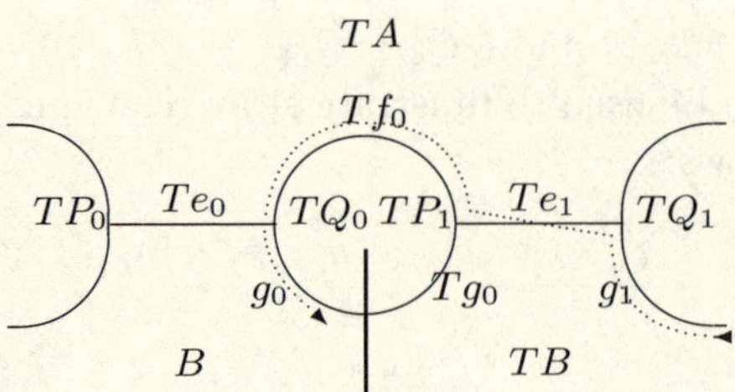

Fig. 10.4. The cycle α_1

The only non-zero intersection numbers occur among the $T^j\alpha_i$. If two of these are to intersect they must share at least a vertex. Checking these, we find that the only others sharing a vertex with α_1 are $\alpha_0, T\alpha_0, T^{-1}\alpha_1, T\alpha_1, T^{-1}\alpha_2$ and α_2. As the intersection form is skew-symmetric, it suffices to calculate $\alpha_1.T\alpha_1$, $\alpha_1.\alpha_0$ and $\alpha_1.T\alpha_0$. All others can now be obtained using skew-symmetry and the actions of T and U. To perform the calculations, we need to deform the cycles in question to be transverse: the intersection numbers are then immediate from the pictures. In these, the curve α_1 is as above; the second curve in each case is represented by a heavy dotted line. The orientation convention is that the intersection number represented by [symbol] is $+1$.

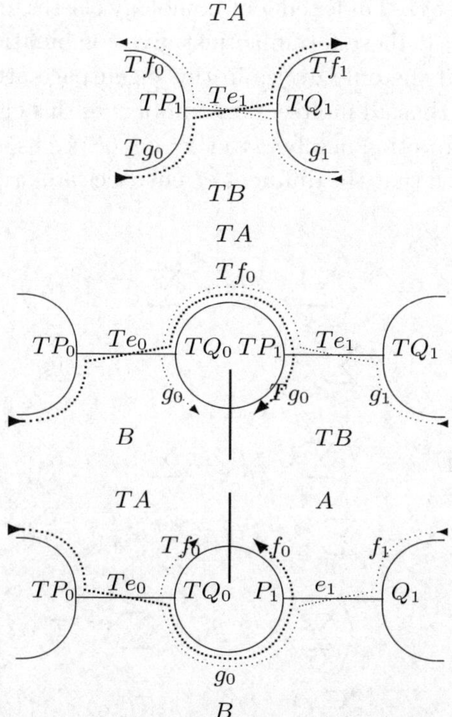

Fig. 10.5. First, the intersection $\alpha_1.T\alpha_1$; then the intersection $\alpha_1.T\alpha_0$; finally, the intersection $\alpha_1.\alpha_0$.

Lemma 10.7.3 *The only non-zero intersection numbers among the* $T^j\alpha_i$ *are:*

$$\begin{array}{rr} T^j\alpha_i.T^{j+1}\alpha_i = 1 & T^{j+1}\alpha_i.T^j\alpha_i = -1 \\ T^j\alpha_{i+1}.T^j\alpha_i = 1 & T^j\alpha_i.T^j\alpha_{i+1} = -1 \\ T^{j+1}\alpha_i.T^j\alpha_{i+1} = 1 & T^j\alpha_{i+1}.T^{j+1}\alpha_i = -1 \end{array}$$

We are now ready to begin the calculation of the equivariant signatures. The eigenspace corresponding to the character with $\chi(T) = \zeta$, where $\zeta^r = 1$, is spanned (if $\zeta \neq 1$) by the elements

$$\beta_i := \sum \zeta^{-j} T^j \alpha_i.$$

We have $U\beta_i = \beta_{i+1}$. We also have the relations

$$\sum \beta_i = \sum \zeta^{-j} T^j \sum \alpha_i = \left(\sum \zeta^{-j} T^j\right)(1 - T)\partial A,$$

$$\sum T^{-i}\beta_i = \sum \zeta^{-j} T^j \sum (T^{-i}\alpha_i) = \left(\sum \zeta^{-j} T^j\right)(1 - T^{-1})\partial B.$$

Thus there are just $r-2$ independent homology classes in this eigenspace.

In the case $\zeta = 1$, these formulae just give combinations of boundary cycles, and indeed the only cycles in this eigenspace are homologous to boundary cycles; thus all intersection numbers in this eigenspace vanish.

We extend intersection numbers to $V := H_1(F; \mathbb{C})$ as a skew-Hermitian form. We have seen that the different eigenspaces are orthogonal. Within V_ζ we have

$$\begin{aligned}
\langle \beta_i, \beta_i \rangle &= \left\langle \sum_{j,k} \zeta^{-j} T^j \alpha_i, \sum \zeta^{-k} T^k \alpha_i \right\rangle \\
&= \sum_{j,k} \zeta^{j-k} \langle T^j \alpha_i, T^k \alpha_i \rangle = r(\zeta^{-1} - \zeta), \\
\langle \beta_i, \beta_{i+1} \rangle &= \left\langle \sum_{j,k} \zeta^{-j} T^j \alpha_i, \sum \zeta^{-k} T^k \alpha_{i+1} \right\rangle \\
&= \sum_{j,k} \zeta^{j-k} \langle T^j \alpha_i, T^k \alpha_{i+1} \rangle = r(\zeta - 1), \\
\langle \beta_{i+1}, \beta_i \rangle &= \left\langle \sum_{j,k} \zeta^{-j} T^j \alpha_{i+1}, \sum \zeta^{-k} T^k \alpha_i \right\rangle \\
&= \sum_{j,k} \zeta^{j-k} \langle T^j \alpha_{i+1}, T^k \alpha_i \rangle = r(1 - \zeta^{-1}),
\end{aligned}$$

and $\langle \beta_i, \beta_k \rangle = 0$ if $i - k \neq 0, \pm 1$.

These products form a skew-Hermitian matrix B. Write $\zeta = e^{2i\theta_0}$, where since $\zeta \neq 1$ we may take $0 < \theta_0 < \pi$. Note that $\sin\theta_0 > 0$. We have

$$b_{i,i} = r(\zeta^{-1} - \zeta) = -2ir \sin 2\theta_0;$$

$$b_{i,i+1} = r(\zeta - 1) = 2ire^{i\theta_0} \sin\theta_0;$$

$$b_{i+1,i} = r(1 - \zeta^{-1}) = 2ire^{-i\theta_0} \sin\theta_0.$$

Now multiply by i to convert to a Hermitian matrix, and divide through by the positive factor $2r\sin\theta_0$. This produces the matrix $A(\theta_0)$ where, for any θ with $0 < \theta < \pi$, $A(\theta)$ is defined by

$$a_{i,i} = 2\cos\theta,\ a_{i,i+1} = -e^{i\theta},\ a_{i+1,i} = -e^{-i\theta};\ a_{i,j} = 0 \text{ otherwise.} \tag{10.13}$$

We already know, from the relations between the β_i, that $A(\theta_0)$ has rank at most $r-2$: indeed, for any θ the sum of the elements in each row vanishes; and if $\sin r\theta = 0$, $\sum e^{-2ki\theta} a_{j,k} = 0$ for each j.

One of the easiest ways to find the signature of a symmetric or Hermitian matrix comes from regarding the matrix as that of a Hermitian form, and applying the Gram-Schmidt orthogonalisation process to a basis of the underlying vector space. Provided the successive principal minors $p_k := \det(a_{i,j})_{1\leq i,j\leq k}$ do not vanish, this reduces the Hermitian form to $\sum_{k=1}^{r}(p_k/p_{k-1})\overline{z_k}z_k$, where we set $p_0 := 1$. It is then sufficient to determine the signs of the terms p_k to find the signature. For the matrix (10.13), since we know that $p_r = 0$ we can restrict to the subspace spanned by the first $(r-1)$ vectors; we will see that also $p_{r-1} = 0$ but $p_{r-2} \neq 0$, so the restriction to the subspace spanned by the first $(r-2)$ vectors is non-degenerate and we need the signature on that subspace. In fact, we have:

Lemma 10.7.4 *The principal minors of $A(\theta)$ (with $\sin\theta \neq 0$) are given, for $0 \leq k < r$, by*

$$p_k = \sin(k+1)\theta/\sin\theta.$$

We have already noted that, for any θ, $A(\theta)$ is singular, so $p_r = 0$.

Proof The result holds for $k = 0$ trivially and for $k = 1$ since $p_1 = a_{1,1} = 2\cos\theta$. We proceed by induction. Expanding the determinant by the first row gives

first, $a_{1,1}$ multiplied by a determinant which, in view of the circulant nature of (10.13), is identical to p_{k-1};

second, $-a_{1,2}$, multiplied by a determinant with only one entry in its first column; expanding by this column then gives $-a_{1,2}a_{2,1}$ multiplied by a determinant which can be identified with p_{k-2}.

Thus $p_k = 2\cos\theta p_{k-1} - p_{k-2}$. Now calculate, using the induction hypothesis:

$$p_k = (2\cos\theta\sin k\theta - \sin(k-1)\theta)\,/\sin\theta = \sin(k+1)\theta/\sin\theta.$$

□

Since $\zeta^r = 1$, $2r\theta_0$ is a multiple of 2π, so $\sin r\theta_0 = 0$ and $p_{r-1} = 0$ as expected. Since $\sin\theta_0 \neq 0$, it also follows that $p_{r-2} \neq 0$.

Suppose θ is such that $\sin k\theta \neq 0$ for $1 \leq k \leq r-1$. Then the principal minors $p_0, p_1, \ldots, p_{r-2}$ are all non-zero, and we can calculate the signature of the Hermitian form $\Phi(\theta)$ defined by the leading $(r-2)\times(r-2)$ minor as above. The number of negative terms is equal to the number of values of k with $1 \leq k \leq r-2$ for which p_k/p_{k-1}, i.e. $\sin(k+1)\theta/\sin k\theta$ is negative. Suppose $(s-1)\pi < (r-1)\theta < s\pi$. Then as ψ increases from 0 to $r\theta$ the sign of $\sin\psi$, originally positive, changes $s-1$ times, at the multiples of π. If ψ is restricted to be a multiple of θ then since $0 < \theta < \pi$, it will encounter each of these intervals in which the sign of $\sin\psi$ is constant. There are thus still just $s-1$ changes of sign, i.e. $s-1$ values of k for which $\sin(k+1)\theta/\sin k\theta$ is negative. Hence the signature of $\Phi(\theta)$, i.e. the number of positive terms less the number of negative terms, is $(r-s-1)-(s-1) = r-2s$.

Now for all θ with $\frac{s}{r-1}\pi < \theta < \frac{s+1}{r-1}\pi$ the form $\Phi(\theta)$ is non-degenerate since $p_{r-2} \neq 0$. Hence as θ varies in this range, the signature of $\Phi(\theta)$ is constant. The above formula thus remains valid even if there are values $1 \le k \le r-1$ with $\sin k\theta = 0$. Taking $\theta = \theta_0$, since $\zeta^r = 1$, $2r\theta_0$ is an integer multiple of 2π, say $r\theta_0 = \pi s$, and then $(s-1)\pi < (r-1)\theta_0 < s\pi$. Thus the signature is $r-2s = r - 2\frac{r\theta_0}{\pi}$. This proves the first assertion of

Theorem 10.7.5 *For the manifold F constructed above, the equivariant signature belonging to the character $\chi(T) = e^{2\pi i s/r}$ is $r-2s$ if $0 < s < r$ and 0 if $s = 0$.*

We have $\sigma_{T^a}(F) = ir\cot\frac{\pi a}{r}$ if a is not divisible by r and 0 if it is.

Proof To obtain the second formula, we calculate. We have

$$\sigma_{T^a}(F) = \sum_{s=1}^{r-1} \chi_s(T^a)\sigma_{\chi_s}(F) = \sum_{s=1}^{r-1} e^{2\pi i a s/r}(r-2s).$$

Write $z := e^{\pi i a/r}$: then $z^{2r} = 1$. To evaluate $\sum_{s=1}^{r-1}(r-2s)z^{2s}$, if $z^2 = 1$ the terms cancel and we have 0; otherwise we multiply by $z - z^{-1}$. After some cancellation, this gives $2z(\sum_0^{r-1} z^{2s}) - r(z + z^{2r-1})$. Here $\sum_0^{r-1} z^{2s} = (1-z^{2r})/(1-z^2) = 0$. Hence the whole sum reduces to $-r(z+z^{-1})/(z-z^{-1})$ or, substituting for z, to $-r(2\cos\frac{\pi a}{r})/(2i\sin\frac{\pi a}{r}) = ir\cot\frac{\pi a}{r}$. □

Note first, that we said we would divide by i, and second that – as follows from general considerations – $\sigma_{x^{-1}}(F) = -\sigma_x(F)$ for all x.

We proceed to the proof of Proposition 10.6.3.

The structure of G-1-manifolds can be analysed as follows. First decompose the set of components into G-orbits: it is enough to work on a single orbit. Choose a component C lying in the orbit; let H be the stabiliser of this component. Then H acts freely on C preserving orientation, hence we can identify C with a circle so that H acts by rotations. Thus H is cyclic; if H has order r, let T be the generator of H which acts by a rotation through an angle $2\pi/r$ in the positive sense. We can identify the union of components in the chosen orbit with $G \times_H C$.

We constructed above an H-2-manifold F with $\partial F = NC$. It follows that $G \times_H F$ is a $G-2-$manifold with $\partial(G \times_H F) = N(G \times_H C)$. We complexify $H_1(F)$ to $H_1(F; \mathbb{C})$ and extend the intersection form to be skew-Hermitian; multiply the form by i to make it Hermitian; and choose an H-invariant orthogonal splitting $H_1(F : \mathbb{C}) = V_+ \oplus V_0 \oplus V_-$, where the form is positive definite on V_+, zero on V_0 and negative definite on V_-. This induces a G-invariant orthogonal splitting of $H_1(G \times_H F; \mathbb{C}) = \mathbb{C}G \otimes_{\mathbb{C}H} H_1(F; \mathbb{C})$:

$$H_1(G \times_H F; \mathbb{C}) = (\mathbb{C}G \otimes_{\mathbb{C}H} V_+) \oplus (\mathbb{C}G \otimes_{\mathbb{C}H} V_0) \oplus (\mathbb{C}G \otimes_{\mathbb{C}H} V_-).$$

Now if $g \in G \setminus H$, g acts by permuting the summands of $\mathbb{C}G \otimes_{\mathbb{C}H} V_\pm$ corresponding to the components of $G \times_H F$; since g leaves none of the components invariant, its trace on $\mathbb{C}G \otimes_{\mathbb{C}H} V_\pm$ is zero. On the other hand, since G is abelian, if $g \in H$, it acts in the same way on each component of $G \times_H F$; so its trace on $\mathbb{C}G \otimes_{\mathbb{C}H} V_\pm$ is $|G : H|$ times its trace on $V_\pm$.

By Theorem 10.7.5, for the character χ_s of H with $\chi_s(T) = e^{2\pi i s/r}$ (where $0 \leq s < r$), we have $\sigma_\chi(F) = r - 2s$ if $s \neq 0$ and 0 if $s = 0$. Thus for any character χ of G, $\sigma_\chi(G \times_H F) = |G : H|(r-2s)$ if $\chi_s(T) = e^{2\pi i s/r}$ (with $0 < s < r$), but vanishes if χ is trivial on H. Dividing this by $|G|$, the Proposition follows.

10.8 Notes

Section 10.1 The Seifert form was first introduced in [163]. See e.g. Kauffman [96] for a modern treatment in book form.

There are several other approaches to the detailed analysis of the Milnor fibration. Many accounts define a 'variation map' $H_n(F, \partial F) \to H_n(F)$ (which is essentially equivalent to the Seifert form); a key result is then that this is an isomorphism. One can also use a morsification, which leads to a distinguished class of $\mathbb{Z}$-bases of the group $H_n(F)$, and hence of the matrices representing the monodromy and intersection numbers. With respect to a distinguished basis, the matrix of the Seifert form satisfies $S(e_i, e_j) = 0$ if $i > j$ and -1 (or $+1$, depending on sign conventions) if $i = j$. Although this is explicit, it is not easy to infer general properties of the Seifert form. This additional information is also available in higher dimensions, where the special arguments applicable when $n = 1$ do not apply. See, for example, Lamotke [107] and Looijenga [121].

Section 10.2 The Monodromy Theorem 10.2.1 holds for isolated singular points in higher dimensions. Several proofs were given c 1970; perhaps the first in an analytic context by Brieskorn [22]. The first in a textbook appears in Looijenga [121]. The approach here mainly follows a paper by du Bois and Michel [54]. It is shown in that paper that the weight filtration as we have defined it coincides with the weight filtration of Steenbrink [171] defined as part of the mixed Hodge structure on H. A'Campo [3] was the first to discuss many of the questions addressed in this section.

Section 10.3 The rational canonical form is described in many algebra textbooks, e.g. [122].

The main calculation of Theorem 10.3.2 is due to A'Campo [5]. A variant of the characteristic polynomial which is often used, since it sometimes extends to the infinite dimensional case, is to replace $\det(tI - h)$ by $\det(I - th)$; the result is then known as the *zeta function.* An alternative form of the result for the characteristic polynomial of the monodromy is also due to A'Campo [5]: see Exercise 10.9.5.

Lemma 10.3.7 is taken from [55], who attribute it to Neumann [139]: their version of the argument uses the 'semistable normalisation', a closed surface $\widehat{F}_i$ (they denote it D_i) obtained by glueing a disc to each boundary component of F_i. Then $\widehat{F}_i$ is a ramified cover of E_i.

An alternative formulation of the work on monodromy (over $\mathbb{R}$ or $\mathbb{C}$) is to use the spectrum, which is the list of eigenvalues of the monodromy, repeated according to their multiplicity. To this one can attach additional data. We do not discuss this here, but refer to papers of Steenbrink et al., see e.g. [171], [159].

Theorem 10.3.10 is due to Lê [108]. A simpler argument was given by A'Campo [3], who also gave a counterexample (see Exercise 10.9.2) to semi-simplicity with 2 branches.

Some of the methods for calculating monodromy are applied to the global situation of polynomial maps $\mathbb{C}^2 \to \mathbb{C}$ in Michel and Weber [128].

Section 10.4 Our presentation of isometric triples largely follows Milnor [131].

An alternative approach is to work more directly with the *Blanchfield pairing*, named after Blanchfield [19], which is defined as follows. Since M is fibred, its infinite cyclic cover is homeomorphic to $F \times \mathbb{R}$, so we can identify $H_1(M; K[t, t^{-1}])$ with H, and consider as a module over $K[t, t^{-1}]$ on which $(1 - t)$ is invertible: the *Blanchfield module.*

Now write L for the quotient field $K(t)$. Then

$$H_1(M;L) \cong H^1(M;L) \cong H_2(M;L) = 0,$$

so the middle map in the exact sequence

$$H^1(M;L) \to H^1(M;L/K[t,t^{-1}]) \to H^2(M;K[t,t^{-1}]) \to H^2(M;L)$$

is an isomorphism, while since L is an injective $K[t,t^{-1}]$-module,

$$H^1(M;L/K[t,t^{-1}]) \cong \mathrm{Hom}\,(H_1(M;K[t,t^{-1}]),L/K[t,t^{-1}]).$$

We thus have an isomorphism

$$H = H_1(M;K[t,t^{-1}]) \cong \mathrm{Hom}\,(H,L/K[t,t^{-1}]),$$

which yields a non-singular pairing $H \times H \to L/K[t,t^{-1}])$. This pairing is easily shown to be symmetric, and is called the Blanchfield pairing: it is essentially equivalent to the isometric structure.

The relationship between the two versions is worked out in Ranicki [153] Section 32, which also gives the relation between different versions of the multisignature. For another account, closer to the concerns of this book, see Neumann [138].

Section 10.5 There are numerous textbook references for the theory of quadratic forms over fields and over various types of rings; rather fewer for hermitian forms. See, for example, Milnor & Husemoller [133], Lam [106], Scharlau [158], Knus [103].

The problem of calculating all the invariants of the Seifert form was treated by du Bois and Hunault [57]. Their result yields only a representative matrix, not an explicit formula for the invariants; however, their result is sufficiently explicit to allow computation in examples. They also obtain results over $\mathbb{Z}$.

An example, viz. the forms $g_{r,s+8}$ and $g_{s,r+8}$ where $g_{r,s}$ denotes

$$(x-y)\left((y^2-x^3)^2 - x^{s+6} - 4yx^{\frac{1}{2}(s+9)}\right) \times \left((x^2-y^5)^2 - y^{r+10} - 4xy^{\frac{1}{2}(r+15)}\right),$$

was given by du Bois and Michel in [55] of two germs with Witt equivalent Seifert forms over $\mathbb{Q}$, but not equisingular; this was refined in [56] to show that (for the same examples) even the isomorphism class of the Seifert form over $\mathbb{Z}$ does not determine the equisingularity type. In this example, the polar quotients are identical in the two cases; they give another example, where the polar quotients differ. That the monodromy fails to determine the equisingularity type was already shown by A'Campo [3]; for the Seifert form over $\mathbb{R}$ examples were given in Schrauwen et al. [159]. From a different viewpoint, Kaenders [94] showed that the restriction to R of the Seifert form determines the intersection multiplicities (compare the discussion in Lemma 10.2.3).

Section 10.6 The first introduction of signatures as link invariants was given by Murasugi [137]; they were generalised by Milnor [130] and by Tristram [180] to versions close to ours.

Our calculation of the signature borrows from the account of Neumann [139]. A very different presentation is given in [159]: checking that their spectral pairs are equivalent to our calculation of the multisignature is a non-trivial exercise. The G-signature theorem is in [16]. The invariant ρ was introduced by the author [189] with essentially the argument given here, and by Atiyah and Singer [16], see also [15].

Section 10.7 Our argument is close to the one sketched in Kauffman [96].

10.9 Exercises

Exercise 10.9.1 Suppose given a singularity with Seifert matrix

$$\begin{pmatrix} 1 & 1 & 0 & 0 \\ 0 & 1 & 1 & 0 \\ 0 & 0 & 1 & 1 \\ 0 & 0 & 0 & 1 \end{pmatrix}.$$

Calculate the matrices A and X, and hence the characteristic polynomial of X. Which type of singularity must it be?

Exercise 10.9.2 Show that the monodromy of the singularity $(x^3 - y^2)(x^2 - y^3) = 0$ has just one Jordan block of size 2, corresponding to the eigenvalue -1.

Exercise 10.9.3 Show that the triple point singularities whose monodromy is not semi-simple are those expressible in the form $A+B$ where A has multiplicity 2, B is smooth, and A and B have an even number of infinitely near points (each of multiplicity 2 on A) in common. Show also that for each of these, the characteristic polynomial of h on H/W_1 is $t+1$.

Exercise 10.9.4 Calculate the characteristic polynomials of the monodromy on $H_1(F)$ and on W_1 for the following examples:
(i) $(y^2 - x^3)(y^3 - x^4) = 0$,
(ii) $(y^2 - x^3)(y^4 - x^3) = 0$.

Exercise 10.9.5 Write S_m for the set of smooth points on the exceptional divisor of a resolution of the given singularity such that the parameter M_i for the component E_i of E is equal to m. Establish the following:
(i) the numbers $\chi(S_m)$ do not depend on the choice of resolution;
(ii) the characteristic polynomial of h_* on H_C is

$$(t-1)\prod_{m\geq 1}(t^m - 1)^{-\chi(S_m)}.$$

(iii) the trace of h_*^k acting on H_C is $1 - \sum_{m|k} m\chi(S_m)$;
(iv) we have $\mu = 1 - \sum_m m\chi(S_m)$.

Exercise 10.9.6 Calculate the following in terms of the numbers b of blowings up to form the good resolution $\pi : T \to S$ with tree $\Gamma_R(C)$, and r of branches of C:

The number $\#\mathcal{V}$ of vertices, the number $\#\mathcal{E}$ of edges and the Euler characteristic χ for each of $\Gamma_R(C)$ and $\Gamma_R^+(C)$.

Describe $\pi^{-1}(C)$ as formed from a certain number of copies of 2-spheres and 2-discs by identifying a certain number of pairs of points, and hence calculate its Euler characteristic.

The multiplicity of 1 as root of the characteristic polynomials of h on $H_1(F)$ and on W_1.

Exercise 10.9.7 Show that the characteristic polynomials of the monodromy for (i) $x(x^5 + y^6) = 0$ and (ii) $(x^2 + y^3)(x^{10} + y^3) = 0$ are the same. Show also that the signatures do not all agree.

Exercise 10.9.8 Show that the examples in the preceding exercise are distinguished by their 2-variable Alexander polynomials.

Exercise 10.9.9 Let C_1 denote the union of the germs $y^2 = x^{15}$ and $x = y^{3/2}(1+y^{5/4})$, and C_2 the union of $y^2 = x^{19}$ and $x = y^{3/2}(1+y^{3/4})$. Calculate the signatures, and deduce that the Seifert forms of C_1 and C_2 over $\mathbb{R}$ are isomorphic.

Exercise 10.9.10 Let $p(t) = t^2 - 2ct + 1$ where $c = \cos 2\phi$ and $0 < \phi < \pi/2$; let H be the $\mathbb{R}[t]$ module generated by e and defined by $p(t)^2 e = 0$. Show that a bilinear form on H defining an isometric structure is determined by $\alpha := \langle e, te\rangle$ and $\beta := \langle se, te\rangle$. Show that the signature of the form defined by Lemma 10.4.4 on the eigenspace $t = e^{2i\phi}$ is equal to the sign of β.

Exercise 10.9.11 Calculate the signature of the form $\xi S - \overline{\xi} S^*$ for the form in the preceding exercise when $\xi = e^{i\phi}$.

11

Ideals and clusters

In Chapter 8 we took a geometric approach to the combinatorics associated to a curve singularity, studying functions on the resolution tree. In this chapter we give a more algebraic presentation. This gives interesting information about the set of ideals in the local ring $\mathcal{O}_0 := \mathcal{O}_{x,y}$ of O. We obtain a relation between these ideals and 'clusters' of infinitely near points, which can be formulated as a Galois correspondence between these.

This has two applications. One is a procedure (Enriques' 'unloading algorithm') leading from a numerical definition of an ideal to the effective numerical parameters defining it. The other is a lead in to the study of integral closures of ideals: we establish the surprisingly close connection between integrally closed ideals and exceptional cycles.

We briefly address the question of determinacy, that is, finding for each reduced $f \in \mathcal{O}_0$ the least integer n such that the terms of degree $\leq n$ in the power series expansion of f are sufficient to determine the equisingularity type of the curve C_f.

In the final section we briefly discuss properties of plane curve singularities from the viewpoint of the local ring $\mathcal{O}_C$, which is that taken in modern algebraic geometry.

11.1 Blowing up ideals

We study ideals I in the ring $\mathcal{O}_0$, (which can be identified with $\mathbb{C}\{x, y\}$) of germs at O of holomorphic functions on the plane T_0. We begin by showing how I gives rise to an ideal in the local ring of the surface obtained by blowing up at a point, and establishing some basic results relating I to these blown up ideals. A sequence of blowings up which leads only to trivial ideals is called a resolution of I. The sequence of multiplicities of the blown up ideals gives numbers which we use to define an exceptional cycle in the resolution.

We use the following notation. We write $I \lhd R$ to denote the fact that I is an ideal in the ring R (all our rings are commutative). The ideal generated by elements $a_1, \ldots, a_k$ is denoted by $\langle a_1, \ldots, a_k \rangle$. The product $I_1 I_2$ of two ideals I_1 and I_2 is the ideal generated by the products $i_1 i_2$ with $i_1 \in I_1$ and $i_2 \in I_2$.

We first relate the order of a function $f \in \mathcal{O}_0$ to ideal theory. Write $\mathfrak{m}$ for the set of germs of functions f with $f(O) = 0$: clearly this is an ideal, and evaluating at O gives an isomorphism $\mathcal{O}_0/\mathfrak{m} \cong \mathbb{C}$, so $\mathfrak{m}$ is a maximal ideal. Indeed, as any $f \notin \mathfrak{m}$ has an inverse in $\mathcal{O}_0$, it is the unique maximal ideal so (as we noted earlier) $\mathcal{O}_0$ is a local ring. Applying Lemma 1.4.3 with $k = n$ $(= 2)$, we see that any $f \in \mathfrak{m}$ can be expressed in the form $xg(x, y) + yh(x, y)$ with $g, h \in \mathcal{O}_0$, i.e. $\mathfrak{m} = \langle x, y \rangle$. More generally,

Lemma 11.1.1 *If $f \in \mathcal{O}_0$ has order at least r, f can be expressed as $\sum_0^r x^{r-i} y^i g_i(x, y)$, with $g_i \in \mathcal{O}_0$. In particular, f has order $\geq r$ if and only if $f \in \mathfrak{m}^r$.*

Proof We proceed by induction on r. The result is established for $r = 1$; assume it holds for r and that f has order $\geq r + 1$. The terms or order r in the expansion of $f = \sum_0^r x^{r-i} y^i g_i(x, y)$ must be $\sum_0^r x^{r-i} y^i g_i(0, 0)$. These vanish by hypothesis, so each $g_i(O) = 0$. Applying Lemma 1.4.3 to the g_i now completes the induction step.

This proves the direct implication in the final assertion of the lemma, and the converse is immediate. □

Although the ring $\mathcal{O}_0$ is infinite dimensional, for most of our arguments we can reduce to finite dimension calculations. We make this systematic by defining the space J^N of N-jets to be the quotient $\mathcal{O}_0/\mathfrak{m}^{N+1}$, and the N-jet $j^N f$ of $f \in \mathcal{O}$ to be the class of f in the quotient. It follows from the lemma that this quotient has a vector space basis $\{x^r y^s \mid 0 \leq r, s, r + s \leq N\}$ and that $j^N f$ can be identified with the sum of the terms of degree at most N in the Taylor expansion of f at O.

Recall from Theorem 2.2.5 that unique factorisation holds in $\mathcal{O}_0$. For any ideal I, write $h(I)$ for the highest common factor of all elements of I, and define $I_{\mathrm{red}} := \{f/h(I) \mid f \in I\}$. Then, since division is unique, I_{red} is an ideal, and the elements of I_{red} have no common factor. The ideal I is equal to the product $\langle h(I) \rangle I_{\mathrm{red}}$. We sometimes refer to this as the *canonical decomposition* of I.

Lemma 11.1.2

(i) *If $J \lhd \mathcal{O}_0$ is an ideal whose elements have no common factor, and $f \in J$, then we can find $g \in J$ such that f and g have no common factor.*

(ii) *The following conditions on an ideal $J \lhd \mathcal{O}_0$ are equivalent:*

(a) the elements of J have no common factor,

(b) $dim(\mathcal{O}_0/J) < \infty$,

(c) for some k, we have $J \supseteq \mathfrak{m}^k$.

Proof Express $f = \prod_{i=1}^r f_i^{a_i}$ as a product of irreducible factors. By hypothesis, for each i there exist elements $g_i \in J$ such that g_i is not divisible by f_i. For any $\lambda = (\lambda_1, \ldots, \lambda_r) \in \mathbb{C}^r$ form the linear combination $g(\lambda) := \sum_i \lambda_i g_i$. The set of λ such that $g(\lambda)$ is divisible by f_i is a vector subspace of $\mathbb{C}^r$ which is not equal to the whole space. Since no proper finite union of subspaces of $\mathbb{C}^r$ can be equal to the whole space, we can choose λ so that $g = g(\lambda)$ has no factor in common with f.

First assume (a). Choose f, g as in (i). Since $f, g \in J$, $\langle f, g\rangle \subseteq J$, so $\dim(\mathcal{O}_0/J) \leq \dim(\mathcal{O}_0/\langle f, g\rangle)$. It will thus suffice to prove the latter finite. Now as f and g have no common factor, the curves C_f and C_g have no component in common. Then (b) follows from the fact that, by Lemma 1.2.1(iv), $\dim(\mathcal{O}_0/\langle f, g\rangle) = C_f.C_g$. Finiteness can also be proved more directly: reduce to a single branch, and prove that non-trivial ideals in $\mathcal{O}_B$ have finite codimension.

Now assume (b), and consider the sequence $a_k := \dim(\mathcal{O}_0/(J + \mathfrak{m}^k))$. The numbers a_k are non-negative integers, and $a_k \geq a_{k+1}$ for each k. Since these numbers cannot decrease below zero, we must have $a_k = a_{k+1}$ for some k. Thus $J + \mathfrak{m}^k = J + \mathfrak{m}^{k+1}$. So for each monomial $m_i := x^i y^{k-i}$ $(0 \leq i \leq k)$ we can write $m_i = j_i + \sum c_{i,j} m_j$ for some elements $j_i \in J$ and $c_{i,j} \in \mathfrak{m}$. Solving these equations for the m_i using Cràmer's rule gives equations expressing Δm_i as a linear combination of the elements j_i, and hence an element of J, where Δ is the determinant of the matrix $I - (c_{i,j})$. But expanding this determinant, we see that it equals 1 added to an element of $\mathfrak{m}$. Hence Δ is an invertible element of $\mathcal{O}_0$, so each m_i itself belongs to J. Since the m_i generate the ideal $\mathfrak{m}^k$, it follows that $\mathfrak{m}^k \subseteq J$.

If (c) holds, then J contains x^k and y^k, which have no common factor. □

An ideal $I \lhd \mathcal{O}_0$ satisfying the conditions (ii) is said to be $\mathfrak{m}$-primary. Thus the canonical decomposition expresses I as the product of a principal ideal and an $\mathfrak{m}$-primary ideal. If I is primary, then $I \supseteq \mathfrak{m}^N$ for some N, and whether or not $f \in I$ is determined by the $(N-1)$-jet of f.

The *multiplicity* of I at the origin is defined to be the least order of any element $f \in I$: we will denote it by $m_0(I)$, so $m_0(I) := \min\{m_0(f) \mid f \in I\}$.

Let $\pi_0 : T_1 \to T_0$ be defined by blowing up the origin, E_0 be the exceptional curve, and $O_1 \in E_0$. We define the *total transform* of the ideal I at O_1 to be the ideal $\pi_0^*(I)$ in $\mathcal{O}_{O_1}$ generated by the pre-images $\{f \circ \pi_1 \mid f \in I\}$. For each $f \in I$, $m_0(f) \geq m_0(I)$, so f vanishes to order at least $m_0(I)$ along E_0. Now define the *strict transform* $I^{(1)}$ at O_1 to be the ideal in $\mathcal{O}_{O_1}$ generated by the $z^{-m_0(I)} f \circ \pi_1$ for $f \in I$. In local coordinates, the blow-up is obtained by substituting $(x_1 y_1, y_1)$ for (x, y), and $I^{(1)}$ is generated by the functions $f^{(1)}(x_1, y_1) := y_1^{-m_0(I)} f(x_1 y_1, y_1)$ for $f \in I$. Observe that $J^k(f^{(1)})$ is determined by $J^{k+m_0(I)}(f)$.

For any given ideal I, the terms of order $m_0(I)$ in the expansions of elements $f \in I$ form a linear family of homogeneous polynomials. If the infinitely near point O_1 does not correspond to a common zero of these polynomials, the ideal $I^{(1)}$ contains a function non-vanishing at O_1, hence equals the entire ring $\mathcal{O}_{O_1}$. Thus the strict transform is trivial at all but a finite number of points on E_0.

Now let $\pi : T \to S$ be, as usual, the composite of a series of blowings-up $\pi_i : T_{i+1} \to T_i$ at points O_i. At each O_i we have the total transform of I, which is the total transform by π_{i-1} of the total transform of I at $\pi_{i-1}(O_i) \in T_{i-1}$, and the strict transform $I^{(i)}$, which is the strict transform by π_{i-1} of the strict transform of I at $\pi_{i-1}(O_i) \in T_{i-1}$. We denote the multiplicity of $I^{(i)}$ at O_i by $m_i(I)$.

For I of finite codimension, we say that π is a *resolution of the ideal* I if the strict transform of I at each point of T is an improper ideal (i.e. the whole ring). In general we say that π resolves I if it resolves both $h(I)$ and I_{red}. We next prove the existence of resolutions. Since the same argument will be used for other purposes below, we present the conclusions in a sharp form.

Proposition 11.1.3

(i) *Let $f \in \mathcal{O}_0$ have order n; assume f regular of order n in x, and write $O_1^{(j)}$ for the points of E_0 at which the strict transform $f^{(j)}$ of f vanishes. Then taking strict transforms $g \mapsto y_1^{-n} g(x_1 y_1, y_1)$*

induces an isomorphism

$$\Phi:\ \mathfrak{m}^n/\langle f\rangle \longrightarrow \bigoplus_j \mathcal{O}_{O_1^{(j)}}/\langle f^{(j)}\rangle.$$

(ii) *For any* $\phi \in \bigoplus_j \mathcal{O}_{O_1^{(j)}}$*, the class of* $y_1^{n-1}\phi$ *in* $\bigoplus_j \mathcal{O}_{O_1^{(j)}}/\langle f^{(j)}\rangle$ *is the total transform of some element of* $\mathcal{O}_0/\langle f\rangle$*.*

(iii) *Let* $I \lhd \mathcal{O}_0$ *have order* n*; pick* $f \in I$ *of order* n*, introduce notation as in (i). Then the induced map* $\mathfrak{m}^n/I \to \bigoplus_j \mathcal{O}_{O_1^{(j)}}/I^{(j)}$ *is surjective. In particular, if* $\dim(\mathcal{O}_0/I) < \infty$ *then, for each* $O_1 \in E_0$*,* $\dim(\mathcal{O}_{O_1}/I^{(1)}) < \dim(\mathcal{O}_0/I)$*.*

(iv) *Any ideal has a resolution.*

Proof (i) For each j, the total transform gives a ring homomorphism $\mathcal{O}_0 \to \mathcal{O}_{O_1^{(j)}}$, and hence

$$\mathfrak{m}^n/\langle f\rangle \subset \mathcal{O}_0/\langle f\rangle \to \mathcal{O}_{O_1^{(j)}}/\langle f^{(j)}\rangle.$$

Now dividing by y_1^n is possible and well-defined on the image of $\mathfrak{m}^n$, so gives an additive homomorphism; assembling these over the possible j defines a map Φ.

By the Weierstrass Preparation Theorem 2.2.2 we may suppose, multiplying f by a unit in $\mathcal{O}_0$, that f has the form $x^n + \sum_1^n a_i(y)x^{n-i}$.

By Corollary 2.2.4, any $F \in \mathcal{O}_0$ may be expressed uniquely in the form $F = f(x,y)Q(x,y) + \sum_1^n R_i(y)x^{n-i}$. Then F has order at least n if and only if, for each i, $R_i(y)$ has order at least i. Thus $\mathfrak{m}^n/\langle f\rangle$ is a free module over the ring $\mathcal{O}_y$ of convergent power series in y, with basis $1, x, x^2, \dots, x^{n-1}$.

Now factorise f. We may take the irreducible factors as monic polynomials in x and collect them into groups according to the tangent directions of the corresponding branches, or equivalently, to the points of E_0: thus $f = \prod_j f_j$, with f_j – of degree d_j, say – corresponding to $O_1^{(j)}$. Note that the strict transform $f_i^{(j)}$ of f_i in $\mathcal{O}_{O_1^{(j)}}$ is invertible if $i \neq j$, so $f^{(j)}$ equals $f_j^{(j)}$ multiplied by a unit. It follows as above from the Preparation Theorem that elements of $\mathcal{O}_{O_1^{(j)}}/\langle f_j^{(j)}\rangle$ are uniquely expressible in the form $\sum_1^{d_j} S_i(y_1)x_1^{d_j-i}$. Thus we have a free $\mathcal{O}_{y_1}$-module of rank d_j; moreover, we may identify y with y_1, and the map Φ is a map of free $\mathcal{O}_y$-modules.

We can calculate the matrix of this map, but it is simpler to proceed as follows. As a result of the blowing up, we may regard $\mathfrak{m}^n/\langle f\rangle$ as the quotient of $\mathcal{O}_y[x_1]$ by the ideal generated by the polynomial

$f'(x_1, y_1) = x_1^{-n} f(x_1, x_1 y_1)$ of degree n. This polynomial factorises as $f' = \prod_j f_j^{(j)}$, and Φ can be identified with the direct sum of the quotient maps $\mathcal{O}_y[x_1]/\langle f' \rangle \to \mathcal{O}_{O_1^{(j)}}/\langle f_j^{(j)} \rangle$. But since the $f_j^{(j)}$ are mutually coprime, these are the projections of a direct sum decomposition, so Φ is indeed an isomorphism.

To see (ii), write ϕ in the form $\sum_{i=1}^n S_i(y_1) x_1^{n-i}$. Then

$$y_1^{n-1}\phi = \sum_{i=1}^{n} (x_1 y_1)^{n-i} y_1^{i-1} S_i(y_1),$$

which is the total transform of $\sum_{i=1}^n x^{n-i} y^{i-1} S_i(y) \in \mathcal{O}_0$.

(iii) It follows from the definition of the ideal transforms $I^{(j)}$ that there is an induced map as written: surjectivity now follows from (i). The second assertion follows.

(iv) Since we can resolve curves, it suffices to prove that any ideal I of finite codimension has a resolution. Blowing up O defines an exceptional curve E_0, and there are only finitely many points of E_0 at which the strict transform $I^{(1)}$ is a proper ideal: moreover, each of these ideals has codimension strictly less than $\dim(\mathcal{O}_0/I)$. Now blow up each of these points of E_0, and continue inductively. Since the codimensions decrease strictly, the procedure terminates. □

We associate to the ideal I (and the blow up T) the cycle

$$[I]_E := \sum_i m_i(I)[\widehat{E}_i]$$

in the surface T. For example, $[\langle f \rangle]_E = [f]_E$. The definition is most useful if I has finite codimension and is resolved by T. If T resolves I, a further blow up will just replace $[I]_E$ by its strict transform.

Another important tool at our disposal is the ability to choose elements of a given ideal which are 'general' in an appropriate sense. Again we give a version here that will suffice for later needs. For each k, write $j^k I := \{j^k f \mid f \in I\}$: clearly this is a linear subspace of J^k.

Theorem 11.1.4

(i) *For any ideal I and blow-up T, there exists $f \in I$ such that $m_i(f) = m_i(I)$ for each i $(0 \le i < N)$.*

(ii) *For any ideal I, $[I]_E \in \mathcal{E}$; i.e. the integers $m_i(I)$ satisfy the inequalities*

$$m_i(I) \ge \sum \{m_j(I) \mid O_j \text{ is proximate to } O_i\}.$$

(iii) *For any $\mathfrak{m}$-primary ideal I resolved by T, we can choose $f, g \in I$, both as in (i), and such that T resolves $C_f \cup C_g$.*

Proof By the definition of $m_0(I)$, $j^{m_0(I)}I$ is non-zero, and $m_0(f) = m_0(I)$ if and only if $j^{m_0(I)}(f) \neq 0$.

The k-jet of $f^{(j)}$ at $O_1^{(j)}$ is determined by $j^{k+m_0(I)}(f)$, and we have a linear map of $j^{k+m_0(I)}I$ to this space of k-jets. Taking $k = m_1(I)$, we see that this map has non-zero image, and so for a general $f \in I$, the image of $j^{m_0(I)+m_1(I)}f$ is non-zero, and thus $m_1(f) = m_1(I)$.

The result in general follows by repeating the argument.

Choose f as in (i): let it define the curve C_f. Then, for each i, $m_i(I) = m_i(f) = m_i(C_f)$, and the result follows from (8.1) for C_f.

We begin with a remark about pencils of binary forms. Suppose that ϕ and ψ are both homogeneous of degree n in x and y and have no common factor. Then a general element of the pencil of forms $\lambda\phi + \mu\psi$ has no repeated root, and two distinct elements have no root in common.

To see this, think of the pencil as defining a rational map $P^1(\mathbb{C}) \to P^1(\mathbb{C})$ by $(x : y) \mapsto (\phi(x, y) : \psi(x, y)$. This map has only a finite number of ramification points, so only a finite number of ratios $(\lambda : \mu)$ correspond to repeated roots. And no two elements of the pencil have a root in common (else all would).

Now return to the argument of (i). We have seen that if f, g are sufficiently general, each of C_f and C_g breaks into pieces corresponding to those i with $\delta_i(I) := [I]_E.[\epsilon_i] \neq 0$. The strict transforms of f and g at O_i both have order $\delta_i(I)$; if f and g are general enough then, since T resolves I, their lowest order terms have no factor in common. Hence, by the remark above, if each of f and g is replaced by a general linear combination of f and g, each of the lowest order terms of their strict transforms at O_i has no repeated factor. The result follows. □

The proof of (i) yields rather more.

Corollary 11.1.5 *For any ideal I and blow-up T, there is an integer M determined by I and N such that for $f \in I$, we have $m_i(f) = m_i(I)$ for each i ($0 \leq i < N$) for all $f \in I$ such that $j^M f$ lies in the complement of a finite union of proper linear subspaces of $j^M I$.*

11.2 The valuative closure of an ideal

There is a closure operation on the set of ideals called valuative closure. We define this, establish its formal properties, and show that it is closely related to the construction of the exceptional cycle. We can give an explicit formula for the codimension of a valuatively closed ideal. We also show that the semigroup of valuatively closed ideals under multiplication satisfies unique factorisation, and determine the set of primes for this semigroup.

For any irreducible curve germ B, define $m_B(I) := \min\{m_B(f) \mid f \in I\}$. We now calculate $m_B(I)$ for any branch B.

Lemma 11.2.1

(i) *For a single blow-up, we have*

$$m_B(I) = m_0(B)m_0(I) + m_{B^{(1)}}(I^{(1)}).$$

Also $m_1(\pi_1^*(I)) = m_0(I) + m_1(I^{(1)})$.

(ii) *If I is $\mathfrak{m}$-primary and T is a resolution of I, then for any B*

$$m_B(I) = \sum_i m_i(B)m_i(I) = -[B]_E.[I]_E.$$

More generally, this holds if no point of the strict transform of B lies on the strict transform of $C_{h(I)}$ or has non-trivial strict transform of I_{red}.

Proof For any f, we have $m_B(f) = m_0(B)m_0(f) + m_{B^{(1)}}(f_1)$. Now take the minimum of each side over all $f \in I$. Since the corresponding functions f_1 generate $I^{(1)}$, the first assertion follows. The second is immediate.

now follows inductively. □

In the simplest case, we have $m_B(I) = m_0(B)m_0(I)$ unless the gradient of B divides all initial terms of elements of I.

We define the *valuative closure* of an ideal $I \lhd \mathcal{O}_0$ to be the set of functions $f \in \mathcal{O}_0$ such that for each irreducible curve germ B, we have $m_B(f) \geq m_B(I)$. Then call I *valuatively closed* if it coincides with its valuative closure.

Lemma 11.2.2 *Let I be an $\mathfrak{m}$-primary ideal such that T resolves I; write $[I]_E = \sum_i a_i[E_i]$. Then the following are equivalent:*

(i) *the function f belongs to the valuative closure of I,*
(ii) *we have $[f]_E \geq [I]_E$,*
(iii) *for each i, $f \circ \pi$ vanishes to order at least a_i along E_i.*

Proof First suppose f belongs to the valuative closure. For each i, choose a curvette ϵ_i. Then $m_{\epsilon_i}(f) \geq m_{\epsilon_i}(I)$. Since $m_k(\epsilon_i) = q_{k,i}$, we have

$$[f]_E = \sum_{i,j} m_i(C_f)q_{i,j}[E_j] = \sum_{i,j} m_i(C_f)m_i(\epsilon_j)[E_j] = \sum_j m_{\epsilon_i}(f)[E_j],$$

and similarly, $[I]_E = \sum_j m_{\epsilon_i}(I)[E_j]$. Hence (ii) holds.

Conversely, if (ii) holds, then by Lemma 11.2.1, for any B we have

$$m_B(f) \geq \sum_i m_i(B)m_i(C_f) = \sum_{i,k} \delta_k(B)q_{i,k}m_i(C_f),$$

and since each $\delta_k(B) \geq 0$ and $\sum_i q_{i,k}m_i(C_f) \geq \sum_i q_{i,k}m_i(I)$, this is in turn no less than $\sum_{i,k} \delta_k(B)q_{i,k}m_i(I) = \sum_i m_i(B)m_i(I) = m_B(I)$, so (i) holds.

Condition (ii) is equivalent to (iii) since $a_i = \sum_j q_{i,j}m_j(I)$ and the order of vanishing of f along E_i is $\sum_j m_j(C_f)q_{i,j}$. □

Corollary 11.2.3 *Suppose I is $\mathfrak{m}$-primary; choose a resolution $\pi : T \to S$ of I.*

(i) *There is an integer M such that for $f \in I$, $[I]_E = [f]_E$ unless $j^M f$ lies in a finite union of proper linear subspaces of $j^M I$.*
(ii) *We have $[I]_E = \inf\{[f]_E \mid f \in I\}$, and the bound is attained.*
(iii) *For any function F with $[F]_E \geq [I]_E$ there exist $g \in I$ such that $[g]_E = [I]_E$ and $[F + g]_E = [I]_E$.*

Proof follows from Theorem 11.1.4(i) and Corollary 11.1.5.

follows from Lemma 11.2.2 (ii) and (i).

As to (iii), it suffices if $j^M g$ avoids both the subspaces of (i) and their translates by $j^M F$. But a general point of a real or complex vector space will avoid any given finite union of proper affine subspaces. □

Theorem 11.2.4 *For I an $\mathfrak{m}$-primary ideal with $[I]_E = \sum m_i[\widehat{E}_i]$, we have*

$$\dim(\mathcal{O}_0/I) \geq \sum_i \tfrac{1}{2}m_i(m_i + 1) = -\tfrac{1}{2}[I]_E.([I]_E + [Z]).$$

Equality holds if and only if I is valuatively closed.

Proof We argue by induction on the number of blowings up required to resolve I. Recall that by Proposition 11.1.3(ii), if $I \lhd \mathcal{O}_0$ have order n and $f \in I$ has order n, the induced map $\beta : \mathfrak{m}^n/I \to \bigoplus_j \mathcal{O}_{O_1^{(j)}}/I^{(j)}$ is surjective. Since $\dim(\mathcal{O}_0/\mathfrak{m}^n) = \frac{1}{2}n(n+1)$, we deduce $\dim(\mathcal{O}_0/I) \geq \frac{1}{2}n(n+1) + \sum_j \dim(\mathcal{O}_{O_1^{(j)}}/I^{(j)})$. The first assertion now follows by induction.

Next suppose I is valuatively closed. We claim that in this case, the map β is an isomorphism. For if $f \in \mathfrak{m}^n$ has zero image under β, its image under $g \mapsto x_1^{-n} g(x_1, x_1 y_1)$ belongs, at each point $O_1^{(j)}$, to the corresponding transform of I, so $f^{(j)} \in I^{(j)}$. Hence $[f^{(j)}] \geq [I^{(j)}]$. So $[f] = n[\widehat{E}_0] + \sum_j [f^{(j)}] \geq n[\widehat{E}_0] + \sum_j [I^{(j)}] = [I]$. Since I is valuatively closed, it follows that $f \in I$. Now arguing by induction, as in the first part of the proof, we see that equality holds at each step, and hence in the result.

Conversely, suppose that equality holds. We may assume by induction that each $I^{(j)}$ is valuatively closed. The map β must be an isomorphism. Thus for any f such that $[f]_E \geq [I]_E$, we have $m_0(f) \geq n$ and, calculating as above, $[f^{(j)}] \geq [I^{(j)}]$ for each j. Thus $f^{(j)} \in I^{(j)}$, so the class of f is in Ker (β). Hence $f \in I$. □

We now consider the valuative closure of an arbitrary ideal.

Lemma 11.2.5 *Let $I \lhd \mathcal{O}_0$ have canonical factorisation $\langle h(I) \rangle I_{\text{red}}$.*

(i) *If f belongs to the valuative closure of I, then $h(I)$ divides f.*
(ii) *The valuative closure of I is the product of $\langle h(I) \rangle$ and the valuative closure of I_{red}.*

Proof Factorise $h(I) = \prod f_i^{a_i}$ with the f_i distinct irreducibles. Taking $B = C_{f_i}$, we see that $f_i \,|\, f$. If $a_i > 1$ we must work a little harder.

For a branch B given by $g(x, y) = 0$, consider the sequence B_N of branches given by $g(x, y) + x^N = 0$. For any other branch B', as N increases past the exponent of contact of B and B', the intersection numbers $B'.B_N$ stabilise at the value $B'.B$, whereas $B.B_N$ tends to ∞ with N.

Now let $a_i > 1$ and let B be defined by $f_i = 0$. We have seen that we can write $f = g f_i$. Now for all N, $m_{B_N}(f) \geq m_{B_N}(I)$. Since I is divisible by $\langle f_i^2 \rangle$, removing the factor f_i, we still have $m_{B_N}(g) \geq m_{B_N}(f_i)$. Since this tends to ∞ with N, $m_B(g) = \infty$, so g vanishes on B, and hence is divisible by f_i. We can now repeat the argument.

It is clear that if g is in the valuative closure of $I_{\rm red}$, then $h(I)g$ is in the valuative closure of I. Conversely, by (i) each element of this closure can be written as $h(I)g$. Now for any $f \in I_{\rm red}$ and any branch B, $m_B(h(I)g) \geq m_B(I) = m_B(h(I)) + m_B(I_{\rm red})$, so $m_B(g) \geq m_B(I_{\rm red})$. It follows that g is in the valuative closure of $I_{\rm red}$. □

Lemma 11.2.6

(i) *For any ideals I, I' in $\mathcal{O}_0$, $[II']_E = [I]_E + [I']_E$.*

(ii) *If I and I' are valuatively closed, so is II'.*

Proof By Corollary 11.2.3, $[I]_E = \inf\{[f]_E \,|\, f \in I\}$, and the bound is attained. Now II' is additively generated by products gg' with $g \in I$ and $g' \in I'$. We have $[gg']_E = [g]_E + [g']_E \geq [I]_E + [I']_E$. Hence $[II']_E \geq [I]_E + [I']_E$.

On the other hand, we can choose $f \in I$, $f' \in I'$ such that $[f]_E = [I]_E$ and $[f']_E = [I']_E$. Then $ff' \in II'$ and $[ff']_E = [f]_E + [f']_E = [I]_E + [I']_E$. Hence $[II']_E \leq [I]_E + [I']_E$. Thus equality holds.

(ii) By Lemma 11.2.5 it will suffice to consider the case when I and I' have finite codimension and are resolved by T.

To establish valuative closure, we must show that any function F with $[F]_E \geq [II']_E$ belongs to II'. By Corollary 11.2.3(iii), we see that for some (in fact, most) $g \in II'$ we have not only $[g]_E = [II']_E$ but also $[F + g]_E = [II']_E$. It will thus suffice to show that if $[g]_E = [II']_E$ then $g \in II'$.

By Theorem 2.2.5, we can express $g = \prod_j g_j$ as a product of irreducible factors. Arguing as in Theorem 11.1.4(iii), we see that we may choose g so that each of these corresponds to a curvette in T. Thus for each j, $[g_j]_E$ is of the form $[\epsilon_i]$ for some i. Now as $[I]_E$, $[I']_E \in \mathcal{E}$, we can write $[I]_E = \sum_i a_i[\epsilon_i]$, $[I']_E = \sum_i a'_i[\epsilon_i]$. Since, for each i, there are $a_i + a'_i$ values of j for which $[g_j]_E = [\epsilon_i]$, we can choose a_i among them. Multiply these together for all i, and call the result f, so that $[f]_E = \sum_i a_i[\epsilon_i] = [I]_E$. If f' is the product of the remaining factors of g, then $[f']_E = [I']_E$. But now $g = ff'$ and since I and I' are valuatively closed, we have $f \in I$ and $f' \in I'$. □

This lemma shows that valuatively closed ideals form an abelian semigroup under multiplication. We can give the precise structure of this semigroup.

Proposition 11.2.7 *The semigroup of valuatively closed ideals in $\mathcal{O}_0$ is a free abelian semigroup. The generators are of two types:*

(i) *for each branch B at O, the ideal of functions vanishing on B;*

(ii) *for each infinitely near point of O, corresponding to an exceptional curve E_i in a resolution T, the valuatively closed ideal I_i with $[I_i]_E = [\epsilon_i]$.*

Proof We have a canonical decomposition $I = \langle h(I)\rangle I_{\text{red}}$, and by Lemma 11.2.5, I is valuatively closed if and only if I_{red} is. It is thus sufficient to consider principal ideals and $\mathfrak{m}$-primary ideals separately.

For principal ideals, unique factorisation was established in Theorem 2.2.5, and in Section 2.3 we defined branches to be curves corresponding to irreducible polynomials.

For $\mathfrak{m}$-primary ideals I we showed in Proposition 11.1.3 that I has a resolution $\pi : T \to S$, and in Lemma 11.2.2 that it is determined by the exceptional cycle $[I]_E$ on T, and in Theorem 11.1.4 that this cycle belongs to $\mathcal{E}$. We saw in Proposition 8.2.4 that $\mathcal{E}$ is a free abelian semigroup with generators $[\epsilon_i]$.

Choose functions f, g such that $f = 0$ and $g = 0$ are curvettes meeting E_i at different points. Then the ideal $I := \langle f, g\rangle$ has finite codimension and is resolved by T. Since $[f]_E = [g]_E = [\epsilon_i]$, it follows by Corollary 11.2.3 that $[I]_E = [f]_E = [\epsilon_i]$. By Lemma 11.2.2, the valuative closure $\overline{I}$ of I consists of all functions h with $[h]_E \geq [I]_E$, and $[\overline{I}]_E = [\epsilon_i]$.

Thus any element of $\mathcal{E}$ corresponds to a valuatively closed $\mathfrak{m}$-primary ideal.

The result now follows since the effect of a blow up $\pi : T' \to T$ on $\mathcal{E}(T)$ is to include it in a free abelian semigroup with one extra generator, corresponding to the new exceptional curve, since the strict transform of $[\epsilon_i]$ in T is the cycle $[\epsilon_i]$ in T'. □

11.3 Ideals and clusters

An assignment of non-negative integers k_i to a finite set of infinitely near points O_i is called a *cluster*. We define the notion of a function passing through a cluster, by analogy with vanishing at a set of distinct points of a plane curve, and having multiplicities at least k_i at them. We define the notion of Galois correspondence, and show that 'functions passing through clusters' leads to a Galois correspondence between clusters and subsets of $\mathcal{O}_0$, and hence two closure operations. The closure operation on ideals coincides with valuative closure; we study the induced closure operation on clusters.

We will call T sufficiently large for a cluster K if it contains curves E_i corresponding to each of the points O_i appearing in K. We sometimes write such a cluster as a formal sum $K = \sum_i k_i[O_i]$.

For any (reduced) curve C_f defined by an equation $f = 0$, we have the multiplicities $m_i(C_f)$ of the strict transforms of C at the points $O_i \in T_i$. Then $\{f \,|\, (\forall i) m_i(C_f) \geq k_i\}$ is NOT in general an ideal in $\mathcal{O}_0$. It is convenient to modify this condition as follows.

Given a curve C and a cluster K with multiplicities k_i at O_i, we say that C *passes through* K if the following inductively defined conditions (C_i) hold:

(C_0): $m_0(C) \geq k_0$. Now define the virtual transform $\widetilde{C^1} := \pi_1^*(C) - k_0 E_0$.

(C_i): $m_i(\widetilde{C^i}) \geq k_i$. Now define the next virtual transform $\widetilde{C^{i+1}} := \pi_i^*(\widetilde{C}) - k_i E_i$.

To analyse this condition, we choose a blow up $\pi : T \to S$ which is sufficiently large for K and associate the cluster K with the cycle $[K] := \sum_i k_i[\widehat{E}_i]$ (*not* with $\sum_i k_i[E_i]$).

Lemma 11.3.1

(i) *The curve C passes through the cluster K if and only if $[C]_E \geq [K]$.*

(ii) *For any cluster K, the set of functions $f \in \mathcal{O}_0$ such that C_f passes through K is an ideal $I(K)$.*

Proof It follows by induction that

$$[\widetilde{C}] = \pi^*[C] - \sum_i k_j[\widehat{E}_j] = [C^{(N)}] + [C]_E - [K].$$

The condition that C pass through K is that the coefficient of each $[E_i]$ in this expression is non-negative. But this is equivalent to requiring $[C]_E \geq [K]$.

C_f passes through K if and only if $[f]_E \geq [K]$, i.e. for each i, $f \circ \pi$ vanishes to order at least $\sum_j k_j q_{j,i}$ along E_i. But if this condition holds for f and f', it clearly holds for $f + f'$, and for fg for any $g \in \mathcal{O}_0$. □

This lemma associates to any cluster K an ideal $I(K)$. Conversely, we associate to any $\mathfrak{m}$-primary ideal I the cluster $K(I)$ such that $[K(I)] = \sum_i m_i(I)[\widehat{E}_i]$. Before going more deeply into the properties of these mappings, we give some general theory which adds some perspective. Note that these definitions of $K(I)$ and $I(K)$ are independent of the choice of T, provided T resolves I and all the points of K.

Suppose given two partially ordered sets X and Y, and maps $A : X \to Y$ and $B : Y \to X$. We say these form a *Galois correspondence* if they have the following properties:

(i) If $I \leq I'$ in X, then $A(I) \geq A(I')$.
(ii) If $K \leq K'$ in Y, then $B(K) \geq B(K')$.
(iii) For any $I \in X$, $B(A(I)) \geq I$.
(iv) For any $K \in Y$, $A(B(K)) \geq K$.

Lemma 11.3.2 *Let (X, Y, A, B) be a Galois correspondence. Then $I = B(A(I))$ if and only if $I = B(K)$ for some K; $K = A(B(K))$ if and only if $K = A(I)$ for some I.*

Proof If $I = B(K)$, then $A(I) \geq K$ by (iv), so $B(A(I)) \leq B(K) = I$ by (ii); combining this with (iii) gives $I = B(A(I))$. The converse is trivial. The proof of the second assertion is precisely similar. □

Thus the map $I \to B(A(I))$ is an idempotent map which respects the ordering of X. Such a map is called a *closure operation* on X. We also have a closure operation on Y, and A and B give inverse bijections between the respective sets of closed elements. The question then arises of characterising the closed elements in each case, and of describing the corresponding closure operations.

Before returning to our ideals and clusters, we give two classical examples of Galois correspondences.

Example 11.3.1 Let L be a field of characteristic 0, and G a finite group of automorphisms of L. Let X denote the set of subsets of G, ordered by inclusion, and Y the set of subsets of L, also ordered by inclusion. For any $I \in X$, i.e. $I \subseteq G$, define $A(I)$ to be the set of elements of L fixed under each automorphism belonging to I. For each $K \in Y$, i.e. $K \subseteq L$, define $B(K)$ to be the set of automorphisms in G leaving fixed each element of K.

In this case, the closed elements of X are the subgroups of G, and if L^G denotes the set (in fact, a field) of elements of L fixed under the whole of G, the closed elements of Y are the fields K such that $L^G \subseteq K \subseteq L$.

The second example is closer to our present interests.

Example 11.3.2 Let X denote the set of subsets of the polynomial algebra $\mathbb{C}[x_1, \ldots, x_n]$, and Y denote the set of subsets of complex n-space $\mathbb{C}^n$. For any $I \in X$, i.e. $I \subseteq \mathbb{C}[x_1, \ldots, x_n]$, define $A(I)$ to be the set of points in $\mathbb{C}^n$ at which each $f \in I$ vanishes. For each $K \subseteq \mathbb{C}^n$, define $B(K)$ to be the set of polynomials in $\mathbb{C}[x_1, \ldots, x_n]$ vanishing at each element of K.

In this case, by Hilbert's zero theorem, the closed elements of X are the radical ideals of $\mathbb{C}[x_1, \ldots, x_n]$; the closed elements of Y are called (affine) algebraic varieties. The closure operation is that defining the Zariski topology.

More generally, we may take any commutative ring R in place of $\mathbb{C}[x_1, \ldots, x_n]$, and let X be the set of subsets of R, Spec(R) the set of prime ideals of R, Y the set of subsets of X. Then for $I \in X$, $A(I) := \{P \in \mathrm{Spec}(R) \,|\, I \subseteq P\}$ and for $K \in Y$, $B(K) := \bigcap K \subseteq X$. This defines the Zariski topology on Spec(R).

We now define a Galois correspondence. We order ideals by inclusion, and order clusters using the ordering $\leq$ on the corresponding cycles $[K]$.

Proposition 11.3.3 *For each $\mathfrak{m}$-primary ideal I, choose a resolution T of I and define the cluster $K(I)$ by $[K(I)] = [I]_E$. For each cluster K choose a blow up T which is sufficiently large for K and define the ideal $I(K) = \{f \,|\, [f]_E \geq [K]\}$. Then the mappings $I \to K(I)$ and $K \to I(K)$ between clusters and $\mathfrak{m}$-primary ideals are well defined and give a Galois correspondence.*

Proof Since neither the definition of $K(I)$ nor that of $I(K)$ is affected by a blow up of T, the mappings are well-defined, and it is sufficient to work in a fixed blow up T, chosen sufficiently large for the ideals and clusters in a given calculation. These definitions coincide with those introduced above since $[K(I)] = \sum_i m_i(I)[\widehat{E}_i] = [I]_E$ and by Lemma 11.3.1(i), $I(K) = \{f \,|\, [f]_E \geq [K]\}$.

If $K \geq K'$, then by definition $[K] \geq [K']$, so if $f \in I(K)$ we have $[C(f)]_E \geq [K] \geq [K']$, so $f \in I(K')$.

If $I \subseteq I'$, then by Corollary 11.2.3 we can choose $f \in I$ with $[I]_E = [f]_E$, and as $f \in I'$, $[f]_E \geq [I']_E$. Thus $[K(I)] = [I]_E \geq [I']_E = [K(I')]$.

We have $f \in I(K(I))$ if and only if $[f]_E \geq [K(I)] = [I]_E$. But this certainly holds for $f \in I$.

Finally, as $f \in I(K) \Leftrightarrow [f]_E \geq [K]$, and $[I(K)]_E = \inf\{[f]_E \,|\, f \in I(K)\}$, we have $[K(I(K))] = [I(K)]_E \geq [K]$. □

Thus the map $I \to I(K(I))$ is a closure operation on the set of $\mathfrak{m}$-primary ideals and $K \to K(I(K))$ is a closure operation on the set of clusters, and the two way correspondence induces a bijection between the respective sets of closed elements. The questions arise of characterising the closed elements in each case, and of describing the corresponding closure operations. For ideals a first result is immediate; we will say more in the next section.

Corollary 11.3.4 *If I is an $\mathfrak{m}$-primary ideal, its closure for the above Galois correspondence coincides with its valuative closure. Thus I is Galois closed if and only if it is valuatively closed.*

Proof We saw in the above proof that $f \in I(K(I))$ if and only if $[f]_E \geq [I]_E$. But by Lemma 11.2.2, f belongs to the valuative closure of I if and only if $[f]_E \geq [I]_E$. This proves the result. □

We now discuss the closure of clusters. Say that the cluster $K = \sum_i k_i[O_i]$ satisfies the *proximity inequalities* if (compare (8.1))

$$k_i \geq \sum \{k_j | O_j \text{ proximate to } O_i\}$$

Theorem 11.3.5 *The following are equivalent:*

(i) *The cluster K satisfies the proximity inequalities, i.e. $[K] \in \mathcal{E}$.*
(ii) *The divisor $[K]$ satisfies $[K].[E_i] \leq 0$ for each i.*
(iii) *There exists a reduced curve C such that $[K] = [C]_E$.*
(iv) *The cluster K is closed for the above Galois correspondence.*

Proof The equivalence of (i) and (ii) follows from the definition of $\mathcal{E}$, and equivalence of (ii) and (iii) from Proposition 8.2.4.

Finally, (iv) holds if and only if there is a valuatively closed $\mathfrak{m}$-primary ideal I with $K = K(I)$, hence $[K] = [I]_E$. By Proposition 11.2.7, this is equivalent to $[K] \in \mathcal{E}$. □

The discussion of the closure operation for clusters requires a little preparation.

Lemma 11.3.6

(i) *If $D \in \mathcal{E}$, then $D \geq 0$.*
(ii) *If $D \in \mathcal{E}^+$, then $D \geq \sum E_i$; in particular, $D > 0$.*
(iii) *If $\mathcal{W} \subseteq \mathcal{E}^+$, then* inf $\mathcal{W} \in \mathcal{E}^+$.

Proof This result was established in Proposition 8.2.4 (iii), but we now offer a different argument.

Set $D = \sum a_i E_i = D_+ - D_-$, where $D_+ := \sup\{D, 0\} = \sum_{a_i>0} a_i E_i$, so also $D_- := \sum_{a_i<0}(-a_i)E_i \geq 0$. Since $D \in \mathcal{E}$ and $D_- \geq 0$, $D.D_- \leq 0$. Since $E_i.E_j \geq 0$ for $i \neq j$ and D_+, D_- have no components in common, $D_+.D_- \geq 0$. Thus $D_-.D_- = D_+.D_- - D.D_- \geq 0$. Since the intersection form is negative definite, it follows that $D_- = 0$.

Now $D \geq 0$. If any $a_i = 0$, $D.E_i \geq 0$ while as $-D$ is nef, $D.E_i \leq 0$, so $D.E_i = 0$. Thus for each j such that m_j is adjacent to m_i in the dual tree Γ, we must have $a_j = 0$. By connectivity of Γ, it follows that $D = 0$, contrary to assumption.

Set $W := \inf \mathcal{W} = \sum a_i E_i$, and choose $D^i \in \mathcal{W}$ having a_i as coefficient of E_i. Then $D^i - W \geq 0$ and has zero coefficient of E_i, so $0 \leq E_i.(D^i - W) \leq -E_i.W$ since $D^i \in \mathcal{E}^+$. Thus $W \in \mathcal{E}$. By (ii) we cannot have $W = 0$. □

Proposition 11.3.7 *Write* $[K]^+ := \inf\{x \in \mathcal{E} \,|\, x \geq [K]\}$ *and define* K^* *by* $[K^*] = [K]^+$.

(i) *The cluster* $K^* = K(I(K))$ *is the closure of* K.
(ii) *Set* $[K^*] - [K] := \sum a_i[E_i]$ *and* $b_i := [K^*].[E_i]$, *then* $a_i \geq 0 \geq b_i$ *for each* i.
(iii) *For any* $K \in H$, *if* $K \notin \mathcal{E}$ *choose* i *such that* $[K].[E_i] > 0$ *and add* $[E_i]$ *to* $[K]$. *Iterate this procedure. Then after a finite number of steps you arrive at* K^*.

Proof By Lemma 11.3.6 the infimum belongs to $\mathcal{E}$, so we have defined a unique cluster K^*.

By Proposition 11.3.3 we have $[I(K)]_E \geq [K]$, and by Theorem 11.1.4, $[I(K)]_E \in \mathcal{E}$. Hence $[I(K)]_E \geq [K^*]$, and so $I(K) \subseteq I(K^*)$. Since $K \leq K^*$, $I(K) \supseteq I(K^*)$. Hence $I(K) = I(K^*)$. But as $[K^*] \in \mathcal{E}$, K^* satisfies the proximity inequalities, so is a closed cluster. Thus $K^* = K(I(K^*)) = K(I(K))$.

(ii) follows from the definition of K^*.

(iii) Since $[K].[E_i] > 0 \geq [K^*].[E_i]$, $([K^*] - [K]).[E_i] < 0$, and so $a_i > 0$. Hence $[K^*] \geq [K] + [E_i]$. Hence iterating the procedure always yields cycles $\leq [K^*]$. The procedure thus terminates with K^*, after $\sum_i a_i$ steps. □

The algorithm of (iii) for the closure operation $K \to K(I(K))$ is known as Enriques' unloading algorithm.

The Galois correspondence gives a bijection between closed clusters and (valuatively) closed $\mathfrak{m}$-primary ideals. It follows from Lemma 11.2.6 that addition of (closed) clusters corresponds to multiplication of ideals. We saw in Proposition 11.2.7 that we have a free semigroup, with generators corresponding to infinitely near points, or to classes of the form $[\epsilon_i]$ in some resolution. Alternatively we may say that unique factorisation holds in $\mathcal{E}$, and the primes are the elements $[\epsilon_i]$.

The following result will be useful in the next section.

Lemma 11.3.8

(i) *Let* $[X] = \sum_i a_i[E_i]$ *be a cycle on such that, for each* i, *either* $a_i \geq 0$ *or* $[X].[E_i] \leq 0$. *Then* $[X] \geq 0$.

(ii) *Let* I *be a valuatively closed* $\mathfrak{m}$*-primary ideal; set* $[I]_E = \sum_i \alpha_i[\epsilon_i]$. *Assume* f *is such that for each* i *either* $\alpha_i = 0$ *or the coefficient of* $[E_i]$ *in* $[f]_E$ *is at least equal to its coefficient in* $[I]_E$. *Then* $f \in I$.

Proof Set $[X^+] := \sup([X], 0) = \sum_{a_i \geq 0} a_i[E_i]$, and $[X] = [X^+] - [X^-]$. Then if $a_i \geq 0$, the coefficient of $[E_i]$ in $[X^-]$ vanishes hence, since $[E_i].[E_j] \geq 0$ for $i \neq j$, $[X^-].[E_i] \geq 0$.

If $a_i < 0$ then, by hypothesis, $[X].[E_i] \leq 0$. Since the coefficient of $[E_i]$ in $[X^+]$ vanishes, we have $[X^+].[E_i] \geq 0$. Hence $[X^-].[E_i] = [X^+].[E_i] - [X].[E_i] \geq 0$.

Thus $[-X^-].[E_i] \leq 0$ for all i, so $[-X^-] \in \mathcal{E}$. By Proposition 8.2.4, $[X] \geq 0$.

Set $[X] := [f]_E - [I]_E = \sum_i a_i[E_i]$. Then the hypothesis is equivalent to saying that, for each i, either $a_i \geq 0$ or $[I]_E.[E_i] = 0$ so that $[X].[E_i] = [f].[E_i] = -\delta_i(C_f) \leq 0$. Hence by (i), $[X] \geq 0$. Thus $[f]_E \geq [I]_E$, and now, since I is valuatively closed, it follows that $f \in I$. □

A direct construction of valuatively closed ideals is as follows.

Lemma 11.3.9 *Let* B *be an irreducible curve, and* $v > 0$ *an integer. Then* $I(B, v) := \{f \in \mathcal{O}_0 \,|\, m_B(f) \geq v\}$ *is a valuatively closed ideal. Moreover*

$$[I(B, v)]_E = \inf\{X \in \mathcal{E} \,|\, -(X.[B]_E) \geq v\}.$$

Proof It is clear that $I(B, v)$ is an ideal. To prove it valuatively closed, note that $f \in I(B, v)$ if and only if $[B].[f] \geq v$. By Lemma 8.2.5, this is equivalent to $-([B]_E.[f]_E) \geq v$. This implies $[f]_E \geq \inf\{X \in \mathcal{E} \,|\, -(X.[B]_E) \geq v\}$. By Lemma 11.3.6(iii), the infimum is an element of $\mathcal{E}$, hence is of the form $[K]$, for some closed cluster K.

We have thus shown that $f \in I(B, v)$ if and only if $[f]_E \geq [K]$, which by Lemma 11.3.1 is equivalent to $f \in I(K)$. Thus $I(B, v)$ coincides with the valuatively closed ideal $I(K)$. □

We now give an explicit form for generators of the semigroup of closed ideals. Consider a curvette ϵ_j; let the m_i be its successive multiplicities attached to the curves E_i of the T. Write $M_j :=|, \sum_i m_i^2 = -[\epsilon_j].[\epsilon_j]$.

Lemma 11.3.10 *The ideal $I(\epsilon_j, M_j)$ has cycle $[I(\epsilon_j, M_j)]_E = [\epsilon_j]_E = [\epsilon_j]$.*

Proof For any f, we have $m_{\epsilon_j}(f) \geq \sum_i m_i(\epsilon_j) m_i(C_f)$, with equality if the strict transforms of ϵ_j and C_f are disjoint.

If the strict transforms are not disjoint, it follows by backwards induction from the proximity inequalities for C_f that $m_i(C_f) \geq m_i(\epsilon_j)$ for all i.

If the strict transforms are disjoint, and $m_0(C_f) < m_0(\epsilon_j)$, it follows by forwards induction from the proximity inequalities for C_f that $m_i(C_f) < m_i(\epsilon_j)$ for all i, and so $\sum_i m_i(\epsilon_j) m_i(C_f) < \sum_i m_i(\epsilon_j)^2 = M_j$.

Thus in either case, it follows from $f \in I(\epsilon_j, M)$ that $m_0(C_f) \geq m_0(\epsilon_j)$. That this is the least value follows since if C_f is another curvette corresponding to a different point of E_j, then $m_0(C_f) = m_0(\epsilon_j)$ and $[f].[\epsilon_j] = M$.

Now blow up and iterate the argument. It follows by induction that $m_i(I(\epsilon_j, M)) = m_i(\epsilon_j)$, so

$$[I(\epsilon_j, M)]_E = \sum m_i(I(\epsilon_j, M))[\widehat{E}_i] = \sum m_i(\epsilon_j)[\widehat{E}_i] = [\epsilon_j]_E.$$

□

In fact, all valuatively closed $\mathfrak{m}$-primary ideals can be constructed by this procedure. Express $[I]_E = \sum a_i[E_i] = \sum_i \alpha_i[\epsilon_i]$, and choose a curvette B_i for each i such that $\alpha_i > 0$. Then $m_{B_i}(I) = -[B_i]_E.[I]_E = -\epsilon_i.[I]_E = a_i$. If, for each of these curvettes, $m_{B_i}(f) \geq m_{B_i}(I)$ then, by Lemma 11.3.8, $f \in I$. Hence $I = \bigcap_{\alpha_i > 0} I(B_i, a_i)$.

11.4 Integrally closed ideals

We begin with the notion of integral closure of rings, and give a characterisation and simple examples. This leads to the definition of integral closures of ideals. The main result in the section is that the integral closure of an ideal in $\mathcal{O}_0$ coincides with its valuative closure. We prove this via a result which is of considerable interest in its own right, giving a sufficient condition for a function h to lie in the ideal generated by two given functions $\langle f, g\rangle$.

Let R be a subring of the (commutative) ring S. An element $x \in S$ is said to be *integral over* R if it satisfies an equation $x^k + \sum_1^k a_i x^{k-i} = 0$ such that each $a_i \in R$. The set of elements of S integral over R is called the *integral closure* of R in S.

Lemma 11.4.1

(i) *Let M be a faithful S-module which is finitely generated over R, and let $s \in S$ be such that $sM \subseteq M$. Then s is integral over R.*

(ii) *The element $s \in S$ is integral over R if and only if the subring of S generated by R and s is finitely generated as R-module.*

(iii) *The integral closure of R in S is a subring.*

Proof Choose a finite set $\{m_i\}$ of generators of M as R-module. Then we can find coefficients $a_{i,j} \in R$, forming a matrix A, such that $sm_i = \sum_j a_{i,j} m_j$. Write Δ for the determinant of the matrix $sI - A$. Then $\Delta m_i = 0$ for each i, so $\Delta = 0$. But expanding the determinant gives an equation showing that s is integral over R.

By (i), the condition is sufficient for integral closure. For the converse, note that if s satisfies an integral relation of degree k, then $R[s]$ is generated as R-module by $1, s, s^2, \ldots, s^{k-1}$.

Let $s, t \in S$ be integral over R, satisfying equations of respective degrees k and l. Then the R-submodule M of S spanned by $\{s^i t^j \mid 0 \leq i < k, 0 \leq j < l\}$ is closed under multiplication by s and t, hence is a subring. For any element $m \in M$, $mM \subseteq M$, so by (i) m is integral over R. Thus the set of elements of S integral over R is closed under addition, subtraction and multiplication, hence is a subring. □

The ring R is said to be *integrally closed in* S if any element of S integral over R belongs to R.

This definition arises in algebraic number theory, where one can take R to be the ring $\mathbb{Z}$ of integers and S a field which is a finite extension of $\mathbb{Q}$: the elements of the integral closure of $\mathbb{Z}$ in S are called the integers of S.

Example 11.4.1 Take S to be the field $\mathbb{Q}[\sqrt{5}]$. Then $\tau := \frac{1}{2}(1+\sqrt{5})$ is integral over $\mathbb{Z}$ since $\tau^2 = \tau + 1$. One can show (Exercise 11.8.8) that the integral closure of $\mathbb{Z}$ in S consists of the elements $a + b\tau$ with $a, b \in \mathbb{Z}$.

An example more pertinent to the topics in this book is as follows. Let B be an irreducible curve defined by an equation $f(x, y) = 0$, and with a parametrisation $\gamma : \mathbb{C} \to \mathbb{C}^2$. We have defined the local ring $\mathcal{O}_B$ of B to be the quotient $\mathcal{O}_0/\langle f \rangle$, and seen that γ induces an injective homomorphism $\gamma^* : \mathcal{O}_B \to \mathcal{O}_t$: let us identify $\mathcal{O}_B$ with its image under γ^*.

Lemma 11.4.2 *The ring $\mathcal{O}_t$ is integral over $\mathcal{O}_B$, and coincides with the integral closure $\overline{\mathcal{O}}_B$ of $\mathcal{O}_B$ in the field $\mathcal{O}\{\{t, t^{-1}\}\}$.*

Proof It follows from Proposition 4.3.1 that a finite set of powers of t forms a basis for $\mathcal{O}_t/\mathcal{O}_B$ over $\mathbb{C}$, so $\mathcal{O}_t$ is finitely generated as $\mathcal{O}_B$-module. It follows that every element of $\mathcal{O}_t$ is integral over $\mathcal{O}_B$.

On the other hand, if a Laurent series $\alpha = \sum_{-N}^{\infty} a_r t^r$ with $N > 0$ and $a_{-N} \neq 0$ is in the integral closure of $\mathcal{O}_B$, then so is $t^{N-1}\alpha$ and hence (subtracting an element of $\mathcal{O}_t$) so is $a_{-N}t^{-1}$, hence also t^{-1}. But this is false: there is no identity $t^{-k} + \sum_1^k a_i t^{i-k} = 0$ with the a_i all in $\mathcal{O}_t$ for otherwise (multiplying by t^{k-1}) it would follow that $t^{-1} \in \mathcal{O}_t$, which is not the case. □

A similar discussion applies to the case of a curve C with several branches B_j. In this case, if $\{t_j\}$ denote parameters on the several branches, we obtain an identification of $\overline{\mathcal{O}}_B$ with $\bigoplus_j \mathbb{C}\{t_j\}$. This coincides with the local ring of the normalisation of the curve. In general, if R is an integral domain with quotient field K, the integral closure of R in K is called the *normalisation* of R.

We turn to integral closures of ideals. Let R be any commutative ring and $I \lhd R$. An element $x \in R$ is said to be *integral over* I if it satisfies an equation $x^k + \sum_1^k a_i x^{k-i} = 0$ such that each $a_i \in I^i$. The set $\overline{I}$ of elements integral over I is called the *integral closure* of I.

There is a characterisation of integral closure of ideals analogous to that for rings. To simplify the exposition we assume – as is the case in the examples of interest to us – that the (commutative) ring R is without zero-divisors and is noetherian (i.e. all ideals are finitely generated).

Lemma 11.4.3

(i) *Let I be an ideal of R. Then x is in the integral closure of I if and only if there is an ideal M of R such that $xM \subseteq IM$.*

(ii) *The integral closure of I is an ideal in R.*

Proof The argument is very similar to the ring case. Choose a finite set $\{m_i\}$ of generators of M as R-module. Then we can find coefficients $a_{i,j} \in I$, forming a matrix A, such that $xm_i = \sum_j a_{i,j}m_j$. Write Δ for the determinant of the matrix $sI - A$. This must vanish; expanding it gives an equation showing that x is integral over I.

Conversely, if x is integral over I, take $M := \sum_1^k x^{k-i}I^{i-1}$. This is finitely generated since I is, and $xM \subseteq IM$ since, by the dependence relation, $x^k \in \sum_1^k x^{k-i}I^i$.

Given x, x' both integral over I, with corresponding ideals M, M', the product ideal MM' satisfies $x(MM') = (xM)M' \subseteq (IM)M' = I(MM')$ and similarly for x'. Hence any of the elements $x \pm x'$ or rx (with $r \in R$) satisfy the condition $y(MM') \subseteq I(MM')$, so are again integral over I. □

The ideal I is said to be *integrally closed* if any element integral over I belongs to I, or equivalently, if $I = \overline{I}$.

Example 11.4.2 Let $I = \langle x^4, y^8\rangle$. Then the integral closure of I contains $a = x^3y^2$, since $a^4 = (x^4)^3(y^8) \in I^4$; $b = x^2y^4$, since $b^4 = (x^4)^2(y^8)^2 \in I^4$; and $c = xy^6$, since $c^4 = (x^4)(y^8)^3 \in I^4$. It can be shown (Exercise 11.8.10) that $\overline{I}$ is generated by these elements together with I, so $\overline{I} = \langle x^4, x^3y^2, x^2y^4, xy^6, y^8\rangle$.

Another example, which we use below, is as follows.

Lemma 11.4.4

(i) *Let R be a unique factorisation domain and P a principal ideal in R. Then P is integrally closed.*

(ii) *For any ideal $I \lhd R$, $\overline{PI} = P\overline{I}$.*

Proof Choose a generator f of P, and let $f = \prod_i p_i^{a_i}$ be a prime factorisation of f. We wish to show that any element $r \in R$ which is integral over P belongs to P, i.e. is divisible by f: thus it is enough to show that, for each i, r is divisible by $p_i^{a_i}$.

By hypothesis there is an equation $r^n + \sum_{j=1}^n A_j r^{n-j} = 0$ where $A_j \in P^j$, so we can write $A_j = c_j p_i^{ja_i}$ for some $c_j \in R$. Let $p_i^{b_i}$ be the highest power of p_i which divides r. If $b_i < a_i$ then all terms of the equation except one, namely r^n, are divisible by $p_i^{nb_i+1}$, a contradiction. Hence indeed r is divisible by $p_i^{a_i}$.

Let r be integral over PI, and satisfy $r^n + \sum_{j=1}^n A_j r^{n-j} = 0$, where $A_j \in (PI)^j$ may be written $f^j B_j$ with $B_j \in I^j$. By (i), r is divisible by f, say $r = sf$, and dividing through by f^n now gives $s^n + \sum_{j=1}^n B_j s^{n-j} = 0$. Hence $s \in \overline{I}$. □

In several situations, integral closure can be characterised by a valuative criterion. In this section we are aiming to prove the

Theorem 11.4.5 *An ideal $I \lhd \mathcal{O}_0$ is valuatively closed if and only if it is integrally closed.*

An $\mathfrak{m}$-primary ideal is closed for the Galois correspondence of Proposition 11.3.3 if and only if these conditions hold.

We saw in Corollary 11.3.4 that an $\mathfrak{m}$-primary ideal is closed for the Galois correspondence if and only if it is valuatively closed.

Next suppose I valuatively closed, and let $I = \langle h(I)\rangle I_{\text{red}}$ be its canonical factorisation. By Lemma 11.4.4, $\langle h(I)\rangle$ is integrally closed, and $\overline{I} = \langle h(I)\rangle \overline{I_{\text{red}}}$. Thus to prove I integrally closed it will suffice to consider the case when I is $\mathfrak{m}$-primary, so we can write $I = I(K)$.

Let $f \in \overline{I(K)}$. Then there is an equation $x^n + \sum_1^n a_r x^{n-r} = 0$ such that each $a_r \in I(K)^r$. Now if $[K] = \sum_i k_i[\widehat{E}_i]$, we have $g \in I(K)$ if and only if, for each i, $g \circ \pi$ vanishes to order at least k_i along E_i. It follows that if $b \in I(K)^r$, then for each i, $b \circ \pi$ vanishes to order at least rk_i along E_i.

We can now argue as in Lemma 11.4.4. Consider the curves E_i in turn. Suppose x vanishes to order h_i, but to no higher order, along E_i; suppose if possible $h_i < k_i$. Then for $1 \leq r \leq n$ the term $a_r x^{n-r}$ vanishes to order at least $rk_i + (n-r)h_i \geq k_i + (n-1)h_i$ along E_i, hence (the negative of) their sum, viz. x^n, also vanishes to at least this order. But it vanishes to no higher order than nh_i, and $h_i < k_i$: a contradiction. Hence indeed $h_i \geq k_i$ for each i, and so $x \in I(K)$. This proves that $I(K)$ is integrally closed.

Our proof of the converse involves a result of independent interest.

Let $f, g \in \mathbb{C}[x, y]$. Then clearly if $h \in \langle f, g\rangle$, h must vanish at any common zero of f and g. More precisely, at each common zero P, $h \in \langle f, g\rangle\mathcal{O}_P$ in the local ring at P (if the curves $f = 0$ and $g = 0$ are smooth and transverse to each other at P, this is automatic).

It is natural to ask whether the converse holds. If the condition holds and we restrict to the curve C_f, the quotient h/g is holomorphic except perhaps at the points P, and bounded at those points, so defines a regular function on the curve. If also C_f is smooth, at P, such a function is the restriction to C_f of a function h holomorphic near P, and h-bg is divisible by f.

It is also natural to ask for ideals $\langle f, g\rangle \lhd \mathcal{O}_0$ whether, if any infinitely near point common to C_f and C_g also lies on C_h, then it follows that $h \in \langle f, g\rangle$. This is not true (see Exercise 11.8.7), but there is a criterion which we now prove.

Suppose $f, g \in \mathcal{O}_0$ share no common factor. Blow up sufficiently often to separate the strict transforms of C_f and C_g. List the E_i which appear in both $[f]_E$ and $[g]_E$, and define a cluster $K = \sum_i k_i[O_i]$ by setting $k_i := m_i(f) + m_i(g) - 1$ for these values of i, and 0 for other values. (By Exercise 11.8.11, this satisfies the proximity inequalities.)

Theorem 11.4.6 (Noether's $Af + Bg$ theorem) *If h goes through the cluster K, then $h \in \langle f, g\rangle$. Equivalently, $I(K) \subseteq \langle f, g\rangle$.*

Proof We argue by induction on the number of blow ups required to form the tree. Thus we may suppose inductively that the result holds for the situation after a single blow up. So $h_1 \in \langle f_1, g_1\rangle$ at each point $O_1^{(j)}$ of this blow up, and hence in $\bigoplus_j \mathcal{O}_{O_1^{(j)}}$.

Now restrict to the curve C_f and its blow up: write m for the order of C_f at O. We see that the class of h_1 in $\bigoplus_j \mathcal{O}_{O_1^{(j)}}/\langle f_1^{(j)}\rangle$ is of the form $g_1 k_1$ for some k_1. By Proposition 11.1.3(i), the class of $y_1^{m-1}k_1$ is the total transform of some element of $\mathcal{O}_0/\langle f\rangle$, so comes from an element $B \in \mathcal{O}_0$. As $h = y^{m+m_0(g)-1}h_1$ and $g = y^{m_0(g)}g_1$, we see that $h - Bg$ restricts to 0 on C_f, hence is divisible by f. □

We are now ready to complete the

Proof [of Theorem 11.4.5]. Suppose I is valuatively closed; let $I = \langle h(I)\rangle I_{\text{red}}$ be its canonical factorisation. By Proposition 11.2.7, I_{red} is valuatively closed, and by Lemma 11.4.4 $\overline{I} = \langle h(I)\rangle \overline{I_{\text{red}}}$. It thus suffices to consider $\mathfrak{m}$-primary ideals. We will show that the valuative closure of any $\mathfrak{m}$-primary ideal I is contained in the integral closure: thus if I is valuatively closed, it is integrally closed.

By Theorem 11.1.4(iii), we can choose two elements $f, g \in I$, with $[f]_E = [g]_E = [I]_E$, and such that each of C_f and C_g is resolved in T into a union of curvettes, the two having no common point.

Write $K := K(I)$ and J for the valuative closure $I(K)$. By Lemma 11.2.6, $J^2 = I(2K)$. By Theorem 11.4.6, any $h \in J^2$ is of the form $\lambda f + \mu g$.

Thus the restrictions of h and μg to C_f coincide, so for each component B of C_f, $m_B(h) = m_B(\mu) + m_B(g)$. By hypothesis, $m_B(h) \geq 2m_B(I)$

and by construction, $m_B(g) = m_B(I)$, so $m_B(\mu) \geq m_B(I)$, i.e. $\mu \in I(B, m_B(I))$.

Now if $[I]_E$ is expressed in the form $\sum_i \alpha_i[\epsilon_i]$, the components of C_f will include, for each i, α_i different curvettes B_i corresponding to the exceptional curve E_i. Thus, for each i such that $\alpha_i > 0$, we have $m_{B_i}(\mu) \geq m_{B_i}(I) = m_{B_i}(J)$. Since J is valuatively closed, it follows from Lemma 11.3.8, that $\mu \in J$. Similarly, $\lambda \in J$.

Thus $J^2 \subseteq \langle f, g\rangle J \subseteq IJ$. It now follows from Lemma 11.4.3(i), taking J as the module M in that result, that J is contained in the integral closure of I.

11.5 Jets and determinacy

We recall that the K-jet $j^K f$ of a function $f(x, y)$ is essentially the same as the set of terms of degree $\leq K$ in the power series expansion of f. In this section we define several equivalence relations on the set of functions $f \in \mathfrak{m}$. We will see that for each reduced f there is an integer K such that all g with $j^K g = j^K f$ are equivalent to f. We say f is determined by its K-jet, or K-determined, up to the equivalence relation in question. We will discuss how to find the least value of K with this property, usually called the degree of determinacy of f.

If all the functions in an equivalence class $\mathcal{E}$ are K-determined, then we can think of the equivalence class as a subset $J^K(\mathcal{E})$ of the finite dimensional vector space J^K of K-jets. In practice, this subset is an open subset of an algebraic variety, so has a well defined dimension, and hence codimension in J^K; the codimension remains the same if we increase K, so we can define it to be the codimension of $\mathcal{E}$. We will also discuss the calculation of these codimensions.

Two functions $f, g \in \mathfrak{m}$ are said to be *right equivalent* if there is a local diffeomorphism $h : (\mathbb{C}^2, O) \to (\mathbb{C}^2, O)$ such that $h \circ f = g$. We shall say that f and g are $\mathcal{K}$-equivalent if there is a local diffeomorphism $h : (\mathbb{C}^2, O) \to (\mathbb{C}^2, O)$ such that $h(C_f) = C_g$. We already defined equisingularity in Chapter 4. It is clear that if f and g are right equivalent, they are $\mathcal{K}$-equivalent, and that if f and g are $\mathcal{K}$-equivalent, then C_f and C_g are equisingular.

Right equivalence is of considerable importance, but is better studied by different techniques to those developed in this book. We state the main conclusions here: see the Notes for references. The results are expressed in terms of the Jacobian ideal $J(f) := \langle \partial f/\partial x, \partial f/\partial y\rangle$. Since by

Theorem 6.5.6, provided f is reduced, the curves $\partial f/\partial x = 0$, $\partial f/\partial y = 0$ have (finite) intersection number $\mu(f)$, the elements of $J(f)$ have no common factor, so by Lemma 11.1.2 $J(f)$, and hence also $\mathfrak{m}J(f)$, is $\mathfrak{m}$-primary and thus of finite codimension.

Proposition 11.5.1 *If f is k-determined (for right equivalence), then $\mathfrak{m}J(f) \supseteq \mathfrak{m}^{k+1}$. Conversely, if $\mathfrak{m}J(f) \supseteq \mathfrak{m}^k$ (or even if $\mathfrak{m}^2 J(f) \supseteq \mathfrak{m}^{k+1}$), then f is k-determined.*

In particular, any reduced germ is finitely determined for right equivalence, and hence also for $\mathcal{K}$-equivalence and for equisingularity.

The relation $\mathcal{K}$-equivalence is often called 'contact equivalence', and is also frequently termed 'analytic equivalence'. There is an analogous result here also, proved by the same method.

Proposition 11.5.2 *If f is k-determined (for $\mathcal{K}$-equivalence), then $\mathfrak{m}J(f) + \langle f \rangle \supseteq \mathfrak{m}^{k+1}$. Conversely, if $\mathfrak{m}J(f) + \langle f \rangle \supseteq \mathfrak{m}^k$, then f is k-determined.*

There is a neat criterion for $\mathcal{K}$-equivalence.

Lemma 11.5.3 *The reduced germs f, g are $\mathcal{K}$-equivalent if and only if the algebras $\mathcal{O}_{C_f}$ and $\mathcal{O}_{C_g}$ are isomorphic.*

Proof The direct implication is immediate. For the converse, first observe that C_f is smooth at O if and only if $f \notin \mathfrak{m}^2$ if and only if the quotient of $\mathcal{O}_{C_f}$ by the square of its maximal ideal has dimension 2: otherwise this quotient has dimension 3. It thus suffices to consider the case $f \in \mathfrak{m}^2$. Now let H be an isomorphism. Consider the diagram

$$\begin{array}{ccc} \mathcal{O}_0 & \dashrightarrow & \mathcal{O}_0 \\ \downarrow & & \downarrow \\ \mathcal{O}_{C_f} & \xrightarrow{H} & \mathcal{O}_{C_g} \end{array}$$

Take $x \in \mathcal{O}_0$; then the image by H of its class in $\mathcal{O}_{C_f}$ lifts to an element $X \in \mathcal{O}_0$; similarly, define Y. Observe that the constant term in each of X and Y vanishes. Define a holomorphic map $h : \mathbb{C}^2 \to \mathbb{C}^2$ near O by $h(x, y) = (X(x, y), Y(x, y))$.

Since we have excluded the smooth case, the arrows in the diagram all induce isomorphisms modulo the square of the maximal ideal. Hence the linear terms in X and Y are linearly independent linear combinations of x and y. Thus the map h is a local diffeomorphism. It induces the homomorphism $\tilde{H} : \mathcal{O}_0 \to \mathcal{O}_0$ such that any power series in x and y goes to the corresponding power series in X and Y (one can verify that this always yields a convergent power series), and hence the given isomorphism H. □

We have seen that two curves are equisingular if they have resolution trees which are isomorphic in an appropriate sense. As in the case of ideals, each step in the resolution is determined by an appropriate jet of the original defining function. Thus for each reduced f, the equisingularity class of f is indeed determined by the M-jet, for M sufficiently large.

Let $\pi : T \to S$ be the minimal good resolution of C_f. We use exceptional cycles in T.

Lemma 11.5.4 *We have $n[\widehat{E}_0] \geq [f]_E$ if and only if every $g \in \mathfrak{m}^n$ passes through the cluster $[f]_E$.*

Proof We have $g \in \mathfrak{m}^n$ if and only if the order of g is at least n; by definition this is equivalent to g passing through the cluster $n[O_0]$. By Lemma 11.3.1(i), this is in turn equivalent to $[g]_E \geq n[\widehat{E}_0]$, i.e. to $[g]_E \geq n[\widehat{E}_0]$. (Note that $n[\widehat{E}_0] \in \mathcal{E}$.)

Now by Lemma 11.3.1(i) again, g passes through $[f]_E$ if and only if $[g]_E \geq [f]_E$. So every $g \in \mathfrak{m}^n$ passes through $[f]_E$ if and only if $[g]_E \geq n[\widehat{E}_0]$ implies $[g]_E \geq [f]_E$, or equivalently, $n[\widehat{E}_0] \geq [f]_E$. □

We can sharpen this result as follows. By (8.6), and the fact that the $[\epsilon_i]$ are the negative dual base to the $[E_i]$, for any g, $M_i(C_g) = -[g]_E.[\epsilon_i]$. Thus if $g \in \mathfrak{m}^n$ so that $[g]_E \geq n[\widehat{E}_0]$, we have $M_i(g) \geq -n[\widehat{E}_0].[\epsilon_i] = nm_i$ by Corollary 8.2.2 (i). If also $nm_i > M_i(f)$, it follows that $M_i(g) > M_i(f)$, i.e the order of vanishing of g along E_i strictly exceeds that of f. Thus when we form strict transforms (so the strict transform of f along E_i vanishes just where E_i meets the strict transform of C_f), the difference of the strict transforms of f and $f+g$ vanishes along E_i. Thus

Lemma 11.5.5 *If $g \in \mathfrak{m}^n$ and $nm_i > M_i(f)$, then the strict transforms of C_f and C_{f+g} meet E_i in the same points.*

This focusses our attention on the quotients $M_i(f)/m_i$. Recall that, by Proposition 9.4.3, the maximum of the quotients $\frac{M_i(C)}{m_i}$ as V_i runs through rupture points of $\Gamma_R(C)$ is the maximum polar quotient $Q(C)$. The following is the main result of this section.

Theorem 11.5.6 *The degree of determinacy of f for equisingularity, i.e. the least integer K such that, for all $g \in \mathfrak{m}^K$, $f + g = 0$ is equisingular to $f = 0$, coincides with the least integer $\lceil Q(C_f) \rceil$ greater than $Q(C_f)$.*

If $n > Q(C_f)$ then $n > M_i(C)/m_i$ for all rupture points V_i on $\Gamma_R(C_f)$. If the inequality holds for other vertices as well, then by Lemma 11.5.4, every $g \in \mathfrak{m}^n$ passes through the cluster $[f]_E$. Further, by Lemma 11.5.5, the strict transforms of C_f and C_{f+g} meet E_i in the same points. The desired equisingularity now follows.

However, it is not necessarily the case that $Q(C) \geq M_i(C)/m_i$ for all vertices V_i in a minimal good resolution graph. We leave it to the reader to verify that a singularity of type E_7 gives an example. The following partial result will allow us to proceed.

Lemma 11.5.7 *If $\Gamma_R(C)$ is a minimal good resolution graph for C, then $\lceil Q(C) \rceil \geq M_i(C)/m_i$ for* all *vertices V_i of $\Gamma_R(C)$.*

Proof Suppose V_i a vertex of $\Gamma_R(C)$ such that $M_i(C)/m_i > Q(C)$. Since, by Proposition 8.3.2, the quotients $M_i(C)/m_i$ are strictly increasing as V_i moves along the core from V_0 to arrowhead vertices, V_i must lie on the path connecting an arrowhead vertex W to the nearest rupture point. It suffices to consider the case when V_t is adjacent to W.

Then the valence $v_t = 2$ and E_t meets the union of other components of $\pi^{-1}(C)$ in just two points: $E_t \cap B$ and another point P. We must have $a_t > 1$, else there would have been no need for the blow up that produced E_t, so the resolution would not be minimal. Thus at the stage of the resolution process where E_t is first created, P must lie on the strict transform of C, as well as on E_{t-1}. We then continually blow up P producing a sequence of exceptional curves $E_{t+1}, \ldots, E_{t+j}$ such that at the end of this sequence the strict transform of C no longer passes through P. Now if the strict transform of C meets E_{t+j} at a point other than its intersections with E_t and E_{t+j-1}, V_{t+j} is already a rupture vertex. Otherwise we need to blow up again; this has the effect on the dual graph of subdividing the edge $V_{t+j-1}V_{t+j}$, and to continue to do so until we create a rupture vertex. In this case there is a rupture point V_r between V_{t+j-1} and V_{t+j} in the graph.

The curve C is the union of the branch B and its complement, C', say. By Proposition 8.3.2 again, $M_i(C')/m_i$ is constant along the path V_rV_t. The behaviour of $M_i(B)$ and m_i along the path between the final rupture point of $\Gamma_R(B)$ and W is given by (8.12) and (8.11): for each blow up that interpolates a vertex next to W, you add 1 to $M_i(B)$ while m_i is unchanged; for a blow up subdividing an edge V_iV_j at V_I, we have $M_I(B) = M_i(B) + M_j(B)$ and $m_I = m_i + m_j$.

Write, for short, $M := M_{t-1}(B)$, $m := m_{t-1}$ and $M' := M_t(C')$. Then $M_t(B) = M+1$, $m_t = m$, and subdividing successively gives $m_{t+r} = (r+1)m$, $M_{t+r}(B) = (r+1)(M+1)-1$ for $1 \leq r \leq j$. Suppose first that V_{t+j} is a rupture point. Then $M_{t+j}(C') = (j+1)M'$, so $mM_{t+j}(C)/m_{t+j} = M'+M+1-\frac{1}{j+1}$. Since $m\lceil Q(C)\rceil \geq mM_{t+j}(C)/m_{t+j}$ and is an integer, we must have $m\lceil Q(C)\rceil \geq M' + M + 1$, which establishes the result in this case.

Otherwise we require a further series of blowings up, which interpolate points between V_{t+j-1} and V_{t+j} on the graph, till we arrive at a rupture point V_r. Applying (8.12) and (8.11) appropriately, we find that for some positive integers p and q we have $m_r = p(jm) + q((j+1)m)$ and $M_r(B) = p(j(M+1)-1) + q((j+1)(M+1)-1)$. Again we have

$$\begin{aligned} M_r(C') &= m_r M_i(C')/m_i = (p(jm) + q((j+1)m))M'/m \\ &= (pj + q(j+1))M'. \end{aligned}$$

We thus obtain

$$m\lceil Q(C)\rceil \geq M' + M + 1 - \frac{p+q}{pj + q(j+1)},$$

and again it follows that $m\lceil Q(C)\rceil \geq M' + M + 1$. □

It follows from monotonicity that equality can only hold for the vertex V_t adjacent to the arrowhead vertex.

Proof [of Theorem 11.5.6] It follows from Lemma 11.5.5 that for $g \in \mathfrak{m}^n$ and $n > Q(C_f)$, the strict transform of g vanishes to higher order than that of f on all E_i corresponding to rupture vertices V_i, and thus on all vertices lying between V_0 and any of these. Thus $M_i(g) > M_i(f)$ and so $M_i(f+g) = M_i(f)$ in all these cases, so the minimal good resolution of $f+g$ contains all these exceptional curves with the same numerical invariants as before.

For the other vertices V_i, by Lemma 11.5.7 we have $M_i(g) \geq M_i(f)$ and hence $M_i(f+g) \geq M_i(f)$. Moreover, by the remark following the lemma, if $M_i(f+g) > M_i(f)$, the vertex V_i is adjacent to an arrowhead vertex. Call V_i exceptional if this occurs.

Now let us compare the numerical functions associated to f and to $f+g$ on the minimal good resolution graph $\Gamma_R(C_f)$. Suppose there is an exceptional vertex V_t. Then V_t is adjacent to an arrowhead vertex W and to just one other vertex, V_r say. It is not possible that V_r also is exceptional. Apply (8.12) at V_r to both C_f and C_{f+g}:

$$a_r M_r(C) = \delta_r(C) + \sum\{M_j(C) \mid V_j \text{ adjacent to } V_r\}.$$

The left hand side is the same in both cases; we have $M_j(C_{f+g}) \geq M_j(C_f)$ for each j, with strict inequality if $j = t$. Hence $\delta_r(C_{f+g}) < \delta_r(C_f)$. In particular, $\delta_r(C_f) > 1$, so V_r is adjacent to an arrowhead vertex W', say.

By Lemma 11.5.5, the strict transforms of C_f and C_{f+g} meet E_r in the same points. The point $E_r \cap B'$ is a point of transverse intersection, so the restriction to E_r of the strict transform of f vanishes simply at this point; hence the same holds for $f+g$. Thus C_{f+g} also has a branch through this point. Hence a decrease in δ_r is not possible.

We have thus established that $g \in \mathfrak{m}^n$ implies that $[f+g]_E = [f]_E$. We know that C_f is a union of curvettes corresponding to vertices V of the dual tree: there are $\delta_i(f)$ curvettes at V_i. It now follows that C_{f+g} has the same description, so C_f and C_{f+g} are equisingular.

Now suppose $n \leq Q(C_f)$: in fact, first consider the case $n < Q(C_f)$. By Theorem 11.1.4, we can find $g \in \mathfrak{m}^n$ such that $[g]_E = [\mathfrak{m}^n]_E = n[\widehat{E}_0]$. Hence, by the previous calculation, for some rupture point V_r we have $M_r(g) < M_r(f)$. It follows that $M_r(f+g) = M_r(g) < M_r(f)$, and so $f+g$ cannot be equisingular to f.

The case of equality is more delicate. Again we may suppose $[g]_E = n[\widehat{E}_0]$, and so that for some rupture point V_r we have $M_r(g) = M_r(f)$. Consider the restriction to E_r of the strict transforms of $f + \lambda g$ as $\lambda \in \mathbb{C}$ varies: these all give homogeneous polynomials of degree $m_r(f)$ in projective coordinates (u, v) on E_r. Write ϕ, ψ for the polynomials corresponding to f and g, θ for their highest common factor and $\alpha := \phi/\theta$, $\beta := \psi/\theta$; set $d := \deg \alpha = \deg \beta$. Then the polynomials $\alpha + \lambda\beta$ have no common factor; for a general value of λ there is no repeated factor, but if $d \geq 2$ for at least one value there will be a repeated factor.

Thus if $d \geq 2$ the set of multiplicities of the roots of $\alpha + \lambda\beta$, hence of $\phi + \lambda\psi$ is not the same for all λ, so not all the $f + \lambda g$ are equisingular. If $d = 1$ and $\deg \theta \geq 1$, we have exceptional values of λ where the root of $\alpha + \lambda\beta$ coincides with a root of θ, and the argument concludes as before. If $d = 0$ there is a value of λ with $\alpha + \lambda\beta = 0$, and thus $M_r(f + \lambda g) > M_r(f)$. Finally if $d = 1$ and θ is a constant, $m_r(f) = 1$; ϕ has just one simple root, so V_r is not a rupture point of $\Gamma_R(C_f)$, contrary to hypothesis. □

We now turn to the calculation of codimensions of equivalence classes: we begin with equisingularity.

Given a reduced function f, choose a good resolution, and let I be the (valuatively closed) ideal of functions g with $[g]_E \geq [f]_E$. Then $[I]_E = [f]_E$. By Corollary 11.2.3 (i), there is an integer M such that, for $g \in I$, $[I]_E = [g]_E$ unless $j^M g$ lies in a finite union of proper linear subspaces of $j^M I$. If we choose M large enough so that $\mathfrak{m}^{M-1} \subseteq I$ then moreover $j^M g \in j^M I$ implies $g \in I$. Hence for all g such that $j^M g \in j^M I$ and avoids the linear subspaces, not only is $g \in I$ but $[g]_E = [I]_E = [f]_E$, and so g is equisingular to f. Now in Theorem 11.2.4, we calculated the codimension of a valuatively closed $\mathfrak{m}$-primary ideal I as $\sum_i \frac{1}{2} m_i(m_i + 1) = -\frac{1}{2}[I]_E.([I]_E + [Z])$, where $[I]_E = \sum m_i[\widehat{E}_i]$.

Write $\mathrm{sat}(f)$ for the number of satellite points occurring in the sequence of infinitely near points obtained in the resolution of f. We next show

Proposition 11.5.8 *The codimension of the equisingularity class of functions containing f is equal to* $\sum_i(\frac{1}{2}m_i(f)(m_i(f)+1)-1)+\mathrm{sat}(f)+1$.

Proof A function f' is equisingular to f if and only if their minimal good resolutions are isomorphic. If this holds, the above procedure gives an ideal I' which bears the same relation to f' as I does to f; moreover, as the construction is the same, we can choose the same number M for both; also the codimensions of I and I' are equal.

It follows that the codimension of the equisingularity class is equal to the difference of the codimension of I and the number of parameters required (in addition to its combinatorial description) to determine I.

We saw in Proposition 11.3.3 that fixing the valuatively closed ideal I is equivalent to fixing the corresponding cluster of infinitely near points. These points are chosen in sequence: at each stage after the first (O_0 is unique) we either choose a point lying on a previously constructed exceptional curve – and here we have one degree of freedom – or we have a satellite point, uniquely determined as the intersection of two exceptional curves. Thus each infinitely near point giving a term in the sum $[I]_E = \sum m_i[\widehat{E}_i]$, except O_0 and the satellite points, contributes 1 to the number of parameters required to determine I. Hence the desired codimension is equal to the codimension of I diminished by $\sum_i 1 - (\mathrm{sat}(F) + 1)$. This gives the result. □

For right equivalence and $\mathcal{K}$-equivalence we again just quote the results.

Proposition 11.5.9 *Suppose f K-determined. Then the set of K-jets of functions right-equivalent (respectively, $\mathcal{K}$-equivalent) to f is a smooth manifold with tangent space the image in J^K of $\mathfrak{m}J(f)$ (respectively, $\mathfrak{m}J(f) + \langle f \rangle$).*

The codimension of the right equivalence class of f is $2 + \mu(f)$.

As to the last statement we recall that by Theorem 6.5.6, $\mu(f)$ is equal to the local intersection number of $\partial f/\partial x = 0$ and $\partial f/\partial y = 0$, and by Lemma 1.2.1 (iv), this is in turn equal to $\dim(\mathcal{O}_2/J(f))$. On the other hand, $J(f)/\mathfrak{m}J(f)$ admits the classes of $\partial f/\partial x$ and $\partial f/\partial y$ as a $\mathbb{C}$-basis.

There is no simple formula for the codimension of the $\mathcal{K}$-equivalence class of f; in fact, this is not an equisingularity invariant. We leave the reader to check that these codimensions differ for the two equisingular curves given in Exercise 4.7.22.

An equisingularity class is the union of the right equivalence classes contained in it, and μ is the same for all of these. Hence the number of parameters required to define the right equivalence class (in a given equisingularity class) – which is usually called the number of moduli, or modality – is the difference of the codimensions. Comparing the formula of Proposition 11.5.8 with that for μ in Theorem 6.5.9, we obtain $(\sum_i m_i(m_i - 1) - r + 1) + 2 - \left(\sum_i(\frac{1}{2}m_i(m_i + 1) - 1) + \mathrm{sat} + 1\right)$, which reduces to $\sum_i \frac{1}{2}(m_i - 1)(m_i - 2) - r - \mathrm{sat} + 2$.

11.6 Local rings and differentials

We conclude this book with a rather brief treatment of a more abstract approach to the theory of singularities of plane curves, which focusses on the local ring $\mathcal{O}_C$. This appears here since it concentrates on ideals, but the topics are those of Section 4.

Let C be a plane curve, $\gamma : \tilde{C} \to C$ a normalisation. For each point $P \in C$, $\gamma^{-1}(P)$ is a finite subset J_P of $\tilde{C}$ (we omit reference to P for the next few paragraphs). Each $j \in J$ determines a branch B_j of C at

P. If t_j is a local coordinate on $\tilde{C}$ at j, composing with γ gives a good local parametrisation for B_j.

The ring $\mathcal{O}_{B_j}$ is a subring of finite codimension in $\mathcal{O}_j := \mathbb{C}\{t_j\}$, which by Lemma 11.4.2 is its integral closure in the quotient field $k_j := \mathbb{C}\{t_j\}[t_j^{-1}]$. Likewise, $\mathcal{O}_J := \bigoplus_j \mathbb{C}\{t_j\}$ is the integral closure of $\mathcal{O}_C$ in their common ring of quotients $k_J := \oplus_j k_j = \bigoplus_j \mathbb{C}\{t_j\}[t_j^{-1}]$.

Conversely, a subring R whose integral closure in k_J is $\mathcal{O}_J$ corresponds to the germ of a plane curve if and only if there exist $x, y \in R$ such that R is a free $\mathbb{C}\{x\}$-module with basis $\{1, y, y^2, \ldots, y^{m-1}\}$ (if $|J| = 1$ it is enough for these elements to generate the module); for then there is a (unique) expression of y^m in the form $\sum_0^{m-1} a_r(x)y^r$, and setting $f(x,y) := y^m - \sum_0^{m-1} a_r(x)y^r$ gives an equation for C.

The *conductor* is the set $\mathfrak{c} := \{f \in \mathcal{O}_J \,|\, f\mathcal{O}_J \subseteq \mathcal{O}_C\}$, i.e. the annihilator of $\mathcal{O}_J/\mathcal{O}_C$. It follows from the definition that $\mathfrak{c} \subseteq \mathcal{O}_C$ and that $\mathcal{O}_J.\mathfrak{c} \subseteq \mathfrak{c}$, so $\mathfrak{c}$ is an ideal both in $\mathcal{O}_J$ and in $\mathcal{O}_C$. We can calculate the conductor explicitly.

Lemma 11.6.1 *If C is a single branch B_j, then $\mathfrak{c}$ is the ideal in $\mathcal{O}_j$ generated by $t_j^{\mu(B_j)}$. In general, $\mathfrak{c}$ is the direct sum of the ideals generated by $t_j^{a_j}$, where $a_j = \mu(B_j) + \sum_{i \neq j} B_i.B_j$.*

Proof For a single branch B_j any ideal in $\mathcal{O}_j$ is generated by a power of t_j. We have the semigroup $S_j = S(B_j)$, and defined $N(S_j)$ to be the greatest integer not lying in the semigroup.Thus all powers t_j^a with $a > N(S_j)$ belong to $\mathcal{O}_{B_j}$, and hence since by Proposition 6.3.2 $\mu(B_j) = N(S_j) + 1$, $t_j^{\mu(B_j)}$ generates $\mathfrak{c}$.

In general any ideal in $\mathcal{O}_J$ is a direct sum of ideals in the several $\mathcal{O}_j$. Let $f_j = 0$ be a reduced equation for B_j. Then an element of $\mathcal{O}_C$ which belongs to the summand $\mathcal{O}_j \subset \mathcal{O}_J$ is the restriction to C of a function vanishing on the B_i for $i \neq j$, and hence of the form $g = g_j \prod_{i \neq j} f_i$ with $g_j \in \mathcal{O}_0$. If also $g \in \mathfrak{c}$, then for any $a \geq 0$, the class of gt_j^a in $\mathcal{O}_j$ has the same form, so $g_j t_j^a \in \mathcal{O}_{B_j}$. Hence g_j belongs to the conductor of $\mathcal{O}_{B_j}$ in $\mathcal{O}_j$, so by the above is divisible by $t_j^{\mu(B_j)}$. Since the order in $\mathcal{O}_j$ of the class of f_i is, by definition, equal to the intersection number $B_i.B_j$, it follows that $\mathfrak{c} \cap \mathcal{O}_j \subseteq \langle t_j^{a_j} \rangle$, where a_j is as above. The converse follows more easily from the same calculation. □

Corollary 11.6.2 *The codimension* $\dim(\mathcal{O}_J/\mathfrak{c}) = 2\delta(C)$, *and* $\dim(\mathcal{O}_C/\mathfrak{c}) = \delta(C)$.

Proof It follows from the lemma that $\dim(\mathcal{O}_J/\mathfrak{c}) = \sum_j a_j = \sum_i \mu(B_i) + 2\sum_{i<j} B_i.B_j$. The result follows since by Proposition 6.3.2 $\mu(B_j) = 2\delta(B_j)$ and it follows from Theorem 6.5.9 that for any union we have $\delta(C \cup C') = \delta(C) + \delta(C') + 2C.C'$. □

The *module of (meromorphic) differentials* at j is defined to be the free k_j-module $\overline{\Omega}_j$ generated by a symbol dt_j. The *residue homomorphism* Res : $\overline{\Omega}_j \to \mathbb{C}$ is defined by $\mathrm{Res}(\sum_{-\infty}^{\infty} a_n t_j^n dt) = a_{-1}$. This can be interpreted as $\mathrm{Res}(\omega) = \frac{1}{2\pi i}\oint d\omega$, where the integral is taken along a small loop round 0, so is independent of the choice of uniformising parameter t_j. Any element $f(t_j) \in k_j$, has a differential $f'(t_j)dt_j$, and we have an exact sequence $0 \to \mathbb{C} \to k_j \xrightarrow{d} \overline{\Omega}_j \xrightarrow{\mathrm{Res}} \mathbb{C} \to 0$. The composite pairing $k_j \times \overline{\Omega}_j \xrightarrow{\mathrm{mult}} \overline{\Omega}_j \xrightarrow{\mathrm{Res}} \mathbb{C}$ is non-degenerate. Summing over j we obtain $\overline{\Omega}_J := \oplus_j \overline{\Omega}_j$ and a non-degenerate pairing $k_J \times \overline{\Omega}_J \to \mathbb{C}$.

We define the module of *Rosenlicht differentials* (also known as the *dualising module*) to be the annihilator

$$\Omega_{B_j} := \{\omega \in \overline{\Omega}_j \mid (\forall f \in \mathcal{O}_{B_j})\mathrm{Res}(f\omega) = 0\}$$

of $\mathcal{O}_{B_j}$ under the above pairing. Then since the pairing is non-degenerate, $\mathcal{O}_{B_j}$ is the annihilator of Ω_{B_j}. Observe that the annihilator of $\mathcal{O}_j$ is the free $\mathcal{O}_j$-module Ω_j with basis dt_j. Hence Res induces a dual pairing of the finite dimensional quotients $\mathcal{O}_j/\mathcal{O}_{B_j}$ and Ω_{B_j}/Ω_j. Similarly let Ω_C be the annihilator of $\mathcal{O}_C$ and $\Omega_J = \oplus_j \Omega_j$. We have a dual pairing of $\mathcal{O}_J/\mathcal{O}_C$ and Ω_C/Ω_J.

The restriction of d gives a surjection $\mathcal{O}_J \to \Omega_J$ with kernel (consisting of constants) of rank r. This kernel meets $\mathcal{O}_C$ in a 1-dimensional subspace, so we have an induced exact sequence

$$0 \to \mathbb{C}^{r-1} \to \mathcal{O}_J/\mathcal{O}_C \to \Omega_J/d\mathcal{O}_C \to 0;$$

in particular, $\delta = \dim(\mathcal{O}_J/\mathcal{O}_C) = r - 1 + \dim(\Omega_J/d\mathcal{O}_C)$. Hence $\dim(\Omega_C/d\mathcal{O}_C) = \dim(\Omega_C/\Omega_J) + \dim(\Omega_J/d\mathcal{O}_C) = \delta + \delta - (r-1) = \mu$.

The above assertions are easy to verify. A deeper fact is that Ω_C is a principal $\mathcal{O}_C$-module. Suppose coordinates chosen so that neither y nor $\partial f/\partial x$ vanishes on any branch of C at O. Then Ω_C is generated by $\omega = dy/(\partial f/\partial x)$. We do not give the proof, which depends on duality in dimension 2: in fact the assertion comes from the calculation $\mathrm{Res}(dx \wedge dy/f) = dy/(\partial f/\partial x)$. Observe that since $df = \frac{\partial f}{\partial x}dx + \frac{\partial f}{\partial y}dy$ vanishes on C, we can also write $\omega = -dx/(\partial f/\partial y)$.

Multiplying by ω induces an isomorphism of $\mathfrak{c}$ on Ω_J. It suffices to check the components separately. We can write $f = \prod f_j$, where $f_j = 0$

is a reduced equation for B_j. Then the restriction to B_j of $\partial f/\partial x$ is equal to the product of the restrictions of $\prod_{i\neq j} f_i$ and $\partial f_j/\partial x$. We have just noted that the former has order $\sum_{i\neq j} B_i.B_j$. The order of the latter is the intersection number of $f_j = 0$ and $\partial f_j/\partial x = 0$; arguing as in the proof of Theorem 6.5.8, this is the sum of the intersection number of $y = 0$ with $\partial f_j/\partial x = 0$ and the intersection number of $\partial f_j/\partial y = 0$ with $\partial f_j/\partial x = 0$, which is equal to $\mu(B_j)$. By the same argument, the intersection number of $y = 0$ with $\partial f_j/\partial x = 0$ is 1 less than the intersection number of $y = 0$ with B_j, which is equal to the order of dy/dt_j. Hence, as claimed, the restriction of ω to B_j has the form $\phi_j dt_j$ where ϕ has order $-\mu(B_j) - \sum_{i\neq j} B_i.B_j = -a_j$.

It follows that multiplying by ω induces an isomorphism of $\mathcal{O}_C/\mathfrak{c}$ on Ω_C/Ω_J. We already know that each has the same dimension, $\delta(C)$, as $\mathcal{O}_J/\mathcal{O}_C$.

For a single branch B, Ω_B has a basis of elements whose orders are the negatives of those not in $S(B)$. Since Ω_B is a principal $\mathcal{O}_B$-module, it follows that $S(B)$ is a dual semigroup. The above isomorphisms generalise this duality to curves with several branches.

Example 11.6.1 Let $f(x, y) = y^2 - x^5$. Parametrise C by (t^2, t^5). Then $\mathcal{O}_C$ has a basis consisting of all powers of t except t, t^3; so these two elements form a basis for $\mathcal{O}_J/\mathcal{O}_C$. Now Ω_J is the free module over $\mathcal{O}_J = \mathbb{C}\{t\}$ with basis dt, and Ω_C/Ω_J has basis $t^{-2}dt, t^{-4}dt$.

We have $\omega = d(t^5)/(-5x^4) = (5t^4 dt)/(-5t^8) = -t^{-4}dt$, and can see directly that this freely generates Ω_C as $\mathcal{O}_C$-module. The conductor consists of elements α with $\alpha\mathcal{O}_C \subset \mathcal{O}_J$, hence has a basis of powers t^j with $j \geq 4$, so coincides with $\mathfrak{m}^4$. For this example, we have $\delta(C) = 2$ and $N(S(B)) = 3$.

Example 11.6.2 Consider the D_7 singularity $f = xy^2 - x^6$. We have two branches, with parametrisations (t_1^2, t_1^5) (as in the previous example) and $(0, t_2)$. The ring $\mathcal{O}_C$ is the subring of $\mathcal{O}_J$ of pairs $(\sum_0^\infty a_i t_1^i, \sum_0^\infty b_i t_1^i)$ with $a_0 = b_0$, $a_1 = a_3 = 0$ and $a_5 = b_1$, so $\dim(\mathcal{O}_J/\mathcal{O}_C) = 4$. A basis for Ω_C/Ω_J is thus $t_1^{-1}dt_1 - t_2^{-1}dt_2, t_1^{-2}dt_1, t_1^{-4}dt_1$, $t_1^{-6}dt_1 - t_2^{-2}dt_2$, and indeed we have $\omega = (5t_1^4 dt_1 + dt_2)/(-5t_1^{10} + t_2^2) = -t_1^{-6}dt_1 + t_2^{-2}dt_2$.

The conductor $\mathfrak{c}$ has basis t_1^i for $i \geq 6$ and t_2^j for $j \geq 2$; so $\dim(\mathcal{O}_J/\mathfrak{c}) = 8$. Here we have $\delta(C) = 4$.

The *local Jacobian* of C at P is defined to be the quotient group $(\mathcal{O}_J)^\times/(\mathcal{O}_C)^\times$. This group has the same dimension δ as $\mathcal{O}_J/\mathcal{O}_C$, and is abstractly isomorphic to $(\mathbb{C}^\times)^{r-1} \times (\mathbb{C}^+)^k$, and hence also to $(S^1)^{r-1} \times (\mathbb{R}^+)^\mu$, where $r = |J|$ is the number of branches and $k = \delta(C) - r + 1 = \frac{1}{2}(\mu(C) + 1 - r)$. The construction of the Jacobian variety $J(\tilde{C})$ of the non-singular curve $\tilde{C}$ in the large is standard in algebraic geometry. There is also a notion of Jacobian of a singular curve C , defined e.g. in [85], which, like the other, is an abelian group. There is a natural exact sequence $0 \to \mathcal{O}_C^\times \to \mathcal{O}_{\tilde{J}}^\times \to J(C) \to J(\tilde{C}) \to 0$, thus the kernel of the (split) surjection $J(C) \to J(\tilde{C})$ can be identified with the direct product over singular points P of C of the local Jacobians.

11.7 Notes

Sections 11.1,11.2 Many of the arguments are adapted from Casas [35]. Readers familiar with sheaves will recognise that blowing up an ideal I should most naturally be regarded as giving a sheaf of ideals along $\pi^{-1}(O)$, and that some of the arguments of Section 11.1 involve paraphrases of this notion.

Section 11.3 The concept of cluster and theory of clusters and unloading is due to Enriques [66]; he uses 'weighted cluster' for what we have called 'cluster'. We have borrowed some of the treatment from Lejeune [115] (also for Chapter 8). A detailed treatment is also given by Casas in [35], where the technique of using clusters is taken further than in this book: see, for example, the use of adjoints and conductors in Section 4.8 loc.cit.

Our account is close to that given for a more general situation in Campillo and González–Sprinberg [30]; compare also Lipman [119] and Cutkosky [42].

Section 11.4 In this context, integrally closed ideals are usually called 'complete ideals'. The theory of complete ideals, in particular the unique factorisation theorem, is due to Zariski [204].

The basic case of Noether's theorem is that if curves C_1, C_2 in P^2, given by homogeneous equations $f_1 = 0$, $f_2 = 0$ of respective degrees d_1, d_2 have $d_1 d_2$ distinct points of intersection, then any curve through all those points has equation of the form $A_1 f_1 + A_2 f_2 = 0$. A direct proof is given in [165]. When the intersection points are not distinct, the result does not always hold for a curve passing through all infinitely near points common to C_1 and C_2. The above proof of the $Af + Bg$ theorem is based on [35] Section 4.9, esp. pp 154–156.

The proof of Theorem 11.4.5 seems to be new. (Both Lejeune and Casas refer to Lipman [119] at a key point.)

A beautiful account of integral closure and its applications in analytic geometry (of higher dimensions) is given by Lejeune and Teissier in [116]. They work for the most part in the context of sheaves of ideals in coherent sheaves of rings, but the situation is formally similar. In particular, they generalise the characterisation by valuative closure to a numerical condition, and show also that if the ideal (sheaf) has generators $\{g_i\}$, then f belongs to its integral closure if and only if, in some neighbourhood of the point in question and for some constant C, the inequality $|f(y)| \leq C \max_i |g_i(y)|$ holds.

They also show that f belongs to the integral closure of an ideal I if and only if, for every curve Γ, the restriction of f to Γ belongs to the ideal induced by I on Γ. In the situation of this chapter, we established this in Lemma 11.2.2.

Section 11.5 The first determinacy results are due to Kuo [104]; the ideas were further developed in [105]. Our approach to determinacy (with respect to equisingularity) follows Casas [35]7.5.1. A less elementary but more attractive approach is given in Lejeune and Teissier [116] and Teissier [174]. Teissier shows, for example,

Theorem 11.7.1 *The least parameters θ_1, θ_2 for which inequalities of the respective forms $||\nabla f|| \geq C|f|^\theta$, $||\nabla f|| \geq C||z||^\theta$ hold on some neighbourhood of O are given by $\theta_1 = 1 - Q^{-1}$, $\theta_2 = Q - 1$.*

A geometrical approach to this which does not, however, amount to a proof is as follows. For any point $z = (x, y)$ near $O \in \mathbb{C}^2$, choose α, β (not both 0) such that $\alpha \partial f/\partial x + \beta \partial f/\partial y$ vanishes at z, so that z lies on the polar curve P corresponding to the tangent direction $\beta x = \alpha y$. Assume this is not tangent to C at O, so the decomposition theorem Theorem 4.5.2 applies, and write Γ for the component of the polar through z.

Changing coordinates, we may suppose $\alpha = 0$. Let t be a good parameter on Γ, and $z = \gamma(t) = (x(t), y(t))$ the corresponding parametrisation. Along Γ, $||z||$ has the same order as $t^{m(\Gamma)}$; by (iii) of Lemma 9.9.1, $|x|$ has the same order. Also along Γ, since $\partial f/\partial y = 0$ we have $||\nabla f|| = |\partial f/\partial x|$, and arguing as in Lemma 6.5.7, $f(\gamma(t)) = \int_0^t (\partial f/\partial x)(dx/dt)dt$. Hence along Γ the order of $f(\gamma(t))$ is the sum of those of x and $\partial f/\partial x$. Now the order of $f(\gamma(t))$ is the intersection number M of Γ and C; thus the order of $||\nabla f||$ is $M - m$, and the estimates can now all be expressed in terms of the polar quotient $\frac{M}{m(\Gamma)}$.

Teissier's actual argument uses the same basic idea, but he considers families of polars, so works in higher dimensional varieties, and uses their normalisation and calculations of multiplicities of divisors. He also shows that for Q the maximum polar quotient in his sense, $N > Q$ if and only if every $g \in f + \mathfrak{m}^{N+1}$ is topologically equivalent to f if and only if the linear family $\lambda f + (1 - \lambda)g$ is equisingular in his sense.

Most other books on singularity theory take right equivalence as the basic notion. See, for example, the introductory text of Dimca [53] or Arnold et al. [11]. A survey of a large number of determinacy theorems, more from this viewpoint, was given by the author [191], with a corrected and updated account in [192].

The equisingularity class of f can also be regarded as a smooth submanifold of $\mathcal{O}_2$: this was established by Wahl [185], who also showed that its tangent space is an ideal in $\mathcal{O}_2$, which he described. This submanifold is foliated by the right equivalence classes: the precise classification up to right equivalence involves describing the parameters required for this. The listing of equisingularity classes (also in more variables) with low modality is due to Arnold (several papers, reprinted in [10], and much also reproduced in [11]). In particular, the cases of modality 0 are those named A_k, D_k, E_k above; those of modality 1 or 2 fall into a small number of cases and naturally occurring families; these too have been extensively investigated. Those of modality 3 were listed in [196].

Section 11.6 The results in this section are well known: see e.g. [64]Chapter 21, [85]. This explains that the source of the duality is the Gorenstein property, which follows in our case from the fact that our curve is a plane curve.

For writing this account I have used Piene [149], which holds in all characteristics, and Montaldi and van Straten [134]. The formula $e_P = 2\delta_P + \kappa_P$ in characteristic p is an easy deduction from these considerations.

11.8 Exercises

Exercise 11.8.1 Show that the sequence $\{[\epsilon_n]\}$ is monotone increasing in the sense that if O_k is proximate to O_j, then $[\epsilon_j] < [\epsilon_k]$. Deduce that $\inf\{z \in \mathcal{E} \mid z \neq 0\} = [\epsilon_0]$. (Hint: use Exercise 8.7.2).

Exercise 11.8.2 Let C be the curve $y^3 = x^5$. Determine the closures of the clusters O_2, $2O_2$, $3O_2$, $4O_2$, $5O_2$, and $6O_2$ by performing the unloading algorithm or otherwise.

Exercise 11.8.3 Define the formal codimension of a cluster K, following Theorem 11.1.7, to be $-\frac{1}{2}[K].([K]+[Z])$. Show that the formal codimension does not increase at any step of the unloading process.

Exercise 11.8.4 For r from 1 to 6, calculate the formal codimension of the cluster rO_2 of Exercise 11.8.2 (as defined in Exercise 11.8.3). Calculate also the codimension of the ideal I with $[I]$ the closure $(rO_2)^*$.

Exercise 11.8.5 Let B be the curve $y^3 = x^5$. The closed ideal $I(B, v)$ is defined to be $\{f \in \mathcal{O}_0 \,|\, ord_B(f) \geq v\}$. Thus $[I(B, v)] = \inf\{X \in \mathcal{E} \,|\, X.[\epsilon_3] \geq v\}$. Express $[I(B, v)]$ as a linear combination of the $[\epsilon_i]$, for all values of v.

Exercise 11.8.6 Show that, with the notation of Proposition 11.2.7, if $a_i > 0$ then $b_i < A_i$ (where $A_i := -[E_i].[E_i]$ is denoted a_i elsewhere, alas).

Write H_Q for $H \otimes \mathbb{Q}$ and $\mathcal{E}_Q$ for $\mathcal{E} \otimes \mathbb{Q}$. Extend Lemma 11.2.6 to show that any $x \in \mathcal{E}_Q$ satisfies $x \geq 0$ and that, for any $\mathcal{W} \subseteq \mathcal{E}$, the infimum $\inf \mathcal{W}$ exists and belongs to $\mathcal{E}$.

Show that, for any $x \in H_Q$, the infimum $x^+ := \inf\{y \in \mathcal{E}_Q \,|\, y \geq x\}$ exists and is attained. Show that if $x^+ - x := \sum a_i[E_i]$ and $b_i := x^+.[E_i]$ then, for each i, $a_i \geq 0 \geq b_i$. Show also that $a_i b_i = 0$ for each i.

Exercise 11.8.7 Give an example of f, g, h such that any infinitely near point common to C_f and C_g also lies on C_h, but $h \notin \langle f, g\rangle$.

Exercise 11.8.8 Show that the integral closure of $\mathbb{Z}$ in $\mathbb{Q}[\sqrt{5}]$ consists of the elements $a + b\tau$ with $a, b \in \mathbb{Z}$, where $\tau := \frac{1}{2}(1+\sqrt{5})$.

Exercise 11.8.9 For a curve C with several branches B_j, where $\{t_j\}$ denote parameters on the several branches, obtain an identification of $\overline{\mathcal{O}}_B$ with $\oplus_j \mathbb{C}\{t_j\}$.

Exercise 11.8.10 Show that the integral closure of $I := \langle x^4, y^8\rangle$ is given by $\overline{I} = \langle x^4, x^3y^2, x^2y^4, xy^6, y^8\rangle$. More generally, let I be any ideal generated by monomials. Show that $\overline{I}$ is generated by the monomials above the Newton polyhedron.

Exercise 11.8.11 Suppose $f, g \in \mathcal{O}_0$ share no common factor. Blow up sufficiently often to separate the strict transforms of C_f and C_g. Define $k_i := m_i(f) + m_i(g) - 1$ if $m_i(f) > 0$ and $m_i(g) > 0$, and 0 otherwise. Prove that this satisfies the proximity inequalities.

Exercise 11.8.12 Find the degree of determinacy of the singularity $(y^3+x^4)(y^2-x^2) = 0$. Determine the codimension of its equisingularity class, and the modality.

Exercise 11.8.13 For each $p \geq 3$, find the degree of determinacy of the singularity $y^3 = x^p$. Determine also the codimension of the equisingularity class containing this singularity, and the modality.

Exercise 11.8.14 For each of the curves $y^3 = x^4$, $xy^3 = x^3$ determine the ring $\mathcal{O}_C$, the module Ω_C of Rosenlicht differentials and the conductor $\mathfrak{c}$.

References

[1] Abhyankar, S. S., Resolution of singularities and modular Galois theory, *Bull. Amer. Math. Soc.* **38** (2001) 131–170.

[2] Abhyankar, S. S. and A. Assi, Factoring the Jacobian, in *Contemporary Math.* **266**, Amer. Math. Soc., 2000, pp. 1–10.

[3] A'Campo, N., Sur la monodromie des singularités isolées d'hypersurfaces complexes, *Invent. Math.* **20** (1973) 147–169.

[4] A'Campo, N., Le nombre de Lefschetz d'une monodromie, *Indag. Math.* **35** (1973) 113–118.

[5] A'Campo, N., La fonction zêta d'une monodromie, *Comment. Math. Helvet.* **50** (1975) 233–248.

[6] A'Campo, N., Le groupe de monodromie du déploiement des singularités isolées de courbes planes I, *Math. Ann.* **213** (1975) 1–32.

[7] A'Campo, N., Monodromy of real isolated singularities, *Topology* **42** (2003) 1229–1240.

[8] Alexander, J. W., Topological invariants of knots and links, *Trans. Amer. Math. Soc.* **30** (1928) 275–306.

[9] Armstrong, M. A., *Basic Topology*, Undergraduate Texts in Math., Springer-Verlag, 1983.

[10] Arnold, V. I., *Singularity Theory*, London Math. Soc. Lecture Notes **53**, Cambridge University Press, 1981.

[11] Arnold, V. I., S. M. Gusein-Zade and A. N. Varchenko, *Singularities of Differentiable Maps I*, Birkhäuser, 1985.

[12] Artal-Bartolo, E., Sur les couples de Zariski, *Jour. Alg. Geom.* **3** (1994) 223–247.

[13] Artin, M., On isolated rational singularities of surfaces, *Amer. Jour. Math.* **88** (1966) 129–136.

[14] Assi, A., Partial derivatives of a meromorphic plane curve, preprint, Angers, 1998.

[15] Atiyah, M. F., V. K. Patodi and I. M. Singer, Spectral asymmetry and Riemannian geometry II, *Math. Proc. Camb. Phil. Soc.* **78** (1975) 405–432.

[16] Atiyah, M. F. and I. M. Singer, The index of elliptic operators III, *Ann. of Math.* **87** (1968) 546–604.

[17] Barth, W., C. Peters and A. Van de Ven, *Compact Complex Surfaces*, Ergebnisse der Math. (3) **4**, Springer-Verlag, 1984.

[18] Benedetti, R. and J.-J. Risler, *Real Algebraic and Semi-algebraic Sets*, Hermann, 1990.

[19] Blanchfield, R. C., Intersection theory of manifolds with operators with applications in knot theory, *Ann. of Math.* **65** (1957) 340–356.

[20] Brasselet, J.-P., From Chern classes to Milnor classes – a history of characteristic classes for singular varieties, in *Singularities – Sapporo 1998* (ed. J.-P. Brasselet and T. Suwa), Advanced Studies in Pure Math. **29**, Kinokuniya, 2000, pp. 31–52.

[21] Brauner, K., Zur Geometrie der Funktionen zweier komplexen Veränderlichen II–IV, *Abh. Math. Sem. Hamburg* **6** (1928) 1–54.

[22] Brieskorn, E., Die Monodromie der isolierten Singularitäten von Hyperflächen, *Manuscripta Math.* **2** (1970) 103–160.

[23] Brieskorn, E. and H. Knörrer, *Plane Algebraic Curves*, Birkhäuser, 1981.

[24] Burau, W., Kennzeichnung der Schlauchknoten, *Abh. Math. Sem. Hamburg* **9** (1932) 125–133.

[25] Burau, W., Kennzeichnung der Schlauchverkettungen, *Abh. Math. Sem. Hamburg* **10** (1934) 285–297.

[26] Burde, G. and H. Zieschang, *Knots*, de Gruyter, 1985; second edition, 2002.

[27] Campillo, A., *Algebroid Curves in Positive Characteristic*, Springer Lecture Notes in Math. **813**, Springer-Verlag, 1980.

[28] Campillo, A., F. Delgado and S. M. Gusein-Zade, On the monodromy of a plane curve singularity and the Poincaré series of its ring of functions, *Funct. Anal. App.* **33i** (1999) 56–57.

[29] Campillo, A., F. Delgado and S. M. Gusein-Zade, The extended semigroup of a plane curve singularity, *Proc. Steklov Institute* **221** (1998) 139–156.

[30] Campillo, A. and G. González-Sprinberg, On characteristic cones, clusters and chains of infinitely near points, in *Singularities* (ed. V. I. Arnold, G.-M. Greuel and J. H. M. Steenbrink), Progress in Math. **162**, Birkhäuser, 1998, pp. 251–261.

[31] Casas-Alvero, E., On the singularities of polar curves, *Manuscripta Math.* **43** (1983) 167–190.

[32] Casas-Alvero, E., Infinitely near imposed singularities and singularities of polar curves, *Math. Ann.* **287** (1990) 429–494.

[33] Casas-Alvero, E., Base points of polar curves, *Ann. Inst. Fourier* **41** (1991) 1–10.

[34] Casas-Alvero, E., Singularities of polar curves, *Compositio Math.* **89** (1993) 339–359.

[35] Casas-Alvero, E., *Singularities of Plane Curves*, London Math. Soc. Lecture Notes **276**, Cambridge University Press, 2000.

[36] Casson, A. and S. Bleiler, *Automorphisms of Surfaces after Nielsen and Thurston*, London Math. Soc. Student Texts **9**, Cambridge University Press, 1988.

[37] Cerf, J., *Sur les difféomorphismes de la sphère de dimension trois*, Springer Lecture Notes in Math. **53**, Springer-Verlag, 1968.

[38] Chang, S. H. and Y.-C. Lu, On C^0-sufficiency of complex jets, *Canadian Jour. Math.* **25** (1973) 874–880.

[39] Coddington, E. A. and N. Levinson, *Theory of Ordinary Differential Equations*, McGraw-Hill, 1955.

[40] Coolidge, J. L., *Algebraic Plane Curves*, Oxford University Press, 1920.

[41] Crowell, R. H., The group G'/G'' of a knot group G, *Duke Math. Jour.* **30** (1963) 349–354.

[42] Cutkosky, S. D., Complete ideals in algebra and geometry, *Contemporary Math.* **159** (1994) 27–39.

[43] Degtyarev, A. I., Classification of surfaces of degree four having a non-simple singular point, *Math. USSR Izvestiya* **35** (1990) 607–627.

[44] Degtyarev, A. I., Isotopic classification of complex plane projective curves of degree five, *Leningrad Math. Jour.* **1** (1990) 881–904.

[45] de Jong, A. J. and J. H. M. Steenbrink, Proof of a conjecture of W. Veys, *Indag. Math.* **6** (1995) 99–104.

[46] Delgado de la Mata, F., The semigroup of values of a curve singularity with several branches, *Manuscripta Math.* **59** (1987) 347–374.

[47] Delgado de la Mata, F., An arithmetical factorization for the critical point set of some map germs from $\mathbb{C}^2$ to $\mathbb{C}^2$, in *Singularities, Lille 1991* (ed. J.-P. Brasselet), London Math. Soc. Lecture Notes **201**, Cambridge University Press, 1994, pp. 61–100.

[48] Delgado de la Mata, F., A factorisation theorem for the polar of a curve with two branches, *Comp. Math.* **92** (1994) 327–375.

[49] Deligne, P., La formule de Milnor, exposé XVI in *Groupes de monodromie en géométrie algébrique, II, Séminaire de Géometrie Algébrique du Bois-Marie 1967–69* (SGA 7 II) (P. Deligne *et al.*), Springer Lecture Notes in Math. **340**, Springer-Verlag, 1973.

[50] Denef, J. and F. Loeser, Caractéristiques d'Euler-Poincaré, fonctions zeta locales, et modifications analytiques, *Jour. Amer. Math. Soc.* **5** (1992) 705–720.

[51] Denef, J. and F. Loeser, Motivic Igusa zeta functions, *Jour. Alg. Geom.* **7** (1998) 503–537.

[52] Dieudonné, J., *Foundations of Modern Analysis*, Academic Press, 1960.

[53] Dimca, A., *Topics on Real and Complex Singularities*, Vieweg, 1987.

[54] Du Bois, P. and F. Michel, Filtration par le poids et monodromie entière, *Bull. Soc. Math. France* **120** (1992) 129–167.

[55] Du Bois, P. and F. Michel, Cobordism of algebraic knots via Seifert forms, *Invent. Math.* **111** (1993) 151–169.

[56] Du Bois, P. and F. Michel, The integral Seifert form does not determine the topology of plane curve germs, *Jour. Alg. Geom.* **3** (1994) 1–38.

[57] Du Bois, P. and O. Hunault, Classification des formes de Seifert rationnelles des germes de courbe plane, *Ann. Inst. Fourier* **46** (1996) 371–410.

[58] Du Plessis, A. A. and C. T. C. Wall, Application of the theory of the discriminant to highly singular plane curves, *Math. Proc. Camb. Phil. Soc.* **126** (1999) 259–266.

[59] Du Val, P., On isolated singularities of surfaces which do not affect the conditions of adjunction I, II, III, *Proc. Camb. Phil. Soc.* **30** (1934) 453–459, 460–465, 483–491.

[60] Du Val, P., Reducible exceptional curves, *Amer. Jour. Math.* **58** (1936) 285–289.

[61] Ebeling, W. and C. T. C. Wall, Kodaira singularities and an extension of Arnold's strange duality, *Compositio Math.* **56** (1985) 3–77.

[62] Eggers, H., *Polarinvarianten und die Topologie von Kurvensingularitäten*, Bonner Math. Schrift **147** 1983.

[63] Ehresmann, C., Sur les espaces fibrés différentiables, *Comptes Rendus Acad. Sci. Paris* **224** (1947) 1611–1612.

[64] Eisenbud, D., *Commutative Algebra with a View Toward Algebraic Geometry*, Springer Graduate Text in Math. **150**, Springer-Verlag, 1996.

[65] Eisenbud, D. and W. Neumann, *Three-dimensional Link Theory and Invariants of Plane Curve Singularities*, Ann. of Math. studies **110**, Princeton University Press, 1985.

[66] Enriques, F. and O. Chisini, *Lezioni sulla Teoria Geometrica delle Equazioni e delle Funzioni Algebriche*, Niccolo Zanichelli, 1915, 1918, 1924.

[67] Ernström, L., Topological Radon transforms and the local Euler obstruction, *Duke Math. Jour.* **76** (1994) 1–21.

[68] Epstein, D. B. A., Curves on 2-manifolds and isotopies, *Acta Math.* **115** (1966) 83–107.

[69] Evers, M., Algebraische Verkettungen, doctoral thesis, Universität Köln, 1980.

[70] Fathi, A., F. Laudenbach and V. Poenaru, *Travaux de Thurston sur les Surfaces*, *Astérisque* **66–67** (1979).

[71] Flenner, H. and M. Zaidenberg, On a class of rational cuspidal plane curves, *Manuscripta Math.* **89** (1996) 439–459.

[72] Fox, R. H., A quick trip through knot theory, in *Topology of 3-manifolds and Related Topics* (ed. M. K. Fort, Jr), Prentice-Hall, 1962, pp. 120–167.

[73] Fulton, W., *Algebraic Curves*, Benjamin, 1969.

[74] Fulton, W., *Intersection Theory*, Ergebnisse (3) **2**, Springer-Verlag, 1984.

[75] García Barroso, E., Invariants des singularités de courbes planes et courbure des fibres de Milnor, doctoral thesis, La Laguna, 1996.

[76] García Barroso, E., Sur les courbes polaires d'une courbe plane réduite, *Proc. London Math. Soc.* **81** (2000) 1–28.

[77] García Barroso, E. and P. D. González Pérez, Decomposition in bunches of the critical locus of a quasi-ordinary map, preprint (27 pp.), 2003.

[78] Grauert, H., Über Modifikation und exzeptionelle analytischen Mengen, *Math. Ann.* **146** (1962) 331–368.

[79] Grauert, H. and R. Remmert, *Coherent Analytic Sheaves*, Springer-Verlag, 1984.

[80] Greuel, G.-M., C. Lossen and E. Shustin, Castelnuovo function, zero-dimensional schemes and singular plane curves, *Jour. Alg. Geom.* **9** (2000) 663–710.

[81] Gunning, R. C. and H. Rossi, *Analytic Functions of Several Complex Variables*, Prentice-Hall, 1965.

[82] Gusein-Zade, S. M., Intersection matrices for certain singularities of functions of two variables, *Funct. Anal. Appl.* **8** (1974) 10–13.

[83] Gusein-Zade, S. M., I. Luengo and A. Melle Hernández, Bifurcations and topology of meromorphic germs, in *New Developments in Singularity Theory* (ed. D. Siersma, C. T. C. Wall and V. Zakalyukin), Kluwer, 2001, pp. 279–304.

[84] Halphen, G.-H., Etude sur les points singuliers des courbes algébriques planes, *Oeuvres* **4** 1–93.

[85] Hartshorne, R., *Algebraic Geometry*, Graduate Text in Math. **52**, Springer-Verlag, 1977.

[86] Hefez, A., Non-reflexive curves, *Compositio Math.* **69** (1989) 3–35.

[87] Hefez, A., *Singularidades de Curvas Irredutiveis Planas*, distributed by USP – São Carlos, 2000.

[88] Hironaka, H., Resolution of singularities of an algebraic variety over a field of characteristic zero, *Ann. of Math.* **79** (1964) 109–326.

[89] Hirsch, M. W., *Differential Topology*, Springer-Verlag, 1976.

[90] Hirzebruch, F., The signature of ramified coverings, in *Global Analysis (Papers in Honor of K. Kodaira)* (ed. D. C. Spencer and S. Iyanaga), University of Tokyo Press, 1969, pp. 253–265.

[91] Hirzebruch, F., Hilbert modular surfaces, *l'Enseignement Math.* **19** (1973) 183–282.

[92] Jaco, W. H. and P. B. Shalen, Seifert fibred spaces in 3-manifolds, *Memoirs Amer. Math. Soc.* **220** (1979).

[93] Johannson, K., *Homotopy Equivalences of 3-manifolds with Boundaries*, Springer Lecture Notes in Math. **761**, Springer-Verlag, 1979.

[94] Kaenders, R., The Seifert form of a plane curve singularity determines its intersection multiplicities, *Indag. Math.* **7** (1996) 185–197.

[95] Kähler, E., Verzweigung einer algebraischen Funktion zweier Veränderlichen in der Umgebung einer singulären Stelle, *Math. Zeits.* **30** (1929) 188–206.

[96] Kauffman, L. H., *On Knots*, Ann. of Math. Study **115**, Princeton University Press, 1987.

[97] Kazarjan, M., Thom polynomials for Lagrange, Legendre and isolated hypersurface singularities, *Proc. London Math. Soc.* **86** (2003) 707–734.

[98] Kedlaya, K. S., The algebraic closure of the power series field in positive characteristic, *Proc. Amer. Math. Soc.* **129** (2001) 3461–3470.

[99] Kirwan, F., *Complex Algebraic Curves*, London Math. Soc. Student Text **23**, Cambridge University Press, 1992.

[100] Kleiman, S. L., The enumerative theory of singularities, in *Real and Complex Singularities, Oslo 1976* (ed. P. Holm), Sijthoff and Noordhoff, 1977, pp. 297–396.

[101] Klein, F., Ein neue Relation zwischen den Singularitäten einer algebraischen Kurve, *Math. Ann.* **10** (1876) 199–209.

[102] Kneser, H., Glättungen von Flächenabbildungen, *Math. Ann.* **100** (1928) 609–617.

[103] Knus, M., *Quadratic and Hermitian Forms over Rings*, Grundlehren **294**, Springer-Verlag, 1991.

[104] Kuo, T.-C., Computation of Lojasiewicz exponent of $f(x, y)$, *Comm. Math. Helv.* **49** (1974) 201–213.

[105] Kuo, T.-C. and Y. C. Lu, On analytic function-germs of two complex variables, *Topology* **16** (1977) 299–310.

[106] Lam, T. Y., *The Algebraic Theory of Quadratic Forms*, Benjamin, 1973.

[107] Lamotke, K., Die Homologie isolierter Singularitäten, *Math. Zeits* **143** (1975) 27–44.

[108] Lê, D. T., Sur les noeuds algébriques, *Compositio Math.* **25** (1972) 281–321.

[109] Lê, D. T., La monodromie n'a pas de points fixes, *Jour. Fac. Sci. Univ. Tokyo 1A* **22** (1975) 409–427.

[110] Lê, D. T. and M. Oka, On resolution complexity of plane curves, *Kodai Math. Jour.* **18** (1995) 1–36.

[111] Lê, D. T., F. Michel and C. Weber, Sur le comportement des polaires associées aux germes de courbes planes, *Compositio Math.* **72** (1989) 87–113.

[112] Lê, D. T., F. Michel and C. Weber, Courbes polaires et topologie des courbes planes, *Ann. Sci. Ecole Norm. Sup.* (4) **24** (1991) 141–169.

[113] Lê, D. T. and M. Tosun, Combinatorics of rational singularities, preprint, Université de Provence, 18 pp.

[114] Lejeune-Jalabert, M., Sur l'équivalence des singularités des courbes algébriques planes. Coefficients de Newton, in *Introduction à la théorie des singularités I*, Travaux en Cours **36**, Hermann, 1988, pp. 49–124.

[115] Lejeune-Jalabert, M., Linear systems with infinitely near base points and complete ideals in dimension two, in *Singularity Theory* (Proceedings of College on Singularities at Trieste 1991) (ed. D. T. Lê, K. Saito and B. Teissier), World Scientific, 1995, pp. 345–369.

[116] Lejeune-Jalabert, M. and B. Teissier, *Clôture intégrale des idéaux et equisingularité*, Notes, Université de Grenoble, 1974.

[117] Libgober, A., Alexander invariants of plane algebraic curves, in *Singularities* (ed. P. Orlik), Proc. Symp. in Pure Math. **40** part 2, Amer. Math. Soc., 1983, pp. 135–143.

[118] Lickorish, W. B. R., *An Introduction to Knot Theory*, Graduate Text in Math. **175**, Springer-Verlag, 1997.

[119] Lipman, J., Rational singularities with applications to algebraic surfaces and complete factorisation, *Publ. Math. IHES* **36** (1969) 195–279.

[120] Loeser, F., Fonctions d'Igusa p-adiques et polynomes de Bernstein, *Amer. Jour. Math.* **110** (1988) 1–21.

[121] Looijenga, E. J. N., *Isolated Singular Points on Complete Intersections*, London Math. Soc. Lecture Notes **77**, Cambridge University Press, 1984.

[122] MacLane, S. and G. Birkhoff, *Algebra*, Macmillan, 1967.

[123] MacPherson, R. D., Chern classes for singular algebraic varieties, *Ann. of Math.* **100** (1974) 423–432.

[124] McCrory, C. and A. Parusìnski, Algebraically constructible functions, *Ann. Sci. Ec. Norm. Sup.* **30** (1997) 527–552.

[125] Merle, M., Invariants polaires de courbes planes, *Invent. Math.* **41** (1977) 103–111.

[126] Michel, F. and C. Weber, *Topologie des germes de courbes planes à plusieurs branches*, Notes, Université de Genève, 1985.

[127] Michel, F. and C. Weber, Sur le role de la monodromie entière dans la topologie des singularités, *Ann. Inst. Fourier* **36** (1986) 183–218.

[128] Michel, F. and C. Weber, On the monodromies of a polynomial map from $\mathbb{C}^2$ to $\mathbb{C}$, *Topology* **40** (2001) 1217–1240.

[129] Milnor, J. W., A unique factorisation theorem for 3-manifolds, *Amer. Jour. Math.* **84** (1962) 1–7.

[130] Milnor, J. W., Infinite cyclic coverings, in *Topology of manifolds* (ed. J. G. Hocking), Prindle, Weber & Schmidt, 1968, pp. 115–133.

[131] Milnor, J. W., On isometries of inner product spaces, *Invent. Math.* **8** (1969) 83–97.

[132] Milnor, J. W., *Singular Points of Complex Hypersurfaces*, Princeton University Press, 1968.

[133] Milnor, J. W. and D. H. Husemoller, *Symmetric Bilinear Forms*, Ergebnisse **73**, Springer-Verlag, 1973.

[134] Montaldi, J. and D. van Straten, 1-forms on singular curves and the topology of real curve singularities, *Topology* **29** (1990) 501–510.

[135] Morse, M., *The Calculus of Variations in the Large*, Colloq. Publ. **18**, Amer. Math. Soc., 1934 (and numerous later editions).

[136] Mumford, D., The topology of a normal singularity of an algebraic surface and a criterion for simplicity, *Publ. Math. IHES* **9** (1961) 195–279.

[137] Murasugi, K., On the Minkowski unit of slice links, *Trans. Amer. Math. Soc.* **114** (1965) 377–383.

[138] Neumann, W. D., Signature related invariants of manifolds – I. Monodromy and γ-invariants, *Topology* **18** (1979) 147–172.

[139] Neumann, W. D., Invariants of plane curve singularities, in *Noeuds, Tresses et Singularités (Plans-sur-Bex 1982)* (ed. C. Weber), Monographies de l'Enseignement Mathématique **31**, l'Enseignement Mathématique, 1983, pp. 223–232.

[140] Neumann, W. D. and F. Raymond, Seifert manifolds, plumbing, μ-invariant and orientation-reversing maps, in *Algebraic and Geometric Topology (Proceedings of a Symposium held at Santa Barbara in Honor of R. L. Wilder)* (ed. K. C. Millett), Springer Lecture Notes in Math. **664**, Springer-Verlag, 1978, pp. 163–196.

[141] Neumann, W. D. and G. A. Swarup, Canonical decompositions of 3-manifolds, *Geometry and Topology* **1** (1997) 21–40.

[142] Newton, I., *The Correspondence of Sir Isaac Newton, vol II (1676–1687)*, Cambridge University Press, 1960, pp. 20–42 and 110–163.

[143] Newton, I., On the theory of fluxions and infinite series with an application to the geometry of curved lines, in *The Mathematical Works of Sir Isaac Newton II*, Johnson Reprint Corp, pp. 135–161.

[144] Newton, I., Enumeratio linearum tertii Ordinis, appendix to *Treatise on Optics*, 1706. English translation (with notes) by C. R. M. Talbot, London, 1861.

[145] Nœther, M., Rationale Ausführungen der Operationen in der Theorie der algebraischen Funktionen, *Math. Ann.* **23** (1883) 311–358.

[146] Oka, M., Geometry of plane curves via toroidal resolution, in *Algebraic Geometry and Singularities* (ed. A. Campillo López and L. Narváez Macarro), Progress in Math. **134**, Birkhäuser, 1996, pp. 95–121.

[147] Orlik, P., *Seifert Manifolds*, Springer Lecture Notes in Math. **291**, Springer-Verlag, 1972.

[148] Persson, U., Horikawa surfaces with maximal Picard numbers, *Math. Ann.* **259** (1982) 287–312.

[149] Piene, R., Polar classes of singular varieties, *Ann. Sci. Ec. Norm. Sup.* (4 ser.) **11** (1978) 247–276.

[150] Plücker, J., *Theorie der Algebraischen Kurven*, Bonn, 1839.

[151] Popescu-Pampu, P., Arbres de contact des singularités quasi-ordinaires et graphes d'adjacence pour les 3-variétés réelles, thesis, Université Paris VII, 2001.

[152] Puiseux, M., Recherches sur les fonctions algébriques, *Jour. de Math.* **15** (1850) 365–480.

[153] Ranicki, A., *High Dimensional Knot Theory*, Springer-Verlag, 1998.

[154] Rapaport, E. S., On the commutator subgroup of a knot group, *Ann. of Math.* **71** (1960) 157–162.

[155] Reeve, J. E., A summary of results in the topological classification of plane algebroid singularities, *Rendiconti del Sem. Mat.* (Univ. Politecnico di Torino) **14** (1954–1955) 159–187.

[156] Rimányi, R., Thom polynomials, symmetries and incidences of singularities, *Invent. Math.* **143** (2001) 499–521.

[157] Rückert, W., Zum Eliminationsproblem der Potenzreihenideale, *Math. Ann.* **107** (1933) 259–281.

[158] Scharlau, W., *Quadratic and Hermitian Forms*, Grundlehren **270**, Springer-Verlag, 1985.

[159] Schrauwen, R., J. H. M. Steenbrink and J. Stevens, Spectral pairs and the topology of curve singularities, in *Complex Geometry and Lie Theory (Sundance UT 1989)* (ed. J. A. Carlson, C. H. Clemens and D. R. Morrison), Proc. Symp. in Pure Math. **53**, Amer. Math. Soc., 1991, pp. 305–328.

[160] Schubert, H., Knoten und Vollringe, *Acta Math.* **90** (1953) 131–286.

[161] Scott, G. P., The geometries of 3-manifolds, *Bull. London Math. Soc.* **15** (1983) 401–487.

[162] Seifert, H., Topologie dreidimensionaler gefaserter Räume, *Acta Math.* **60** (1933) 147–238.

[163] Seifert, H., Uber das Geschlecht von Knoten, *Math. Ann.* **110** (1934) 571–592.

[164] Semple, J. G. and G. T. Kneebone, *Algebraic Curves*, Oxford University Press, 1959.

[165] Semple, J. G. and L. Roth, *Introduction to Algebraic Geometry*, Oxford University Press, 1949.

[166] Serre, J.-P., *Corps Locaux*, Hermann, 1962.

[167] Shustin, E., Versal deformations in the space of plane curves of fixed degree, *Funct. Anal. Appl.* **21** (1987) 82–84.

[168] Smith, H. J. S., On the higher singularities of plane curves, *Proc. London Math. Soc.* **6** (1875) 153–182.

[169] Spanier, E., *Algebraic Topology*, McGraw-Hill, 1966.

[170] Späth, H., Der Weierstra*β*sche Vorbereitungssatz, *Jour. reine angew. Math.* **161** (1929) 95–100.

[171] Steenbrink, J. H. M., Mixed Hodge structure on the vanishing cohomology, in *Real and Complex Singularities, Oslo 1976* (ed. P. Holm), Sijthoff and Noordhoff, 1977, pp. 525–563.

[172] Stefănescu, D., On meromorphic formal power series, *Bull. Math. Sci. Math. R. S. Roumanie* (n.s.) **27(75)** (1983) 169–178.

[173] Stickelberger, W., Uber einen Satz des Herrn Noether, *Math. Ann.* **30** (1887) 401–409.

[174] Teissier, B., Variétés polaires I. Invariants polaires des singularités des hypersurfaces, *Invent. Math.* **40** (1977) 267–292.

[175] Teissier, B., Introduction to equisingularity problems, in *Algebraic Geometry (Arcata 1974)* (ed. R. Hartshorne), Proc. Symp. in Pure Math. **29**, Amer. Math. Soc., 1975, pp. 593–642.

[176] Thurston, W. P., Three dimensional manifolds, Kleinian groups, and hyperbolic geometry, *Bull. Amer. Math. Soc.* **6** (1982) 357–382.

[177] Tibar, M., Bouquet decomposition of the Milnor fibre, *Topology* **35** (1996) 227–241.

[178] Torres, G., On the Alexander polynomial, *Ann. of Math.* **57** (1953) 57–89.

[179] Tosun, M., Tyurina components and rational cycles for rational singularities, *Turkish Jour. Math.* **23** (1999) 361–374.

[180] Tristram, A. G., Some cobordism invariants for links, *Proc. Camb. Phil. Soc.* **66** (1969) 251–264.

[181] Urabe, T., Dynkin graphs and combinations of singularities on plane sextic curves, in *Singularities, Proceedings, University of Iowa, 1986* (ed. R. Randell), Contemporary Math. **90** (1989), pp. 295–316.

[182] van der Waerden, B., *Modern Algebra I* (tr. F. Blum, revised English edition), Frederick Ungar, 1949.

[183] Veys, W., Determination of the poles of the topological zeta function for curves, *Manuscripta Math.* **87** (1995) 435–448.

[184] Viro, O., Some integral calculus based on Euler characteristic, in *Topology and Geometry – Rohlin Seminar*, Springer Lecture Notes in Math. **1346**, Springer-Verlag, 1988, pp. 127–138.

[185] Wahl, J., Equisingular deformations of plane algebraic curves, *Amer. Jour. Math.* **96** (1974) 529–577.

[186] Waldhausen, F., Eine Klasse von 3-dimensionalen Mannigfaltigkeiten I, *Invent. Math.* **3** (1967) 308–333.

[187] Waldhausen, F., Eine Klasse von 3-dimensionalen Mannigfaltigkeiten II, *Invent. Math.* **4** (1967) 87–117.

[188] Walker, R. J., *Algebraic Curves*, Princeton University Press, 1950; reprinted by Dover 1962.

[189] Wall, C. T. C., *Surgery on Compact Manifolds*, Academic Press, 1970.

[190] Wall, C. T. C., Introduction to the preparation theorem, in *Proceedings of Liverpool Singularities Symposium I* (ed. C. T. C. Wall), Lecture Notes in Math. **192**, Springer-Verlag, 1971, pp. 90–96.

[191] Wall, C. T. C., Finite determinacy of smooth map-germs, *Bull. London Math. Soc.* **13** (1981) 481–539.

[192] Wall, C. T. C., Classification and stability of singularities of smooth maps, in *Singularity Theory* (Proceedings of College on Singularities at Trieste 1991) (ed. D. T. Lê, K. Saito and B. Teissier), World Scientific, 1995, pp. 920–952.

[193] Wall, C. T. C., Duality of singular plane curves, *Jour. London Math. Soc.* **50** (1994) 265–275.

[194] Wall, C. T. C., Real rational quartic curves, in *Real and Complex Singularities* (ed. W. L. Marar), Pitman Research Notes in Math. **333**, Longman, 1995, pp. 1–32.

[195] Wall, C. T. C., Duality of real projective plane curves: Klein's equation, *Topology* **35** (1996) 355–362.

[196] Wall, C. T. C., Notes on trimodal singularities, *Manuscripta Math.* **100** (1999) 131–157.

[197] Wallace, A., Tangency and duality over arbitrary fields, *Proc. London Math. Soc.* **6** (1956) 321–342.

[198] Weierstrass, K., *Abhandlungen aus dem Funktionenlehre*, Springer, 1886.

[199] Wirtinger, W., Uber die Verzweigungen bei Funktionen von zwei Veränderlichen, *Jahresbericht Deutsch. Math. Verein.* **14** (1905) 517.

[200] Yamamoto, M., Classification of isolated algebraic singularities by their Alexander polynomials, *Topology* **23** (1984) 277–287.

[201] Yang, J.-G., Sextic curves with simple singularities, *Tohoku Math. Jour.* **48** (1996) 203–227.

[202] Zariski, O., On the problem of existence of algebraic functions of two variables possessing a given branch curve, *Amer. Jour. Math.* **51** (1929) 305–328.

[203] Zariski, O., On the topology of algebroid singularities, *Amer. Jour. Math.* **54** (1932) 453–465.

[204] Zariski, O., Polynomial ideals defined by infinitely near base points, *Amer. Jour. Math.* **60** (1938) 151–204.

[205] Zariski, O., Studies in equisingularity I, *Amer. Jour. Math.* **87** (1965) 507–536. (This, and several other cited papers of Zariski, are reprinted in *Collected Papers, Vol. IV: Equisingularity on Algebraic Varieties* (ed. J. Lipman and B. Teissier), MIT Press, 1979.

[206] Zariski, O., Studies in equisingularity II: equisingularity in codimension 1 and characteristic zero, *Amer. Jour. Math.* **87** (1965) 972–1006.

[207] Zariski, O., Studies in equisingularity III: saturation of local rings and equisingularity, *Amer. Jour. Math.* **90** (1968) 961–1023.

[208] Zariski, O., General theory of saturation and of saturated local rings II: saturated local rings of dimension 1, *Amer. Jour. Math.* **93** (1971) 872–964.

[209] Zariski, O., Le problème de modules pour les branches des courbes planes, Lecture Notes (ed. F. Kmety and M. Merle), Ecole Polytechnique, 1973.

[210] Zeuthen, H. G., Sur les différentes formes des courbes planes du quatrième ordre, *Math. Ann.* **7** (1874) 410–432.

[211] Zeuthen, H. G., Note sur les singularités des courbes planes, *Math. Ann.* **10** (1876) 210–220.

Index